T0324693

De Sitter Invariant
Special Relativity

De Sitter Invariant Special Relativity

Mu-Lin Yan
University of Science and Technology of China, China

 **World Scientific**

NEW JERSEY · LONDON · SINGAPORE · BEIJING · SHANGHAI · HONG KONG · TAIPEI · CHENNAI

Published by

World Scientific Publishing Co. Pte. Ltd.

5 Toh Tuck Link, Singapore 596224

USA office: 27 Warren Street, Suite 401-402, Hackensack, NJ 07601

UK office: 57 Shelton Street, Covent Garden, London WC2H 9HE

British Library Cataloguing-in-Publication Data
A catalogue record for this book is available from the British Library.

ISBN 978-981-4618-70-0

In-house Editor: Song Yu

Printed in Singapore

To my friend Professor Guo, Han-Ying (1939–2010)

Preface

The de Sitter Invariant Special Relativity is a Special Relativity in space-time with non-zero Einstein cosmological constant Λ. However, even though the possibility of existence of such an universal parameter Λ has been widely accepted or conjectured for studying new physics, the corresponding de Sitter invariant special relativity achieved 40 years ago seems still unfortunately less known in the physics community as of recent years. In this case, a textbook-type publication (or Lecture Notes) as introduction of the theory of de Sitter invariant Special relatively is needed. This is the motivation for writing this book.

Usually, the main contents of a physics textbook should be descriptions of the well-established theories and facts. This book satisfies this requirement due to the following two facts discussed throughout the book:

(1) The theory of *de Sitter invariant special relativity* has already been complete and self-consistent mathematically. To see this point, we could briefly recall the history. In 1917, A. Einstein suggested introducing a cosmological constant Λ into his gravitational field equation from his view of cosmology at that time. In 1935 P.A.M. Dirac reconsidered the spacetime with Λ (called the de Sitter space) from space symmetry group viewpoint[1]. He wrote that[1]:

"... the de Sitter space (with no local gravitational fields) is associated with a very interesting group, and so the study of the equations of atomic physics in this space is of special interest, from a mathematical point of view."

---

[1]P.A.M. Dirac, *The electron wave equation in de-Sitter space*, Annals of Mathematics, **35**, 657 (1935)

The meaning of the phrase *"equations of atomic physics"* he interpreted is that:[1]

> *"The equations of atomic physics are usually formulated in terms of the space-time of the special theory of relativity."*

This indicates that Dirac noticed at that time that when $\Lambda \neq 0$, the usual Einstein's special relativity based on Poincaré group would become a special relativity based on de Sitter group. Thus, how to realize such a special relativity was a great challenge. In 1974, Q.K. Lu (K.H. Look), Z.L. Zou (C.L. Tsou), H.Y. Guo (H.Y. Kuo)[2] successfully achieved the de Sitter invariant special relativity by means of the classical domain method. The success is that the principle of relativity is well realized in the de Sitter constant curved spacetime in their work.[2] They found out that the basic metric of the spacetime of de Sitter invariant special relativity is Beltrami metric rather than the Minkowski-Lorentz metric of usual Einstein's special relativity. This marked the establishment of the de Sitter invariant special relativity.

(2) The most important achievement in astrophysics and cosmology during past two decades is the certainty of that the present Universe expansion is accelerative.[3] This observational result can be straightforwardly attributed to the existence of a non-zero cosmological constant Λ_{eff} which is the sum of Einstein's cosmology constant Λ and a vacuum energy (or dark energy) density term $8\pi G\rho_{\text{vac}}$. This discovery naturally revived the necessity for $\Lambda \neq 0$, which serves as a precondition of de Sitter special relativity.

Using de Sitter special relativity to study the atomic physics was firstly called for by P.A.M. Dirac about 80 years ago.[1] After the de Sitter invariant special relativity was established, one could respond to such a call. In this book we describe how to use this theory to study Hydrogen atom. We show in the book that the results are very non-trivial and meaningful.

---

[2]K.H. Look (Q.K. Lu), C.L. Tsou (Z.L. Zou) and H.Y. Kuo, (H.Y. Guo), 1974, *"Motion effects and cosmological red-shift phenomena in classical domain space-time"*, Acta Physica Sinica, **23** (1974) 225 (in Chinese). (The interpreting papers in English to that work: see, e.g., H.Y Guo, *et al.*, Phys. Lett., **A331**, 1 (2004); M.L. Yan, *et al.* Commun Theor Phys., **48**, 27 (2007).)

[3]A.G. Riess, *et al.*, *"Observational evidence from supernovae for an accelerating universe and a cosmological constant"*, Astro. J. **116**, 1009(1998). S. Perlmutter *et al.*, *"Measurements of Omega and Lambda from 42 high redshift supernovae"*, Astrophys. J. 517, 565 (1999) [astro-ph/9812133].

These results are useful for answering the puzzles about changes of some universal physics constants in spacetime during the cosmological evolution. Especially, since the variation of fine-construct constant $\alpha \equiv e^2/(\hbar c)$ has been observed and measured in Keck- and VLT-telescopes for a long time, the theoretic predictions could compared with the data. One of the most surprising findings is that the combinations of Keck- and VLT-data for α-variations support the de Sitter invariant special relativistic predictions within error bar. This agreement could help shed light on searching new physics.

Special Relativity is one of the foundations of physics and has impact on a number branches of physics. Undoubtedly, the de Sitter Invariant Special Relativity is one of the candidates for the new physics beyond the Standard Model (SM) of physics, but it still respects the Einstein's philosophy on relativity and connects to relevant branches of physics. Perhaps the best way to describe new conceptions in this new special relativity is to start from some review or recall of the existing descriptions in various branches of the SM of physics, and then compare them with those in the new theory. Many excellent works and textbooks in relevant fields of physics are cited in this book. The corresponding contents are selected as self-contained as possible in order to be reader-friendly.

As an Introduction to the de Sitter invariant special relativity, this book mainly focuses on the foundations and some applications of it, but does not cover every aspects of developments of the theory. The list of references in the field of de Sitter special relativity cited in the book is by no means complete.

Besides writing Chapter 9, Dr. Sze-Shiang Feng spent lots of time checking and revising the typescript of this book. The author is indebted to him for his great help without which the book cannot be accomplished on time.

Mu-Lin Yan[4]

[4]Supported in part by National Natural Science Foundation of China under Grant No. 11375169.

Contents

Preface   vii

1. General Introduction   1

2. Overview of Einstein's Special Relativity (E-SR)   9

   2.1 Inertial system of reference and relativity principle   9
   2.2 Lorentz transformation and Minkowski spacetime   10
   2.3 Landau-Lifshitz action and relativistic mechanics   12
   2.4 Canonical formulism of E-SR mechanics   14
   2.5 Noether's theorem and the motion integrals   15
   2.6 Noether charges of E-SR mechanics   19
   2.7 Quantization of E-SR mechanics   23
   2.8 Localization of E-SR spacetime symmetry and Einstein's
        General Relativity (I)   25
   2.9 Localization of E-SR spacetime symmetry and Einstein's
        General Relativity (II)   37

3. De Sitter Invariant Special Relativity   43

   3.1 Beltrami metrics   43
   3.2 Motion of a particle in the Beltrami spacetime   45
   3.3 Lagrangian mechanics of free particles in the Beltrami
        spacetime   48
   3.4 Spacetime transformation preserving Beltrami metric . . .   51
   3.5 Noether charges of dS-SR mechanics   57

3.6   Hamilton and canonical momenta   63
3.7   Dispersion relation   68
3.8   Minkowski point in Beltrami spacetime   70

4. De Sitter Invariant General Relativity   73

4.1   Localization of Lu-Zou-Guo transformation and action of
       de Sitter invariant General Relativity   74
4.2   Static spherically symmetric solution in empty spacetime   76
4.3   Beltrami–de Sitter–Schwarzschild Solution   79

5. Dynamics of Expansion of the Universe in General Relativity   87

5.1   Robertson–Walker metric   88
5.2   Friedmann equation   91
5.3   The Friedmann models   95
5.4   ΛCDM model   97
5.5   Accelerated expansion   103
5.6   Cosmological constant   104

6. Relativistic Quantum Mechanics for de Sitter Invariant
    Special Relativity   107

6.1   Quantization of the de Sitter invariant special relativistic
       mechanics   107
6.2   Dirac equation in Beltrami spacetime   110
       6.2.1   Introduction   110
       6.2.2   Weyl's approach   111
       6.2.3   Dirac equation in Beltrami spacetime   114

7. Distant Hydrogen Atom in Cosmology   119

7.1   Introduction   119
7.2   Beltrami local inertial coordinate system in light cone of
       Ricci Friedmann–Robertson–Walker universe   122
7.3   Hydrogen atom at distant locally inertial coordinate
       system in light cone of RFW universe   126
7.4   Solutions of ordinary Einstein relativistic Dirac
       equation for Hydrogen atom at QSO   127

7.4.1   Eigenvalues and eigenstates 127

7.4.2 $2s^{1/2}$ and $2p^{1/2}$ states of Hydrogen 130

7.5   De Sitter invariant relativistic Dirac equation for distant Hydrogen atom 132

7.6   De Sitter relativistic quantum mechanics spectrum equation for Hydrogen 136

7.7   Adiabatic approximation solution to de Sitter–Dirac spectra equation and time variation of physical constants 139

7.8   **$2s^{1/2}$–$2p^{1/2}$** splitting in the dS-SR Dirac equation of Hydrogen 142

     7.8.1   Energy levels shifts of Hydrogen in dS-SR QM as perturbation effects of E-SR Dirac equation of atom 142

     7.8.2   **$2s^{1/2}$ – $2p^{1/2}$** splitting caused by **H′** 145

7.9   High-order calculations of $1/R^2$-expansions 150

7.10   Universal constant variations over cosmological time . . . 154

8.   Temporal and Spatial Variation of the Fine Structure Constant   157

8.1   Methods to search for changes of fine-structure constant . 157

8.2   Spacetime variations of fine-structure constant estimated via combining Keck /HIRES- and VLT /UVES-observation data 160

8.3   An overview of implication of spacetime variations of fine-structure constant 163

8.4   Light-cone of Ricci Friedmann–Robertson–Walker universe   169

8.5   Local inertial coordinate system in light-cone of RFW universe with Einstein cosmology constant 171

8.6   Electric Coulomb law at light-cone of RFW universe . . . 177

8.7   Fine structure constant variation in $\{Q^0, Q^1, 0, 0\}$ 181

     8.7.1   Formulation of alpha-variation for case of $\Theta = \pi$, and 0 182

     8.7.2   Comparisons with observations of alpha-varying for cases of $\Theta = 0, \pi$ 190

8.8 α-Varying in whole sky 193

8.9   Summary and comments   196

9. De Sitter Invariance of Generally Covariant Dirac Equation   201

9.1   Gürsey-Lee metric of de Sitter spacetime   201
9.2   Proof of Eq.(9.21)   209
9.3   de Sitter invariance   210

Appendix A   Solutions of Dirac Equation of Hydrogen in
Minkowski Spacetime   213

A.1   Pauli equation in a central potential   213
A.2   Relativistic quantum numbers and spin-angular functions   218
A.3   Dirac equation for Hydrogen   220
A.4   Explicit wave functions for one-electron atom with
principal quantum number $n = 1$ and 2   227
A.5   Mathematica calculation for practice   228

Appendix B   The Confluent Hypergeometric Function   231

Appendix C   Spherical Harmonics   233

Appendix D   Electric Coulomb Law in QSO-Light-Cone Space   235

Appendix E   Adiabatic Approximative Wave Functions in
vacde Sitter–Dirac Equation of Hydrogen   239

Appendix F   Matrix Elements of $\mathbf{H'}$ in $(\mathbf{2s^{1/2}\text{-}2p^{1/2}})$-
Hilbert Space   243

Appendix G   Solutions of Exercises   249

Bibliography   259

Index   263

Chapter 1

General Introduction

Einstein's Special Relativity is one of the cornerstones of modern physics. There is one universal parameter c (i.e., speed of light) in the Einstein's Special Relativity (E-SR), which serves as the maximal velocity of physics. One might be curious about whether there is another universal parameter R that serves as the maximal length in physics besides the universal maximal velocity limit c. The answer is yes. This book intends to describe a special theory of relativity with two universal parameters c and R. Such a theory is called the de Sitter Invariant Special Relativity, or the Special Relativity with Cosmology Constant.

Einstein's Special Relativity is based on the principle of relativity and has global Poincaré-Minkowski space-time symmetry. This theory assumes that the space-time metric is $\eta_{\mu\nu} = \mathrm{diag}\{+, -, -, -\}$, which will be called *the basic metric* of the Einstein's special relativity. The most general space-time transformation to preserve metric $\eta_{\mu\nu}$ is Poincaré group (or inhomogeneous Lorentz group $ISO(3,1)$). It is well known that the Poincaré group is the limit of the de Sitter group with the pseudo-sphere radius $|R| \to \infty$. And the de Sitter spacetime is the maximally symmetric spacetime. It is quite natural to extend Einstein's theory of special relativity to a theory that physics is covariant under de Sitter transformation.

The 4-dimensional de Sitter spacetime continuum is a 4-dimensional pseudo-sphere (or hyperboloid) $\mathcal{S}_\Lambda$ embedded in a 5-dimensional Minkowski spacetime with metric $\eta_{AB} = \mathrm{diag}(1, -1, -1, -1, -1)$ (or $\eta_{AB} = \mathrm{diag}(1, -1, -1, -1, +1)$ for anti de Sitter continuum) (see, e.g., [Roberson (1928)]). The spacetime symmetry is $SO(4,1)$ (or $SO(3,2)$). Since $\mathcal{S}_\Lambda$ is a 4-dimensional hypersphere, it always be possible to use some geodesic

1

projection calculations to induce a following de Sitter spacetime metric:

$$ds^2 = g_{\mu\nu}(x)dx^\mu dx^\nu, \tag{1.1}$$

where $\mu = \{0, 1, 2, 3\}$ and the fifth coordinate disappears. According to the principle of relativistic mechanics, once the spacetime metric fixed, the action describing the motion of free particle (i.e., a moving body which is not acted upon by external forces) in the spacetime is determined by Landau-Lifshitz action [Landau, Lifshitz (1987)] (this statement coincides with the geodesic principle for free particle's motion in the gravitational fields in the general relativity). As a deduction, the moving of free particle will be linear and uniform in the *inertial reference systems*. Instead, this claim provides basis for defining the *inertial reference systems*, which is called inertial moving law for free particles and serves a key step and a starting point to build a special relativity theory. A well-known example is the mechanics of Einstein's special relativity. Its basic metric is $\eta_{\mu\nu}$. The corresponding Landau-Lifshitz action is:

$$S = -mc \int \sqrt{\eta_{\mu\nu}dx^\mu dx^\nu} = -mc^2 \int dt\sqrt{1 - \dot{\mathbf{x}}^2/c^2} \equiv \int dtL. \tag{1.2}$$

The least action principle gives

$$\frac{d}{dt}\frac{\partial L}{\partial \dot{x}^i} = \frac{\partial L}{\partial x^i}, \tag{1.3}$$

and the solution of (1.2) and (1.3) is

$$\ddot{\mathbf{x}} = 0, \ \Rightarrow \ \dot{\mathbf{x}} = constant, \tag{1.4}$$

which indicates that the inertial moving law for free particle holds. This fact is the basis of the Einstein's Special Relativity.

We want to discuss what is *the basic metric* for de Sitter invariant special relativity. Equation (1.1) is of $SO(4, 1)$ (or $SO(3, 2)$) reduced de Sitter spacetime metric. However, it is geodesic projection calculation process dependent. It is crucial and highly non-trivial to find suitable explicit expression of that $g_{\mu\nu}$, which also leads to the inertial motion law for free particles (1.4). In 1970s Lu, Zou and Guo [Lu (1970)][Lu, Zou and Guo (1974)] solved this problem, and such a metric were successfully determined. Namely, under requirement that the inertial moving law for free particle in the de Sitter spacetime holds, it is found that the reduced metric $g_{\mu\nu}$ of (1.1) is so called Beltrami metric $B_{\mu\nu}(x)$:

$$g_{\mu\nu}(x) = B_{\mu\nu}(x). \tag{1.5}$$

(see Eq.(3.6) for the expression of $B_{\mu\nu}(x)$). $B_{\mu\nu}(x)$ serves as the *basic metric* of the de Sitter invariant special relativity. In this way, distinguishing from Einstein's special relativity, a new special relativity theory based on considering the extension from Poincaré to de Sitter symmetry is formulated. The special relativity principle in the de Sitter (and anti de Sitter) spacetime is realized. This is an important discovery. In 2005, the Lagrangian-Hamiltonian formalism for the mechanics of de Sitter invariant special relativity were suggested [Yan, Xiao, Huang and Li (2005)].

In history, in 1935, P.A.M. Dirac called to study the electron wave equation in de Sitter (and Anti-de Sitter) spacetime and to study the atomic physics in that spacetime [Dirac (1935)]. This could be considered the first effort to try to explore an effect of relativistic theory with the de Sitter spacetime symmetry based on the group viewpoint. However, it was not known how to realize the special relativity principle in the constant curved de Sitter (and anti de Sitter) spacetime at that time. This hindered the discovery of such a special relativity with de Sitter spacetime symmetry for long time.

Several remarks on Lu-Zou-Guo's de Sitter invariant special relativity [Lu (1970)][Lu, Zou and Guo (1974)] (and [Yan, Xiao, Huang and Li (2005)]) are as follows:

(1) Since the Einstein's special relativity is well-known, a comparison between the quantities of E-SR and their counterpart in de Sitter invariant special relativity would be beneficial. The following Table 1.1 is a dictionary for the comparison.

(2) To the de Sitter invariant special relativity, the Lagrangian of mechanics for free particle is $L_{dS} = -mc\sqrt{B_{\mu\nu}(x)\dot{x}^\mu \dot{x}^\nu}$ (see Table 1.1). As is well know that various laws of conservation (of momentum, angular momentum, etc) are particular cases of one general theorem: to every one-parameter group of diffeomorphisms of the configuration manifold of a Lagrangian system which preserves the Lagrangian function, there corresponds a first integral of the equations of motion. That theorem is called Noether's theorem. This theorem is valid for both Lagrangian $L_{Eins} = -mc\sqrt{\eta_{\mu\nu}\dot{x}^\mu \dot{x}^\nu}$ and Lagrangian $L_{dS} = -mc\sqrt{B_{\mu\nu}(x)\dot{x}^\mu \dot{x}^\nu}$. Differentiating from L_{Eins}, L_{dS} is time dependent and coordinate dependent (i.e., there is no cyclic coordinates in L_{dS}). At first sight, the

Table 1.1   Comparison of quantities in Einstein's special relativity (E-SR) and their counterpart in de Sitter invariant special relativity (dS-SR).

Quantities	E-SR	dS-SR
Symmetry	Poincaré group	de Sitter group
Universal parameter	c	c and R (or $\Lambda \equiv 3/R^2$)
Basic metric	Lorentz metric $\eta_{\mu\nu}$	Beltrami metric $B_{\mu\nu}$ (see Eq.(3.6))
Action	$S_{Eins} = -mc$ $\times \int \sqrt{\eta_{\mu\nu}dx^\mu dx^\nu}$ $\equiv \int dt L_{Eins},$	$S_{dS} = -mc \int \sqrt{B_{\mu\nu}dx^\mu dx^\nu}$ $\equiv \int dt L_{dS},$
Lagrangian	$L_{Eins} = -mc^2$ $\times \sqrt{1 - \dot{\mathbf{x}}^2/c^2}$ (see Eq.(2.24))	$L_{dS} = -mc^2$ $\times \sqrt{\frac{R^2(c^2-\dot{\mathbf{x}}^2)-\mathbf{x}^2\dot{\mathbf{x}}^2+(\mathbf{x}\cdot\dot{\mathbf{x}})^2+c^2(\mathbf{x}-\dot{\mathbf{x}}t)^2}{R^{-2}c^2(R^2+\mathbf{x}^2-c^2t^2)^2}}$ (see Eq.(3.32))
Inertial motion law	$\delta S_{Eins} = 0, \Rightarrow \ddot{x}^i = 0$	$\delta S_{dS} = 0, \Rightarrow \ddot{x}^i = 0$
Spacetime transformation between inertial systems	$x^\mu \xrightarrow{Lorentz} x'^\mu$ $= M^\mu_\nu x^\nu + a^\mu$ $M :=$ $(M^\mu_\nu) \in SO(1,3),$ (see Eq.(2.4))	$x^\mu \xrightarrow{LZG} x'^\mu = \frac{\pm\sigma(a)^{1/2}}{\sigma(a,x)}(x^\nu - a^\nu)D^\mu_\nu,$ $D^\mu_\nu = L^\mu_\nu + \frac{\eta_{\nu\rho}a^\rho a^\lambda}{R^2(\sigma(a)+\sigma^{1/2}(a))}L^\mu_\lambda,$ $L := (L^\mu_\nu)$ $\sigma(x) = 1 - \frac{1}{R^2}\eta_{\mu\nu}x^\mu x^\nu,$ $\sigma(a,x) = 1 - \frac{\eta_{\mu\nu}}{R^2}a^\mu x^\nu.$ (see Eq.(3.70))

energy-momentum conservation laws will be spoiled by such a time and coordinate-dependent Lagrangian. However, these fundamental conservation laws in physics are protected by the de Sitter spacetime symmetry. This point is rather subtle. In Chapter 3, all Noether charges (including energy E, momentums $\mathbf{p}$, boost charges $\mathbf{K}$, and angler momentums $\mathbf{L}$) for de Sitter invariant special relativity are determined.

(3) As an essential advantage of the Lagrangian-Hamiltonian formulation over other formalisms, both canonical momentum $\pi_i \equiv \partial L/\partial \dot{x}^i$ conjugating to the coordinate x^i and canonical energy $H \equiv \pi_i \dot{x}^i - L$ (or Hamilton) conjugating to the time t for free particles in the inertial reference frame can be determined rationally by the mechanical

principle. We address that the canonical quantities are different from the the corresponding conserved Noether charges in principle. In other words, the conservation of Noether chargers is irrelevant to whether or not the corresponding canonical quantities are conservative. In the de Sitter invariant special relativistic mechanics, the Lagrangian $L_{dS} = -mc\sqrt{B_{\mu\nu}(x)\dot{x}^\mu\dot{x}^\nu}$ is both time and coordinate dependent, and hence its canonical energy and canonical momentums are not conservative. But this issue does not affect the conservation of its Noether chargers. Namely, in de Sitter invariant special relativity, though $\dot{H} \neq 0$ and $\dot{\pi}_i \neq 0$, but $\dot{E} = \dot{p}^i = \dot{K}^i = \dot{L}^i = 0$ still hold true.

(4) The canonical energy (i.e., Hamiltonian) and the canonical momenta are members of frame of the canonical formulation of mechanics. This is the foundation of the quantization of the dynamics. In this way, the relativistic quantum mechanics of de Sitter invariant special relativity is arrived. In the chapter 6, such a quantum mechanics is described.

(5) According to the gauge principle, the ordinary general relativity (or Einstein general relativity) can be constructed by means of localizing the spacetime symmetry group $ISO(3,1)$ of the Einstein's special relativity. This mechanism provides a bridge from special relativity to general relativity. It works also for the de Sitter invariant special relativity which has global spacetime symmetry $SO(4,1)$ (or $SO(3,2)$). In Chapters 4 and 5, we show how to employ this mechanism to construct a general relativity and its characteristics. Namely we will achieve the de Sitter invariant general relativity by localizing the de Sitter invariant special relativity.

(6) A significant feature of the de Sitter invariant general relativity (i.e., GR with Λ) is that *the local inertial coordinate system* of this theory is an inertial frame in the Beltrami spacetime instead of one in the flat Minkowski spacetime as usual. Existence of *local inertial coordinate system* is required by the Equivalence Principle. The principle states that experiments in a sufficiently small falling laboratory, over a sufficiently short time, give results that are indistinguishable from those of the same experiments in an inertial frame in empty space of special relativity (see, e.g., *pp.119* of [Hartle (2003)]). Such a sufficiently small falling laboratory, over a sufficiently short time represents *a*

local inertial coordinates system. This principle suggests that the local properties of curved spacetime should be indistinguishable from those of the spacetime with *inertial metric* of special relativity. A concrete expression of this ideal is the requirement that, given a metric $g_{\alpha\beta}$ in one system of coordinates x^{α}, at each point P of spacetime it is possible to introduce new coordinates x'^{α} such that

$$g'_{\alpha\beta}(x'_P) = \text{inertial metric of Special Relativity (SR) at } x'_P, \quad (1.6)$$

and the connection at x'_P is the Christoffel symbols deduced from $g'_{\alpha\beta}(x'_P)$.

In usual Einstein's general relativity (without Λ), the above expression is

$$g'_{\alpha\beta}(x'_P) = \eta_{\alpha\beta}, \quad and \quad \Gamma^{\lambda}_{\alpha\beta} = 0, \quad (1.7)$$

which satisfies the Einstein equation of Einstein General Relativity (E-GR) in empty space: $G_{\mu\nu} \equiv R_{\mu\nu} - \frac{1}{2}g_{\mu\nu}R = 0$, where definitions of $R\mu\nu, R$ are given in Eqs. (2.182)–(2.191).

In de Sitter invariant General Relativity (dS-GR) (or GR with a Λ), the local inertial coordinate system at x'^{α}_P is characterized by

$$g'_{\alpha\beta}(x'_P) = B_{\alpha\beta}(x'_P), \quad (1.8)$$

$$\Gamma^{\lambda}_{\alpha\beta} = \frac{1}{2}B^{\lambda\rho}(\partial_{\alpha}B_{\rho\beta} + \partial_{\beta}B_{\rho\alpha} - \partial_{\rho}B_{\alpha\beta}). \quad (1.9)$$

which satisfies the Einstein equation of dS-GR in empty spacetime: $G_{\mu\nu} + \Lambda g_{\mu\nu} = 0$ with $\Lambda = 3/R^2$. (Note $\eta_{\mu\nu}$ does not satisfy that equation, i.e., $G_{\mu\nu}(\eta) + \Lambda\eta_{\mu\nu} \neq 0$. So it cannot be the metric of the local inertial system in dS-GR with Λ.)

(7) From cosmological view, particle physical studies on all ordinary atom, nucleus etc in laboratories are accomplished in *a local inertial systems* of a specific point in the Universe continuum which is governed by gravities described by general relativity. Thus, it is natural to ask whether *the local inertial system* is described in the Minkowski metric $\eta_{\mu\nu}$ or in the Beltrami metric $B_{\mu\nu}$. It is meaningful and important to pursue answer to this question. As is well known that the Hydrogen atom's fine structure is a solvable problem in the standard relativistic quantum mechanics, and it has already been extensively studied in terms of ordinary Einstein's special relativistic quantum mechanics

(see, e.g., reference of [Strange (2008)]). Therefore, a further calculation to one-electron atom's levels in terms of the de Sitter invariant special relativistic quantum mechanics would be interesting and helpful [Yan (2012)]. Actually, in principle, comparison between the predictions and the observation data would help determine the metric of *the local inertial coordinate system* in the Universe of the real world. Chapters 7 and 8 devote to solution to the Hydrogen atom by the relativistic quantum mechanics with de Sitter spacetime symmetry.

(8) When $R \to \infty$, the de Sitter invariant special relativity will be reduced to ordinary special relativity. Therefore, the de Sitter invariant SR is full special relativity with maximum spacetime symmetry, and the ordinary SR is a specific limit of it. However, comparing the predictions of the ordinary E-SR with the experimental data (in laboratories), the theory is correct with extremely high accuracy [Zhang (1997)]. This fact indicates the value of $|R|$ must be a very huge length scale. The corrections coming from the de Sitter invariant special relativity should only be visible in some observation experiments with cosmologically large distances, which are compatible with $\sim |R|$. Such effects may emerge in a a couple of situations, such as in the analysis of evidence for dark energy found by the Supernova Cosmology Project[Perlmutter, S (1999)] [Yan, Hu, Huang (2012)], in dealing with the observation of atomic absorbed spectra from the distant galaxy, and the quasi-stellar object (typically at cosmic distance scale) [Webb *et al.* (1999)] [Murphy (2007)] [van Weerdenburg, *et al.* (2011)] [Yan (2012)] etc.

Special relativity is a theory on the basic spacetime symmetry in physics. It is directly related to many fundamental concepts of physics, such as the definitions of energy-momentum, motion integrals, the procedure of quantization leading to the quantum mechanics, and to extend beyond itself to general relativity etc. Therefore any correction to the ordinary E-SR will influence many aspects in physics. In this book, we describe the dS-SR which distinguishes itself from the ordinary E-SR. Several relevant aspects are described from new point of view. They are: Mechanics; Quantum Mechanics; General Relativity; Cosmology; Hydrogen Atom Solutions; Maximal Symmetric Spacetime and so on. All of the corresponding contents are selected as self-contained as possible in order to be reader friendly. Some exercises for readers are presented in Chapter-sections: 3.1; 3.2; 3.4; 3.6; 5.2; 6.2; 7.4; 8.3 and 8.6. The solutions of the exercises are given in Appendix G.

Besides the works mentioned above, there are a number of discussions on various topics related to the de Sitter special relativity in literatures such as [Guo, Huang, Wu, Zhou (2010)], [Guo, Huang, Wu (2008)], [Guo, Huang, Wu (2008)], [Guo, Huang, Tian, Wu, Zhou (2007)], [Chang, Chen, Guan, Huang (2005)],[Huang, Tian, Wu, Xu, Zhou (2012)], [Tian, Guo, Huang, Xu, Zhou (2005)], and so on.

Chapter 2

Overview of Einstein's Special Relativity (E-SR)

We start with a brief description of ordinary Einstein's Special Relativity (to denote it E-SR for short hereafter).

2.1 Inertial system of reference and relativity principle

For the description of processes taking place in nature, one must have a *system of reference*. By a system of reference we mean a system of coordinates serving to indicate the position of a particle in space, as well as clocks fixed in this system serving to indicate the time.

There exist systems of reference in which a freely moving body, i.e., a moving body which is not acted upon by external forces, proceeds with constant velocity. Such reference systems are defined to be *inertial*.

If two reference systems move uniformly relative to each other, and if one of them is an inertial system, then clearly the other is also inertial (in this system the every free motion will be linear and uniform). In this way one can obtain arbitrarily many inertial systems of reference, moving uniformly to one another.

Experiments, e.g., the observations in the Galileo-boat which moves uniformly, show that the so-called *principle of relativity* is valid. According to this principle all the laws of nature are identical in all inertial systems of reference. In other words, the equations expressing the laws of nature are invariant with respect to transformations of coordinates and time from one inertial system to another. This means that the equation describing any law of nature, when written in terms of coordinates and time in different inertial reference systems, has the same form.

2.2 Lorentz transformation and Minkowski spacetime

The null result of Michelson-Morley experiment (1881) indicates that the propagating velocity of light c is the same in different inertial systems. The propagating equation of light waves in an inertial system of $\{x^0 \equiv ct,\ x^1,\ x^2,\ x^3\}$ is as follows

$$\frac{1}{c^2}\frac{\partial^2 \phi(x)}{\partial t^2} - \nabla^2 \phi(x) = 0,$$

$$\text{or}\quad \left(\frac{\partial^2}{\partial(ix^0)^2} + \frac{\partial^2}{\partial(x^1)^2} + \frac{\partial^2}{\partial(x^2)^2} + \frac{\partial^2}{\partial(x^3)^2}\right)\phi(x) = 0,$$

$$\text{or}\quad \eta^{\mu\nu}\frac{\partial}{\partial x^\mu}\frac{\partial}{\partial x^\nu}\phi(x) = 0, \tag{2.1}$$

where the superscript $\{\mu,\nu\} = \{0,\ 1,\ 2,\ 3\}$, the repeated index means summation from 0 to 3 (Einstein convenience), $\eta^{\mu\nu}$ is the contravariant tensor of metric as follows

$$\{\eta^{\mu\nu}\} = \begin{pmatrix} 1 & 0 & 0 & 0 \\ 0 & -1 & 0 & 0 \\ 0 & 0 & -1 & 0 \\ 0 & 0 & 0 & -1 \end{pmatrix}, \tag{2.2}$$

and the covariant metric tensor corresponding to $\eta^{\mu\nu}$ is $\eta_{\mu\nu} = \text{diag}\{1,-1,-1,-1,\}$. The principle of relativity requires that the Eq.(2.1) must be invariant under the spacetime transformations from initial inertial system $K := \{x^0 \equiv ct,\ x^1,\ x^2,\ x^3\}$ into another new inertial system $K' := \{x'^0 \equiv ct',\ x'^1,\ x'^2,\ x'^3\}$. Namely, in the new system, the wave equation reads

$$\eta^{\mu\nu}\frac{\partial}{\partial x'^\mu}\frac{\partial}{\partial x'^\nu}\phi'(x') = 0, \tag{2.3}$$

where $\phi'(x') = \phi(x)$ is a scalar wave function. The $ISO(3,1)$-transformations of $x^\mu \to x'^\mu$ are

$$x^\mu \to x'^\mu = M^\mu{}_\nu x^\nu + a^\mu = \eta^{\mu\nu}M_{\nu\lambda}x^\lambda + a^\mu, \tag{2.4}$$

and hence

$$\frac{\partial}{\partial x^\mu} = \frac{\partial x'^\lambda}{\partial x^\mu}\frac{\partial}{\partial x'^\lambda} = M^\lambda{}_\mu\frac{\partial}{\partial x'^\lambda}, \quad \text{and} \quad \frac{\partial}{\partial x^\nu} = \frac{\partial x'^\rho}{\partial x^\nu}\frac{\partial}{\partial x'^\rho} = M^\rho{}_\nu\frac{\partial}{\partial x'^\rho}, \tag{2.5}$$

where $M^\mu{}_\nu$ are constant (4×4)−tensor elements, a^μ is a constant 4-vector, and the Greek indices are raised up (lowered down) by $\eta^{\mu\nu}$ ($\eta_{\mu\nu}$). Substituting Eq.(2.5) into Eq.(2.1), we have

$$\eta^{\mu\nu} M^\lambda{}_\mu M^\rho{}_\nu \frac{\partial}{\partial x'^\lambda} \frac{\partial}{\partial x'^\rho} \phi'(x') = 0. \tag{2.6}$$

Comparing Eq.(2.6) with Eq.(2.3), we obtain

$$\eta^{\mu\nu} M^\lambda{}_\mu M^\rho{}_\nu = \eta^{\lambda\rho}, \quad \text{or} \quad M^\lambda{}_\mu M^{T\ \mu\rho} = \eta^{\lambda\rho}, \quad \text{or} \quad M^\lambda{}_\mu M^{T\ \mu}{}_\rho = \delta^\lambda_\rho, \tag{2.7}$$

where the notation M^T is the transpose of matrix M. Eq.(2.7) indicates $M \in SO(3,1)$ called Lorentz group. The Eq.(2.4) is called inhomogeneous $ISO(3,1)$ (or Poincaré) transformation of coordinates of space and time. When $a^\mu = 0$, that is called Lorentz transformation. Suppose $M^0{}_0 = M^1{}_1 = \cosh\psi$, $M^0{}_1 = M^1{}_0 = \sinh\psi$, $M^2{}_2 = M^3{}_3 = 1$ and others=0, it is easy to check that Eq.(2.7) is satisfied for such M-matrix elements. Substituting them into Eq.(2.4), the Lorentz transformation reads

$$x'^1 = x^1 \cosh\psi + ct \sinh\psi, \quad ct' = x^1 \sinh\psi + ct \cosh\psi. \tag{2.8}$$

Let us consider the motion of the origin of K−system in the K'−system. Then $x^1 = 0$ and the formulas (2.8) take the form:

$$x'^1 = ct \sinh\psi, \quad ct' = ct \cosh\psi, \tag{2.9}$$

or dividing one by the other,

$$\frac{x'^1}{ct'} = \tanh\psi. \tag{2.10}$$

But x'^1/t' is clearly the velocity $-u$ of the K−system relative to K'. So

$$\tanh\psi = -\frac{u}{c}. \tag{2.11}$$

From this we have

$$\sinh\psi = \frac{-v/c}{\sqrt{1 - \frac{u^2}{c^2}}}, \quad \cosh\psi = \frac{1}{\sqrt{1 - \frac{u^2}{c^2}}}. \tag{2.12}$$

Substituting these relations into Eq.(2.8), we find an explicit expression of the Lorentz transformation:

$$x'^1 = \frac{x^1 - vt}{\sqrt{1 - \frac{u^2}{c^2}}}, \quad x'^2 = x^2, \quad x'^3 = x^3, \quad t' = \frac{t - ux^1/c^2}{\sqrt{1 - \frac{u^2}{c^2}}}, \tag{2.13}$$

or

$$x'^0 = \gamma(x^0 - \beta x^1), \quad x'^1 = \gamma(x^1 - \beta x^0), \quad x'^2 = x^2, \quad x'^3 = x^3, \tag{2.14}$$

where Lorentz factors: $\gamma = 1/\sqrt{1 - u^2/c^2}$, $\beta = u/c$.

The 4-D spacetime with metric $\eta_{\mu\nu}$ is the Minkowski spacetime. Under the spacetime transformation $x^\mu \to x'^\mu$, the tensor of $\eta_{\mu\nu}$ transforms to $\eta'_{\mu\nu}$ according to:

$$\eta'_{\mu\nu} = \frac{\partial x^\lambda}{\partial x'^\mu} \frac{\partial x^\rho}{\partial x'^\nu} \eta_{\lambda\rho}. \tag{2.15}$$

When the transformation is the Lorentz transformation in the Minkowski space, we have

$$x^\lambda = M^\lambda{}_\mu x'^\mu, \quad \text{and} \quad \frac{\partial x^\lambda}{\partial x'^\mu} = M^\lambda{}_\mu, \quad \frac{\partial x^\rho}{\partial x'^\nu} = M^\rho{}_\nu. \tag{2.16}$$

Therefore

$$\eta'_{\mu\nu} = M^\lambda{}_\mu M^\rho{}_\nu \eta_{\lambda\rho} = \eta_{\mu\nu}. \tag{2.17}$$

This indicates that the Lorentz transformation preserves Lorentz metric $\eta_{\mu\nu}$.

When two events in Minkowski space are infinitely close to each other, then the interval ds between them is

$$ds^2 = cdt^2 - (dx^1)^2 - (dx^2)^2 - (dx^3)^2 = \eta_{\mu\nu} dx^\mu dx^\nu. \tag{2.18}$$

Under $ISO(3,1)$−transformation (2.4), $ds \to ds'$. Then we have

$$ds'^2 = \eta_{\mu\nu} dx'^\mu dx'^\nu = \eta_{\mu\nu} M^\mu{}_\lambda M^\nu{}_\rho dx^\lambda dx^\rho = \eta_{\lambda\rho} dx^\lambda dx^\rho = ds^2, \tag{2.19}$$

where Eq.(2.17) has been used. Equation (2.19) shows the invariance of interval in Minkowski space under the Poincaré transformation (2.4).

2.3   Landau-Lifshitz action and relativistic mechanics

The outstanding prerequisite of above discussions is the existence of inertial reference system. Hence, it is necessary to look into the Relativistic Mechanics for free particles, which should meet the inertial motion law and be invariant under Lorentz transformations.

As is well known, the dynamics of each mechanical system is determined by the *principle of least action*, which states that there exists a certain integral S called the *action* which takes a minimum value for actual motion so that its variation δS is zero. By symmetry considerations, Landau and

Lifshitz in their famous book "*The Classical Theory of Fields*" suggested that the Action for a free particle should have the form

$$S = -\alpha \int ds = -\alpha \int \sqrt{\eta_{\mu\nu} dx^\mu dx^\nu}, \tag{2.20}$$

where α is a constant which will be determined by the classical mechanics under low velocity limit. Hereafter, we call S of (2.20) *L-L action*. Noting $ds = \sqrt{c^2 dt^2 - \sum_{i=1}^{3}(dx^i)^2} = cdt\sqrt{1 - v^2/c^2}$ (where $v^2 = (\dot{x}^1)^2 + (\dot{x}^2)^2 + (\dot{x}^3)^2$), we have

$$S = -\alpha c \int \sqrt{1 - \frac{v^2}{c^2}} dt. \tag{2.21}$$

The Lagrangian function $L = L(t, x^i, \dot{x}^i)$ can be defined via the action integral with respect to the time:

$$S = \int L dt. \tag{2.22}$$

Comparing Eq.(2.21) with Eq.(2.22), we obtain

$$L = -\alpha c \sqrt{1 - \frac{v^2}{c^2}}. \tag{2.23}$$

When $v^2 << c^2$, we approximately have $L \simeq -\alpha c + \alpha v^2/(2c)$. Neglecting the constant term and comparing the expression of that L with the usual classical mechanics $L = mv^2/2$, we find that $\alpha = mc$. Then

$$L = -mc^2 \sqrt{1 - \frac{v^2}{c^2}} = -mc^2 \sqrt{1 + \frac{\eta_{ij}\dot{x}^i \dot{x}^j}{c^2}}, \tag{2.24}$$

where Latin index $i = 1, 2, 3$, $v^i = \dot{x}^i$, and $\{\eta_{ij}\} = \text{diag}\{-1, -1, -1\}$. Equation (2.24) is the Lagrangian function for free particle in E-SR.

From $\delta S = \delta \int L dt = 0$, we get the equation of motion (i.e., the Euler-Lagrange equation):

$$\frac{d}{dt} \frac{\partial L}{\partial \dot{x}^i} - \frac{\partial L}{\partial x^i} = 0. \tag{2.25}$$

To the free particle in E-SR, substituting Eq.(2.24) into Eq.(2.25), the equation of motion is:

$$\ddot{x}^i = 0, \tag{2.26}$$

and the solution is

$$x^i = v^i t + constant, \quad \text{with} \quad v^i = constant. \tag{2.27}$$

We conclude that the inertial motion law holds for free particle in E-SR mechanics.

2.4   Canonical formulism of E-SR mechanics

When the Lagrangian function L is known, both canonical momenta π_i conjugating to x^i and canonical energy (i.e., Hamilton) conjugating to t are well defined as follows

$$\pi_i = \frac{\partial L}{\partial \dot{x}^i} = \frac{\partial L}{\partial v^i} = \frac{-m\eta_{ij}v^j}{\sqrt{1 - \frac{v^2}{c^2}}} = \frac{-m}{\sqrt{1 - \frac{v^2}{c^2}}}(\eta_{i\nu}\dot{x}^\nu), \qquad (2.28)$$

$$H = \sum_{i=1}^{3} \frac{\partial L}{\partial \dot{x}^i}\dot{x}^i - L = \frac{mc^2}{\sqrt{1 - \frac{v^2}{c^2}}} = \frac{-m}{\sqrt{1 - \frac{v^2}{c^2}}}(-c\eta_{0\nu}\dot{x}^\nu)$$

$$= \sqrt{-c^2\eta^{ij}\pi_i\pi_j + m^2c^4}, \qquad (2.29)$$

where Eq.(2.24) has been used. And then the equation of motion (2.24) becomes two canonical equations

$$\dot{x}^i = \frac{\partial H}{\partial \pi_i} = \{x^i,\ H\}_{PB}, \qquad (2.30)$$

$$\dot{\pi}_i = -\frac{\partial H}{\partial x^i} = \{\pi_i,\ H\}_{PB}, \qquad (2.31)$$

where the definition of the Poisson Bracket is $\{f,\quad g\}_{PB} := \sum_{i=1}^{3}\left(\frac{\partial f}{\partial x^i}\frac{\partial g}{\partial \pi_i} - \frac{\partial f}{\partial \pi_i}\frac{\partial g}{\partial x^i}\right)$, and therefore

$$\{x^i,\ \pi_j\}_{PB} = \delta^i_j,\ \ \{x^i,\ x^j\}_{PB} = 0,\ \ \{\pi_i,\ \pi_j\}_{PB} = 0. \qquad (2.32)$$

Combining Eq.(2.28) with Eq.(2.29), the covariant canonic four-momentum is

$$\pi_\mu \equiv (\pi_0,\ \pi_i) = \left(-\frac{H}{c},\ \pi_i\right) = \frac{-m}{\sqrt{1 - \frac{v^2}{c^2}}}\eta_{\mu\nu}\dot{x}^\nu$$

$$= -mc\eta_{\mu\nu}\frac{dx^\nu}{ds}, \qquad (2.33)$$

where $ds = \sqrt{\eta_{\mu\nu}dx^\mu dx^\nu} = cdt\sqrt{1 - v^2/c^2}$ were used. Equation (2.33) means that

$$\eta^{\mu\nu}\pi_\mu\pi_\nu = m^2c^2, \qquad (2.34)$$

which is called dispersion relation of E-SR.

Generally, for arbitrary function $\mathcal{J}(t, x, \pi)$, one has

$$\dot{\mathcal{J}}(t, x, \pi) \equiv \frac{d\mathcal{J}}{dt} = \frac{\partial \mathcal{J}}{\partial t} + \{H,\ \mathcal{J}\}_{PB}, \qquad (2.35)$$

where the canonical equations (2.30) and (2.31) were used.

2.5 Noether's theorem and the motion integrals

Various laws of conservation (of momentum, angular momentum, etc.) are particular cases of one general theorem: to every one-parameter group of Lagrangian-preserving diffeomorphisms of the configuration manifold of a Lagrangian system, there corresponds a first integral of the equations of motion. That theorem is called Noether's theorem . For the sake of generality, we use notation $\mathbf{q} \equiv \{q_1, q_2, \cdots q_f\}$ to denotes various generalized coordinates (including Cartesian coordinates x^i, $i = 1, 2, 3$ in SR) $\dot{\mathbf{q}}$ is the generalized velocities in this section.

Noether's theorem: Consider a system whose dynamics at a given instant of time t is described by a Lagrangian function $L(t, \mathbf{q}, \dot{\mathbf{q}})$. The dynamical behavior of the system is determined by the Lagrangian equation of motion. Suppose there is a symmetry group of transformations

$$t \to T, \tag{2.36}$$

$$\mathbf{q} \to \mathbf{Q}, \tag{2.37}$$

which leave the action $S \equiv \int L(t, \mathbf{q}, \dot{\mathbf{q}}) dt$ invariant. Namely,

$$\int L(t, \mathbf{q}, \dot{\mathbf{q}}) dt = \int L(T, \mathbf{Q}, \dot{\mathbf{Q}}) dT. \tag{2.38}$$

The theorem states that this invariance will lead to a constant of motion, which is called also the Noether Charge.

More precisely, under transformation

$$T = T(t, \mathbf{q}, \dot{\mathbf{q}}, \epsilon), \tag{2.39}$$

$$\mathbf{Q} = \mathbf{Q}(t, \mathbf{q}, \dot{\mathbf{q}}, \epsilon), \tag{2.40}$$

where ϵ is a time-independent infinitesimal parameter, and

$$(T)_{\epsilon=0} = t, \tag{2.41}$$

$$(\mathbf{Q})_{\epsilon=0} = \mathbf{q}, \tag{2.42}$$

then function $\dot{\mathbf{Q}}$ in Eq.(2.38) is:

$$\dot{\mathbf{Q}}(t, \mathbf{q}, \dot{\mathbf{q}}, \ddot{\mathbf{q}}, \epsilon) \equiv \frac{d\mathbf{Q}}{dT} = \frac{d\mathbf{Q}/dt}{dT/dt} = \frac{\dot{\mathbf{Q}}}{\dot{T}} = \frac{\dot{\mathbf{Q}}(t, \mathbf{q}, \dot{\mathbf{q}}, \ddot{\mathbf{q}}, \epsilon)}{\dot{T}(t, \mathbf{q}, \dot{\mathbf{q}}, \ddot{\mathbf{q}}, \epsilon)}. \tag{2.43}$$

Rewriting Eq.(2.38) as follows

$$\int [L(T, \mathbf{Q}, \dot{\mathbf{Q}})\dot{T} - L(t, \mathbf{q}, \dot{\mathbf{q}})] dt = 0, \tag{2.44}$$

and noting

$$L(T, \mathbf{Q}, \acute{\mathbf{Q}})\dot{T} - L(t, \mathbf{q}, \dot{\mathbf{q}}) = \left[L(t, \mathbf{q}, \dot{\mathbf{q}})\dot{t} + \epsilon \left(\frac{\partial L(T, \mathbf{Q}, \acute{\mathbf{Q}})\dot{T}}{\partial \epsilon} \right)_{\epsilon=0} \right.$$

$$\left. - L(t, \mathbf{q}, \dot{\mathbf{q}}) - \epsilon \frac{dF}{dt} \right.$$

$$= 0, \tag{2.45}$$

we obtain the following Noether theorem condition

$$\left[\frac{\partial}{\partial \epsilon} \left\{ L\Big(T(t, \mathbf{q}, \dot{\mathbf{q}}, \epsilon), \mathbf{Q}(t, \mathbf{q}, \dot{\mathbf{q}}, \epsilon), \acute{\mathbf{Q}}(t, \mathbf{q}, \dot{\mathbf{q}}, \ddot{\mathbf{q}}, \epsilon) \Big) \dot{T}(t, \mathbf{q}, \dot{\mathbf{q}}, \ddot{\mathbf{q}}, \epsilon) \right\} \right]_{\epsilon=0} = \dot{F}, \tag{2.46}$$

where $\dot{F}$ is the total time derivative of some function $F(t, \mathbf{q}, \dot{\mathbf{q}})$. Therefore the quantity

$$L\xi + \sum_i \frac{\partial L(t, \mathbf{q}, \dot{\mathbf{q}})}{\partial \dot{q}_i} (\eta_i - \dot{q}_i \xi) - F \tag{2.47}$$

where

$$\xi = \left[\frac{\partial T = T(t, \mathbf{q}, \dot{\mathbf{q}}, \epsilon)}{\partial \epsilon} \right]_{\epsilon=0}, \tag{2.48}$$

$$\eta_i = \left[\frac{\partial Q_i(t, \mathbf{q}, \dot{\mathbf{q}}, \epsilon)}{\partial \epsilon} \right]_{\epsilon=0}, \tag{2.49}$$

is a constant of the motion.

Proof: Carrying out the derivative in Eq.(2.46) we obtain

$$\left\{ \left[\sum_i \frac{\partial L(T, \mathbf{Q}, \acute{\mathbf{Q}})}{\partial Q_i} \frac{\partial Q_i(t, \mathbf{q}, \dot{\mathbf{q}}, \epsilon)}{\partial \epsilon} + \sum_i \frac{\partial L(T, \mathbf{Q}, \acute{\mathbf{Q}})}{\partial \acute{Q}_i} \frac{\partial \acute{Q}_i(t, \mathbf{q}, \dot{\mathbf{q}}, \ddot{\mathbf{q}}, \epsilon)}{\partial \epsilon} \right. \right.$$

$$\left. + \frac{\partial L(T, \mathbf{Q}, \acute{\mathbf{Q}})}{\partial T} \frac{\partial T(t, \mathbf{q}, \dot{\mathbf{q}}, \epsilon)}{\partial \epsilon} \right] \dot{T} + L(T, \mathbf{Q}, \acute{\mathbf{Q}}) \frac{\partial T(t, \mathbf{q}, \dot{\mathbf{q}}, \ddot{\mathbf{q}}, \epsilon)}{\partial \epsilon} \right\}_{\epsilon=0} = \dot{F} . \tag{2.50}$$

To evaluate the terms in Eq.(2.50), we first expand T and Q_i in powers of ϵ, obtaining

$$T = (T)_{\epsilon=0} + \left[\frac{\partial T(t, \mathbf{q}, \dot{\mathbf{q}}, \epsilon)}{\partial \epsilon} \right]_{\epsilon=0} \epsilon + \cdots, \tag{2.51}$$

$$Q_i = (Q_i)_{\epsilon=0} + \left[\frac{\partial Q_i(t, \mathbf{q}, \dot{\mathbf{q}}, \epsilon)}{\partial \epsilon} \right]_{\epsilon=0} \epsilon + \cdots, \tag{2.52}$$

Substituting Eqs.(2.41), (2.42), (2.48) and (2.49) in Eqs.(2.40) and (2.41) we obtain

$$T = t + \xi\epsilon + \cdots, \tag{2.53}$$

$$Q_i = q_i + \eta_i\epsilon + \cdots. \tag{2.54}$$

From Eq.(2.53) and Eq.(2.54) we obtain

$$\dot{T} = 1 + \dot{\xi}\epsilon + \cdots, \tag{2.55}$$

$$\dot{Q}_i = \dot{q}_i + \dot{\eta}_i\epsilon + \cdots. \tag{2.56}$$

From Eqs.(2.55) and (2.56) it follows that

$$(\dot{T})_{\epsilon=0} = 1, \quad (\dot{Q}_i)_{\epsilon=0} = \dot{q}, \quad (\acute{Q}_i) = \left(\frac{\dot{Q}_i}{\dot{T}}\right)_{\epsilon=0} = \dot{q}_i,$$

$$\left[\frac{\partial T(t,\mathbf{q},\dot{\mathbf{q}},\epsilon)}{\partial\epsilon}\right]_{\epsilon=0} = \xi,$$

$$\left[\frac{\partial Q_i(t,\mathbf{q},\dot{\mathbf{q}},\epsilon)}{\partial\epsilon}\right]_{\epsilon=0} = \eta_i, \quad \left[\frac{\partial \dot{T}(t,\mathbf{q},\dot{\mathbf{q}},\epsilon)}{\partial\epsilon}\right]_{\epsilon=0} = \dot{\xi},$$

$$\left[\frac{\partial \dot{Q}_i(t,\mathbf{q},\dot{\mathbf{q}},\epsilon)}{\partial\epsilon}\right]_{\epsilon=0} = \dot{\eta}_i,$$

$$\left[\frac{\partial \acute{Q}_i(t,\mathbf{q},\dot{\mathbf{q}},\ddot{\mathbf{q}},\epsilon)}{\partial\epsilon}\right]_{\epsilon=0} = \left\{\frac{\partial}{\partial\epsilon}\left[\frac{\dot{Q}_i(t,\mathbf{q},\dot{\mathbf{q}},\ddot{\mathbf{q}},\epsilon)}{\dot{T}(t,\mathbf{q},\dot{\mathbf{q}},\ddot{\mathbf{q}},\epsilon)}\right]\right\}_{\epsilon=0}$$

$$= \left[\frac{1}{\dot{T}}\frac{\partial \dot{Q}_i(t,\mathbf{q},\dot{\mathbf{q}},\epsilon)}{\partial\epsilon} - \frac{\dot{Q}_i}{\dot{T}^2}\frac{\partial \dot{T}(t,\mathbf{q},\dot{\mathbf{q}},\epsilon)}{\partial\epsilon}\right]_{\epsilon=0}$$

$$= \dot{\eta}_i - \dot{q}_i\dot{\xi}. \tag{2.57}$$

Using Eqs.(2.42), (2.43), (2.57) in Eq.(2.50) and noting also that

$$\left[\frac{\partial L(T,\mathbf{Q},\acute{\mathbf{Q}})}{\partial Q_i}\right]_{\epsilon=0} = \frac{\partial L(t,\mathbf{q},\dot{\mathbf{q}})}{\partial q_i} \tag{2.58}$$

$$\left[\frac{\partial L(T,\mathbf{Q},\acute{\mathbf{Q}})}{\partial T}\right]_{\epsilon=0} = \frac{\partial L(t,\mathbf{q},\dot{\mathbf{q}})}{\partial t} \tag{2.59}$$

$$\left[L(T,\mathbf{Q},\acute{\mathbf{Q}})\right]_{\epsilon=0} = L(t,\mathbf{q},\dot{\mathbf{q}}), \tag{2.60}$$

we obtain

$$\sum_i \frac{\partial L(t, \mathbf{q}, \dot{\mathbf{q}})}{\partial q_i} \eta_i + \sum_i \frac{\partial L(t, \mathbf{q}, \dot{\mathbf{q}})}{\partial \dot{q}_i} (\dot{\eta}_i - \dot{q}_i \dot{\xi}) + \frac{\partial L(t, \mathbf{q}, \dot{\mathbf{q}})}{\partial t} \xi + L(t, \mathbf{q}, \dot{\mathbf{q}}) \dot{\xi} = \dot{F}.$$

(2.61)

In the remainder of the proof we suppress the argument of functions appearing in a partial derivative, since in all cases the argument would be $(t, \mathbf{q}, \dot{\mathbf{q}})$ and there is very little chance of confusion. We now note

$$\frac{\partial L}{\partial t} = \dot{L} - \sum_i \frac{\partial L}{\partial q_i} \dot{q}_i - \sum_i \frac{\partial L}{\partial \dot{q}_i} \ddot{q}_i$$

(2.62)

$$\frac{\partial L}{\partial \dot{q}_i} \dot{\eta}_i = \frac{d}{dt}\left(\frac{\partial L}{\partial q_i} \eta_i\right) - \frac{d}{dt}\left(\frac{\partial L}{\partial \dot{q}_i}\right) \eta_i$$

(2.63)

$$\frac{\partial L}{\partial \dot{q}_i} \dot{q}_i \dot{\xi} = \frac{d}{dt}\left(\frac{\partial L}{\partial \dot{q}_i} \dot{q}_i \xi\right) - \frac{d}{dt}\left(\frac{\partial L}{\partial \dot{q}_i}\right) \dot{q}_i \xi - \frac{\partial L}{\partial \dot{q}_i} \ddot{q}_i \xi$$

(2.64)

Substituting Eqs.(2.62)−(2.64) in Eq.(2.61) we obtain

$$\sum_i \frac{\partial L}{\partial q_i} \eta_i + \sum_i \left[\frac{d}{dt}\left(\frac{\partial L}{\partial q_i} \eta_i\right) - \frac{d}{dt}\left(\frac{\partial L}{\partial \dot{q}_i}\right) \eta_i\right]$$
$$- \sum_i \left[\frac{d}{dt}\left(\frac{\partial L}{\partial \dot{q}_i} \dot{q}_i \xi\right) - \frac{d}{dt}\left(\frac{\partial L}{\partial \dot{q}_i}\right) \dot{q}_i \xi - \frac{\partial L}{\partial \dot{q}_i} \ddot{q}_i \xi\right]$$
$$+ \left[L - \sum_i \left(\frac{\partial L}{\partial q_i} \dot{q}_i - \frac{\partial L}{\partial \dot{q}_i} \ddot{q}_i\right)\right] \xi + L\dot{\xi} = \dot{F}$$

(2.65)

which can be simplified to

$$\sum_i \left[\frac{\partial L}{\partial q_i} - \frac{d}{dt}\left(\frac{\partial L}{\partial \dot{q}_i}\right)\right] [\eta_i - \dot{q}_i \xi] + \frac{d}{dt}\left[L\xi + \sum_i \frac{\partial L}{\partial \dot{q}_i} (\eta_i - \dot{q}_i \xi) - F\right] = 0.$$

(2.66)

From Lagrangian equations

$$\frac{\partial L}{\partial q_i} - \frac{d}{dt}\left(\frac{\partial L}{\partial \dot{q}_i}\right) = 0.$$

(2.67)

we obtain from Eq.(2.66)

$$\frac{d}{dt}\left[L\xi + \sum_i \frac{\partial L}{\partial \dot{q}_i} (\eta_i - \dot{q}_i \xi) - F\right] = 0.$$

(2.68)

It follows that the quantity in square brackets in Eq.(2.68) is a constant of the motion, which is what we seek.

2.6   Noether charges of E-SR mechanics

Now let's come back to E-SR with Lagrangian function of Eq.(2.24):

$$L = -mc^2\sqrt{1 + \frac{\eta_{ij}\dot{x}^i\dot{x}^j}{c^2}}. \qquad (2.69)$$

In the E-SR formulation within an inertial system K, the generalized co-ordinates are $\{q_1, q_2, q_3\} = \{x^1, x^2, x^3\}$ and time is t. In another inertial system K', coordinates and time are $\{Q_1, Q_2, Q_3\} = \{x'^1, x'^2, x'^3\}$ and time $T = t'$. the transformations (2.36) (2.37) preserving the action are replaced by Poincaré group transformations (see Eqs.(2.4) and (2.14)):

$$x^\mu \to x'^\mu = M^\mu{}_\nu x^\nu + a^\mu, \qquad (2.70)$$

Or more explicitly in the case without rotations in space:

$$t \to t' = \gamma(t - \beta x^1/c) + a^0/c, \qquad (2.71)$$

$$x^1 \to x'^1 = \gamma(x^1 - \beta ct) + a^1, \qquad (2.72)$$

$$x^2 \to x'^2 = x^2 + a^2, \qquad (2.73)$$

$$x^3 \to x'^3 = x^3 + a^3. \qquad (2.74)$$

The Noether theorem condition Eq.(2.46) becomes

$$\left[\frac{\partial}{\partial\epsilon}\left\{L\left(t'(t,\mathbf{x},\dot{\mathbf{x}},\epsilon), \mathbf{x}'(t,\mathbf{x},\dot{\mathbf{x}},\epsilon), \dot{\mathbf{x}}'(t,\mathbf{x},\dot{\mathbf{x}},\ddot{\mathbf{x}},\epsilon)\right)\dot{t}'(t,\mathbf{x},\dot{\mathbf{x}},\ddot{\mathbf{x}},\epsilon)\right\}\right]_{\epsilon=0} = \dot{F},$$
$$(2.75)$$

and the corresponding Noether charge Eq.(2.47) becomes:

$$G \equiv -L\xi - \sum_i \frac{\partial L(t,\mathbf{x},\dot{\mathbf{x}})}{\partial\dot{x}^i}(\eta^i - \dot{x}^i\xi) + F \qquad (2.76)$$

where

$$\xi = \left[\frac{\partial t'(t,\mathbf{x},\dot{\mathbf{x}},\epsilon)}{\partial\epsilon}\right]_{\epsilon=0}, \qquad (2.77)$$

$$\eta^i = \left[\frac{\partial x'^i(t,\mathbf{x},\dot{\mathbf{x}},\epsilon)}{\partial\epsilon}\right]_{\epsilon=0}. \qquad (2.78)$$

When $[\frac{\partial}{\partial\epsilon}L(\epsilon)\dot{t}'(\epsilon)]_{\epsilon=0} = 0$, from Eq.(2.75) we have $\dot{F} = 0$. In this case, we take the specific solution $F = 0$ to be available for all Noether charges of mechanics (2.76) in order to obtain an uniform Noether charge's expression. Namely,

$$\dot{F} = 0, \Rightarrow F = 0. \qquad (2.79)$$

This is an additional convention for the uniform and the simplest expressions of Noether charges (2.76).

All Noether charges for E-SR are enumerated as follows:

(1) Energy: The Lagrangian function (2.69) is time-independent. Choosing $\beta = a^1 = a^2 = a^3 = 0$, $a^0 = \epsilon c$ in Eqs.(2.71)–(2.74), we have

$$t' = t + \epsilon, \quad \mathbf{x}' = \mathbf{x}, \quad \text{and then} \quad \frac{\partial}{\partial \epsilon}\left\{L\left(t', \mathbf{x}', \dot{\mathbf{x}}'\right) \dot{t}'\right\} = 0, \quad F = 0,$$

(2.80)

where Noether's theorem (the condition Eq.(2.75) is satisfied) and convention (2.79) were used. For this choice of t' and $\mathbf{x}'$, Eqs.(2.77)–(2.78) imply

$$\xi = 1, \quad \eta^i = 0,$$

(2.81)

Hence, from Eq.(2.76), we obtain the energy as a Noether charge of E-SR:

$$G_{a^0} = -L + \sum_i \frac{\partial L}{\partial \dot{x}^i}\dot{x}^i \equiv E = \text{constant.}$$

(2.82)

Substituting Eq.(2.69) (or (2.24)) into (2.82), we get the energy:

$$E = \frac{mc^2}{\sqrt{1 - \frac{v^2}{c^2}}}.$$

(2.83)

Comparing Eq.(2.83) with Eq.(2.29), we have

$$E = H.$$

(2.84)

This means that the energy Noether charge is equal to the canonic energy (or Hamiltonian) in E-SR.

(2) Momentum: We let $\beta = a^0 = a^2 = a^3 = 0$, $a^1 = \epsilon$ in Eqs.(2.71)–(2.74). Then we have

$$t' = t, \quad x'^1 = x^1 + \epsilon, \quad x'^2 = x^2, \quad x'^3 = x^3, \quad F = 0 \quad (2.85)$$

in Noether's theorem. The condition Eq.(2.75) is satisfied. For this choice of t' and $\mathbf{x}'$, Eqs.(2.77) and (2.78) indicate

$$\xi = 0, \quad \eta^i = \delta^i_1,$$

(2.86)

Hence, from Eq.(2.76), we obtain the momentum Noether charge of E-SR:

$$G_{a^1} = \frac{\partial L}{\partial \dot{x}^1} \equiv p^1 = \text{constant.}$$

(2.87)

Similarly, $G_{a^i} \equiv p^i =$ constant to $i = 2$, 3. Therefore, we have

$$p^i = \frac{m\dot{x}^i}{\sqrt{1 - \frac{v^2}{c^2}}} = \pi_i = \text{constant.} \tag{2.88}$$

(3) Lorentz boost: Taking $\beta = \epsilon$, $a^0 = a^1 = a^2 = a^3 = 0$ in Eqs.(2.71)–(2.74), then we have

$$t' = t - \frac{\epsilon x^1}{c}, \quad (\text{note} : \gamma \equiv 1/\sqrt{1 - \beta^2} \simeq 1 + \mathcal{O}(\epsilon^2))$$

$$x'^1 = x^1 - \epsilon ct, \quad x'^2 = x^2, \quad x'^3 = x^3, \tag{2.89}$$

Firstly we check the condition Eq.(2.75) and determine the function F. From Eq.(2.89) we have

$$\acute{x}'^i \equiv \frac{d\dot{x}'^i/dt}{dt'/dt} = \frac{\dot{x}^i - c\epsilon\delta_1^i}{1 - \frac{\epsilon}{c}\dot{x}^1},$$

and $$L(t', \mathbf{x}', \acute{\mathbf{x}}') = -mc^2\sqrt{1 + \frac{\eta_{ij}\acute{x}'^i\acute{x}'^j}{c^2}}$$

$$\simeq -mc^2\sqrt{1 - \frac{\dot{\mathbf{x}}^2}{c^2}}\left(1 + \frac{\dot{x}^1}{c}\epsilon + \mathcal{O}(\epsilon^2)\right). \tag{2.90}$$

Noting $\dot{t}' = 1 - \frac{\dot{x}^1}{c}\epsilon$, we have

$$L(t', \mathbf{x}', \acute{\mathbf{x}}')\dot{t}' = -mc^2\sqrt{1 - \frac{\dot{\mathbf{x}}^2}{c^2}}\left(1 + \frac{\dot{x}^1}{c}\epsilon\right)\left(1 - \frac{\dot{x}^1}{c}\epsilon\right) + \mathcal{O}(\epsilon^2)$$

$$= -mc^2\sqrt{1 - \frac{\dot{\mathbf{x}}^2}{c^2}} + \mathcal{O}(\epsilon^2),$$

and then $$\left[\frac{\partial}{\partial\epsilon}\{L(t', \mathbf{x}', \acute{\mathbf{x}}')\dot{t}'\}\right]_{\epsilon=0} = \left[\frac{\partial}{\partial\epsilon}\{\mathcal{O}(\epsilon^2)\}\right]_{\epsilon=0} = 0. \tag{2.91}$$

Hence, when $F = 0$, the condition of Eq.(2.75) is satisfied. Next, from the choice of Eq.(2.89), we have

$$\xi = \frac{\partial t'}{\partial\epsilon} = -\frac{x^1}{c}, \quad \eta^1 = \frac{\partial x'^1}{\partial\epsilon} = -ct, \quad \eta^2 = \eta^3 = \dot{x}^2 = \dot{x}^3 = 0. \tag{2.92}$$

And from Eqs.(2.76) and (2.69), we obtain the boost charge:

$$G_{\beta^1} \equiv K^1 = -L\xi - \sum_i \frac{\partial L(t, \mathbf{x}, \dot{\mathbf{x}})}{\partial\dot{x}^i}(\eta^i - \dot{x}^i\xi) + F$$

$$= -\frac{mc}{\sqrt{1 - \beta^2}}(x^1 - t\dot{x}^1), \tag{2.93}$$

where $\beta^2 = (\beta^{(1)})^2 = (\dot{x}^1)^2$. Similarly, one can repeat above calculations for the cases of $\{\beta^{(2)} \neq 0, \ \beta^{(1)} = \beta^{(3)} = 0\}$ (or $\{\beta^{(3)} \neq 0, \ \beta^{(1)} = \beta^{(2)} = 0\}$) and obtains the generic expression for the Lorentz boost Noether charge:

$$K^i = -\frac{mc}{\sqrt{1 - \beta^2}}(x^i - t\dot{x}^i), \quad \text{with } i = 1, 2, 3. \qquad (2.94)$$

(4) Angular momentum: Going back to transformation Eq.(2.70), we see that the Lagrangian (2.69) is invariant under an arbitrary rotation about some axis, which for convenience we choose to be x^3 axis. Then condition (2.75) in Noether theorem will be satisfied if we replace x^i, $\dot{x}^i$ and t in the Lagrangian by

$$t' = t \qquad (2.95)$$

$$x'^1 = x^1 \cos\epsilon - x^2 \sin\epsilon \qquad (2.96)$$

$$x'^2 = x^1 \sin\epsilon + x^2 \cos\epsilon \qquad (2.97)$$

$$x'^3 = x^3 \qquad (2.98)$$

and let $F = 0$ (see convention (2.79)). For this choice

$$\xi = 0 \qquad (2.99)$$

$$\eta^1 = -x^2 \qquad (2.100)$$

$$\eta^2 = x^1 \qquad (2.101)$$

$$\eta^3 = 0. \qquad (2.102)$$

Hence, from Eqs.(2.76) and (2.69), we obtain the third component of angular momentum charge:

$$G_3 \equiv L^3 = -L\xi - \sum_i \frac{\partial L(t, \mathbf{x}, \dot{\mathbf{x}})}{\partial \dot{x}^i}(\eta^i - \dot{x}^i\xi) + F$$

$$= \frac{m}{\sqrt{1 - \beta^2}}(x^1\dot{x}^2 - x^2\dot{x}^1). \qquad (2.103)$$

Similarly, one can repeat above calculations for x^1-axis case and x^2-axis case respectively, and obtains the generic expression for L^i:

$$L^i = \frac{m}{\sqrt{1 - \beta^2}}\epsilon^i{}_{jk}x^j\dot{x}^k, \quad \text{with } i = 1, 2, 3, \text{ and } \epsilon^{123} = +1.$$

$$(2.104)$$

2.7   Quantization of E-SR mechanics

In this section, we discuss the procedure of passing from classical E-SR mechanics to quantum E-SR mechanics. This is called *"quantization"* of E-SR theory.

Lagrangian-Hamiltonian formulation of mechanics (see Section 2.4) is the foundation of quantization. When the classical Poisson Brackets in canonical equations for canonical coordinates and canonical momentum become operator's commutators:

$$\{x, \pi\}_{PB} \Rightarrow \frac{1}{i\hbar}[x, \hat{\pi}], \tag{2.105}$$

the classical mechanics will be quantized. Under (2.105), the Poisson Bracket relations of Eq.(2.32) are passed to the canonical commutation relations:

$$[\hat{\pi}_i, \ x^j] = -i\hbar\delta_i^j, \quad [x^i, \ x^j] = [\hat{\pi}_i, \ \hat{\pi}_j] = 0, \tag{2.106}$$

where x^i, $\hat{\pi}_i$ with $i = 1, 2, 3$ are canonically conjugate operators. From Eqs.(2.30), (2.31) and (2.105), we get Heisenberg equations with quantum Hamilton operator $\hat{H}$:

$$\dot{x}^i = \frac{i}{\hbar}[\hat{H}, \ x^i], \quad \dot{\hat{\pi}}_i = \frac{i}{\hbar}[\hat{H}, \ \hat{\pi}_i] \tag{2.107}$$

which describe the dynamics of the quantum system. From the dispersion relation Eq.(2.34), we have E-SR wave equation in the Schrödinger picture for spinless particle as follows

$$\eta^{\mu\nu}\hat{\pi}_\mu\hat{\pi}_\nu\phi(x) = m^2c^2\phi(x), \tag{2.108}$$

where $\hat{\pi}_0 = -\frac{\hat{H}}{c}$, and $\phi(x)$ is wave function. The Hamiltonian operator $\hat{H} \equiv -c\hat{\pi}_0$ represents the generator of time evolution, i.e.,

$$[t, \ \hat{H}] = -i\hbar, \quad \text{or} \quad [x^0, \ \hat{\pi}_0] = i\hbar. \tag{2.109}$$

It is straightforward by using Eqs.(2.29) and (2.106) to check that

$$[\hat{H}, \ \hat{\pi}_i] = -c[\hat{\pi}_0, \ \hat{\pi}_i] = 0, \quad \text{or} \quad [\hat{\pi}_0, \ \hat{\pi}_i] = 0. \tag{2.110}$$

Combining Eqs.(2.106), (2.109) and (2.110), we have *the 4-dimensional invariant commutation relations* for x^μ and $\hat{\pi}_\nu$:

$$[x^\mu, \hat{\pi}_\nu] = i\hbar\delta_\nu^\mu, \quad [x^\mu, x^\nu] = 0, \quad [\hat{\pi}_\mu, \hat{\pi}_\nu] = 0. \tag{2.111}$$

The simplest solution of Eq.(2.111) is

$$\hat{\pi}_\mu = -i\hbar\partial_\mu. \tag{2.112}$$

Substituting Eq.(2.112) to Eq.(2.108), we obtain

$$\left(\Box + \frac{m^2c^2}{\hbar^2}\right)\phi(x) \equiv \left(\frac{1}{c^2}\frac{\partial^2}{\partial t^2} - \nabla^2 + \frac{m^2c^2}{\hbar^2}\right)\phi(x) = 0, \tag{2.113}$$

which is called the Klein-Gordon equation in Minkowski spacetime.

In the following, we consider wave equation of particles with spin $1/2$, which is called Dirac equation. The Lorentz transformations unite time and space into a single four-dimensional entity. The basic equation underlying relativistic quantum theory must reflect this unity, implying that there must be complete symmetry between the time and space parts of the equation. Clearly the Klein-Gordon equation (2.113) satisfies this constraint, but the Schrödinger equation does not because it is first order in the time and second order in the space derivatives. Let us assume that, like Schrödinger equation, the Dirac equation will be linear in the time derivatives. Therefore it must also be linear in space derivatives. We know from the dispersion relation Eq.(2.34) of classical E-SR that the canonical energy of a free particle $H = -c\pi_0$ is given by

$$H^2 \equiv c^2\pi_0^2 = \vec{\pi}^2c^2 + m^2c^4. \tag{2.114}$$

To obtain a linear equation we write the square root of it in a following way:

$$H = \sqrt{\vec{\pi}^2c^2 + m^2c^4} = \vec{\alpha}\cdot\vec{\pi}c + \beta mc^2. \tag{2.115}$$

Here, $\vec{\alpha}$ and β have to be such that when we square Eq.(2.115) we arrive back at Eq.(2.114). Clearly, β has to be a scalar and $\vec{\alpha}$ has to be a vector so that both terms on right hand side of this equation are scalars. Let's square this explicitly:

$$\begin{aligned}
H^2 &= (\vec{\alpha}\cdot\vec{\pi}c + \beta mc^2)(\vec{\alpha}\cdot\vec{\pi}c + \beta mc^2)\\
&= \alpha_1^2\pi_1^2c^2 + \alpha_2^2\pi_2^2c^2 + \alpha_3^2\pi_3^2c^2 + \vec{\alpha}\cdot\vec{\pi}\beta mc^3 + \beta\vec{\alpha}\cdot\vec{\pi}mc^3 + \beta^2m^2c^4\\
&\quad + c^2((\alpha_1\alpha_2 + \alpha_2\alpha_1)\pi_1\pi_2 + (\alpha_2\alpha_3 + \alpha_3\alpha_2)\pi_2\pi_3 + (\alpha_3\alpha_1 + \alpha_1\alpha_3)\pi_3\pi_1).
\end{aligned} \tag{2.116}$$

For this to be consistent with Eq.(2.114) we require that

$$\alpha_1^2 = \alpha_2^2 = \alpha_3^2 = \beta^2 = \mathbf{I} \tag{2.117}$$

$$\vec{\alpha}\beta + \beta\vec{\alpha} \equiv \{\vec{\alpha}, \beta\} = 0 \tag{2.118}$$

$$\{\alpha_1, \alpha_2\} = \{\alpha_2, \alpha_3\} = \{\alpha_3, \alpha_1\} = 0 \tag{2.119}$$

where $\{\alpha_1, \alpha_2\} = \alpha_1\alpha_2 + \alpha_2\alpha_1$, and etc. No existing numbers can satisfy Eqs.(2.117), (2.118) and (2.119) simultaneously. However, it turns out that there are matrices that do so. Combining such matrices solutions of $\{\vec{\alpha}, \beta\}$ and the operator expressions of π_μ Eq.(2.112) and substituting them to (2.115), we obtain the Dirac equation as follows

$$i\hbar\frac{\partial}{\partial t}\psi(\mathbf{r}, t) = c(-i\hbar\boldsymbol{\alpha}\cdot\nabla + \beta mc)\psi(\mathbf{r}, t) \tag{2.120}$$

where $\vec{\alpha}$, β are 4×4-matrices and $\psi(\mathbf{r}, t)$ is spinor wave function with 4 components. Redefining Dirac matrices as follows

$$\gamma^i = \beta\alpha_i = \begin{pmatrix} 0 & \sigma^i \\ -\sigma^i & 0 \end{pmatrix} \tag{2.121}$$

$$\gamma^0 = \beta = \begin{pmatrix} 0 & 1 \\ 1 & 0 \end{pmatrix} \tag{2.122}$$

where

$$\sigma^1 = \begin{pmatrix} 0 & 1 \\ 1 & 0 \end{pmatrix}, \quad \sigma^2 = \begin{pmatrix} 0 & -i \\ i & 0 \end{pmatrix}, \quad \sigma^3 = \begin{pmatrix} 1 & 0 \\ 0 & -1 \end{pmatrix},$$

we have explicit Lorentz invariant expression of Dirac equation (2.120):

$$(i\hbar\gamma^\mu\partial_\mu - mc)\psi(\mathbf{r}, t) = 0. \tag{2.123}$$

2.8 Localization of E-SR spacetime symmetry and Einstein's General Relativity (I)

In this section and next, we turn to discuss the localization of E-SR spacetime symmetry, and to show that Einstein's General Relativity can be achieved in this way. For this aim, conceptions of gauge fields, gauge $ISO(3,1)$ symmetry and Riemann geometry technique will be employed. We hence interpret them firstly in this section.

Conceptions of gauge fields: To interpret the main idea of gauge theory, we consider the $U(1)$-gauge field theory in Minkowski spacetime, in which the electromagnetic fields are the $U(1)$-gauge fields. To see this point, let us consider the charged scalar fields described by following Lagrange density of complex scalar fields:

$$\mathcal{L}_0(x) = \eta^{\mu\nu}\hbar^2\partial_\mu\phi^*(x)\partial_\nu\phi(x) - m^2c^2\phi^*(x)\phi(x). \qquad (2.124)$$

The action is

$$S_0 \equiv \int d^4x \mathcal{L}_0(x) = \int d^4x \mathcal{L}_0(\phi, \partial_\mu\phi, \phi^*, \partial_\mu\phi^*)(x). \qquad (2.125)$$

From variations of this action with respect to $\phi(x)$ and $\phi^*(x)$ respectively, the equations of motion can be derived:

$$\partial_\mu \frac{\partial\mathcal{L}_0}{\partial(\partial_\mu\phi(x))} - \frac{\partial\mathcal{L}_0}{\partial\phi(x)} = 0, \quad \Rightarrow \left(\partial^2 - \frac{m^2c^2}{\hbar^2}\right)\phi^*(x) = 0, \quad (2.126)$$

$$\partial_\mu \frac{\partial\mathcal{L}_0}{\partial(\partial_\mu\phi^*(x))} - \frac{\partial\mathcal{L}_0}{\partial\phi^*(x)} = 0, \quad \Rightarrow \left(\partial^2 - \frac{m^2c^2}{\hbar^2}\right)\phi(x) = 0. \quad (2.127)$$

Clearly, $\mathcal{L}_0(x)$ is invariant under phase transformation

$$\phi(x) \to e^{i\lambda}\phi(x)|_{\lambda\to\epsilon} = \phi(x) + i\lambda\phi(x), \qquad (2.128)$$

$$\phi^*(x) \to e^{-i\lambda}\phi^*(x)|_{\lambda\to\epsilon} = \phi^*(x) - i\lambda\phi^*(x), \qquad (2.129)$$

where λ is a spacetime independent constant. Therefore we have

$$
\begin{aligned}
0 = \frac{\partial\mathcal{L}_0}{\partial\lambda} &= \left(\frac{\partial\mathcal{L}_0}{\partial\phi(x)}i\phi + \frac{\partial\mathcal{L}_0}{\partial(\partial_\mu\phi(x))}i\partial_\mu\phi\right) \\
&\quad - \left(\frac{\partial\mathcal{L}_0}{\partial\phi^*(x)}i\phi^* + \frac{\partial\mathcal{L}_0}{\partial(\partial_\mu\phi^*(x))}i\partial_\mu\phi^*\right) \\
&= \left[\left(\partial_\mu\frac{\partial\mathcal{L}_0}{\partial(\partial_\mu\phi(x))}\right)i\phi + \frac{\partial\mathcal{L}_0}{\partial(\partial_\mu\phi(x))}i\partial_\mu\phi\right] \\
&\quad - \left[\left(\partial_\mu\frac{\partial\mathcal{L}_0}{\partial(\partial_\mu\phi^*(x))}\right)i\phi^* + \frac{\partial\mathcal{L}_0}{\partial(\partial_\mu\phi^*(x))}i\partial_\mu\phi^*\right] \\
&= \partial_\mu J^\mu(x), \qquad (2.130)
\end{aligned}
$$

where the equations of motion, Eqs.(2.128) and (2.129), have been used, and

$$J^\mu(x) = i\hbar^2((\partial^\mu\phi^*)\phi - (\partial^\mu\phi)\phi^*). \qquad (2.131)$$

Physically, conserved current $J^\mu(x)$ is the electric current density of ϕ-field, and $\partial_\mu J^\mu(x) = 0$ (Eq.(2.130)) means the electric charge of ϕ-particle is conserved. This result can also be thought the consequence of Noether theorem in fields theory due to the global phase symmetry of ϕ-fields in $\mathcal{L}_0$ (2.124). The Noether charge is

$$Q = \int d^3 J^0 \equiv e. \tag{2.132}$$

A further interesting question arises here is what fields are excited by this charge? In other words, besides conserved Noether charges due to the symmetry, could one further construct a field-theoretic dynamics related to invariant properties of $\mathcal{L}_0$? Actually, it has long been realized that the existence of the electromagnetic fields can be related to such invariant properties of the Lagrangian. Namely, the electromagnetic fields are *U(1)-gauge fields.*

To understand this point, let us consider the more general gauge transformations of the forms Eqs.(2.128) and (2.129), but in which the parameter λ becomes arbitrary functions of position, i.e., λ is localized to be $\lambda(x)$. In this case, we have

$$\partial_\mu \phi(x) \to \partial_\mu \phi'(x) = \partial_\mu \left(e^{i\lambda(x)} \phi(x) \right) |_{\lambda \to \epsilon}$$
$$= (1 + i\lambda(x)) \partial_\mu \phi(x) + i\phi(x) \partial_\mu \lambda(x), \tag{2.133}$$

$$\partial_\mu \phi^*(x) \to \partial_\mu \phi'^*(x) = \partial_\mu \left(e^{-i\lambda(x)} \phi^*(x) \right) |_{\lambda \to \epsilon}$$
$$= (1 - i\lambda(x)) \partial_\mu \phi^*(x) - i\phi^*(x) \partial_\mu \lambda(x), \tag{2.134}$$

and the first term of $\mathcal{L}_0(x)$ (see Eq.(2.124)) is no longer invariant, because

$$\eta^{\mu\nu} \hbar^2 \partial_\mu \phi^*(x) \partial_\nu \phi(x)$$
$$\to \eta^{\mu\nu} \hbar^2 \partial_\mu \phi^*(x) \partial_\nu \phi(x) + i\eta^{\mu\nu} \hbar^2 [(\partial_\mu \phi^*(x)) \phi(x)(\partial_\nu \lambda(x))$$
$$- \phi^*(x)(\partial_\nu \phi(x))(\partial_\mu \lambda(x))]. \tag{2.135}$$

Thus, we find out that under localized phase transformations $\mathcal{L}(x)$ is no longer invariant. Namely, when the phase $\lambda \to \lambda(x)$, we find $\delta \mathcal{L}_0(x) = \delta \mathcal{L}_0(\phi, \partial\phi; \phi^*, \partial\phi^*) \neq 0$. We wish to make the Lagrange invariant under the general gauge transformations for which λ is a function of x, then it is necessary to introduce a new field A_μ (we shall call it $U(1)$-gauge field) which transforms according to

$$A_\mu \to A_\mu + \frac{1}{\tilde{e}} \partial_\mu \lambda, \tag{2.136}$$

where $\tilde{e}$ (note, $\tilde{e} = e/(\hbar c)$ in *Gaussian system of units*) is a constant, and to replace $\partial_\mu \phi$, $\partial_\mu \phi^*$ by "covariant derivatives"

$$D_\mu \phi \equiv (\partial_\mu - i\tilde{e}A_\mu)\phi, \qquad D_\mu \phi^* \equiv (\partial_\mu + i\tilde{e}A_\mu)\phi^*. \qquad (2.137)$$

Using Eq.(2.136) and $\phi(x) \to \phi'(x) = \exp(i\lambda(x))\phi(x)$ we can check that

$$D_\mu \phi \to D'_\mu \phi' = e^{i\lambda(x)}\phi, \qquad (2.138)$$

$$D_\mu \phi^* \to D'_\mu \phi'^* = e^{-i\lambda(x)}\phi^*. \qquad (2.139)$$

Thus, when $\mathcal{L}_0(x) = \mathcal{L}_0(\phi, \partial\phi; \phi^*, \partial\phi^*) \to \mathcal{L}_D(x) = \mathcal{L}_D(\phi, D\phi; \phi^*, D\phi^*)$, we have

$$\mathcal{L}_D(x) = \eta^{\mu\nu}\hbar^2 D_\mu \phi^*(x) D_\nu \phi(x) - m^2 c^2 \phi^*(x)\phi(x), \qquad (2.140)$$

which is invariant under gauge phase transformation (2.136) and $\phi(x) \to \phi'(x) = \exp(i\lambda(x))\phi(x)$.

Furthermore, from Eq.(2.137) one finds

$$(D_\mu D_\nu - D_\nu D_\mu)\phi = -i\tilde{e}F_{\mu\nu}\phi,$$

where

$$F_{\mu\nu} = \partial_\mu A_\nu - \partial_\nu A_\mu. \qquad (2.141)$$

Unlike A_μ, the expression $F_{\mu\nu}$ is a invariant quantity under gauge transformation (2.136). It satisfies the cyclic identity (i.e., *Bianchi identity*)

$$\partial_\lambda F_{\mu\nu} + \partial_\mu F_{\nu\lambda} + \partial_\nu F_{\lambda\mu} = 0. \qquad (2.142)$$

It remains to find a free Lagrangian density $\mathcal{L}_G$ for the new gauge field A_μ. Clearly $\mathcal{L}_G$ must be separately invariant, and it is easy to see that this implies it must contain A_μ only through the invariant combination $F_{\mu\nu}$. The simplest such Lagrangian density is

$$\mathcal{L}_G = -\frac{1}{4}F_{\mu\nu}F^{\mu\nu}, \qquad (2.143)$$

where the tensor indices are raised with the Lorentz metric in Minkowski spacetime $\eta_{\mu\nu}$ with diagonal elements $(1, -1, -1, -1)$. From Eqs.(2.140) and (2.143), the total Lagrangian density is

$$\mathcal{L}_{tot}(\phi, A) = \mathcal{L}_D(\phi, A) + \mathcal{L}_G(A). \qquad (2.144)$$

By means of variation $(\delta/\delta A_\mu(x)) \int d^4x \mathcal{L}_{tot}(\phi, A)$, we obtain

$$\partial_\nu F^{\mu\nu} \equiv F^{\mu\nu}{}_{,\nu} = -J^\mu, \qquad (2.145)$$

where $J^\mu = (\delta/\delta A_\mu(x)) \int d^4x \mathcal{L}_D(\phi, A)$. Because of the antisymmetry of $F^{\mu\nu}$, then we find J^μ is of conserved current:

$$\partial_\mu J^\mu = 0. \tag{2.146}$$

Thus the consequence of Noether theorem Eq.(2.130) is proved in the gauge theory. In 1954, Yang and Mills extend $U(1)$ gauge theory to non-Abelian gauge theory [Yang, C.N. (1954)]. Since then, it is realized that the Gauge Field Theory is one of the fundamental principles of modern physics. The modern particle physics build upon non-Abelian gauge field theory (i.e., Yang-Mills field theory) for appropriate internal symmetries. The unification of interactions between the particles can be realized by the gauge theory based on localization of internal symmetries . The resulting gauge dynamics is unique.

localization of $ISO(3,1)$: The spacetime symmetry of Einstein's Special Relativity (E-SR) is the global inhomogeneous Lorentz group (or Poincaré group) $ISO(3,1)$ in the Minkowski spacetime with Lorentz metric $\eta_{\mu\nu}$ (see Eq.(8.26)). The infinitesimal version of the transformation (8.26) is:

$$x^\mu \to x'^\mu = x^\mu + \delta x^\mu$$
$$\delta x^\mu = \epsilon^\mu{}_\nu x^\nu + \epsilon^\mu \tag{2.147}$$

where the group parameters $\epsilon^\mu{}_\nu$ and ϵ^μ are space-time independent constants. The corresponding ten Noethers have calculated in section §(2.6). They are E (see Eq.(2.83)), p^i (see Eq.(2.88)), K^i (see Eq.(2.94)) and L^i (see Eq.(2.104)).

The procedures of localization of global $ISO(3,1)$ are as follows

$$\epsilon^\mu{}_\nu \to \epsilon^\mu{}_\nu(x), \quad \epsilon^\mu \to \epsilon^\mu(x), \tag{2.148}$$

where $\epsilon^\mu{}_\nu(x)$, and $\epsilon^\mu(x)$ are arbitrary tensor function and vector function of x respectively. Under such localization transformations (2.148), the global transformation (2.147) becomes

$$x^\mu \to x'^\mu = x^\mu + \epsilon^\mu{}_\nu(x)x^\nu + \epsilon^\mu(x) \equiv x'^\mu(x^\nu) = x^\mu + \delta x^\mu(x), \tag{2.149}$$

where

$$\delta x^\mu(x) = \epsilon^\mu{}_\nu(x)x^\nu + \epsilon^\mu(x), \tag{2.150}$$

or

$$\delta x^\mu(x) = f^\mu(x), \tag{2.151}$$

where $f^\mu(x) \equiv \epsilon^\mu{}_\nu(x)x^\nu + \epsilon^\mu(x)$ are four arbitrary functions of x. Hence, Eqs.(2.149) with (2.151) represents a general spacetime coordinates transformation, or curvilinear coordinates transformation. Remembering the group parameter number of the global spacetime $ISO(3,1)$ is 10 (i.e., $\epsilon^\mu{}_\nu$ and ϵ^μ), however, it becomes 4 for localized spacetime $ISO(3,1)$ transformations (2.151), which is less than the global $ISO(3,1)$'s. This situation indicates that such gauge theory due to Eq.(2.151) seems to be not a full gauge theory of whole external $ISO(3,1)$ symmetry, but it should be a minimal version of that gauge theory arising from the symmetry localization. In the below, we mainly describe such minimal version theory, and assume the spacetime is torsion-free just like Einstein did in GR.

Expression of (2.148) means that the global $ISO(3,1)$ symmetry of E-SR turns into local $ISO(3,1)$ symmetry, or gauge $ISO(3,1)$ symmetry. Equation (2.149) plays a role of gauge transformation in the gauge theory. According to the principle of gauge theory, the forms of physics laws should be invariant under the gauge transformation (2.149). Thus, starting with the symmetry localization procedure of gauge theory one arrives at one of the hypotheses of GR: *the principle of general relativity*, that states that the forms of physics laws should be the same in all reference systems including both inertial and non-inertial systems. Namely, under the general coordinates transformations Eq.(2.149) the transformation properties of the expressions of physics laws have to be the same.

Riemann geometry: To analyze this issue further, it is necessary to describe the main points of Riemann geometry calculations. A brief (but enough for this book) description on them is as follows (about more detailed illustrations to the calculations, please, e.g., see [Landau, Lifshitz (1987)]):

(1) **Tensors in curvilinear coordinates frame:** From Eq.(2.148), we have

$$dx'^\mu = \frac{\partial x'^\mu}{\partial x^\nu} dx^\nu, \quad \text{when } \det\left|\frac{\partial x'^\mu}{\partial x^\nu}\right| \neq 0, \text{ or } \infty, \quad dx^\mu = \frac{\partial x^\mu}{\partial x'^\nu} dx'^\nu,$$

$$\tag{2.152}$$

and hence dx^μ is a contravariant 4-vector. Generally, the contravariant tensor $A^{\mu\nu\cdots}$ are defined via following transforming properties:

$$A'^{\mu\nu\cdots} = \frac{\partial x'^\mu}{\partial x^\alpha} \frac{\partial x'^\nu}{\partial x^\beta} \cdots A^{\alpha\beta\cdots}, \tag{2.153}$$

and the covariant tensor $A_{\mu\nu\cdots}$'s transforming property is as follows:

$$A'_{\mu\nu\cdots} = \frac{\partial x^\alpha}{\partial x'^\mu} \frac{\partial x^\beta}{\partial x'^\nu} \cdots A_{\alpha\beta\cdots}. \tag{2.154}$$

(2) **Metric tensor:** The square of the line element in curvilinear coordinates is a quadratic form in the differentials dx^μ:

$$ds^2 = g_{\mu\nu} dx^\mu dx^\nu, \tag{2.155}$$

where the $g_{\mu\nu}$ are functions of the coordinates; $g_{\mu\nu}$ is symmetric in the indices μ and ν, i.e., $g_{\mu\nu} = g_{\nu\mu}$. Since the (contracted) product of $g_{\mu\nu}$ and the contravariant tensor $dx^\mu dx^\nu$ is a scalar, the $g_{\mu\nu}$ form a covariant tensor; it called the *metric tensor*. The contravariant metric tensor is tensor $g^{\nu\lambda}$ reciprocal to tensor $g_{\mu\nu}$, that is

$$g_{\mu\nu} g^{\nu\lambda} = \delta_\mu^\lambda. \tag{2.156}$$

The connection between covariant tensor and contravariant is as follows

$$A^\mu = g^{\mu\nu} A_\nu, \qquad A_\mu = g_{\mu\nu} A^\nu. \tag{2.157}$$

(Note, notation A_μ we used hereafter in this section represents a generic covariant vector in curvilinear coordinates systems, rather than a specific electromagnetic potential vector in gauge theory.)

(3) **Completely antisymmetric unit pseudo-tensor $\epsilon^{\mu\nu\lambda\rho}$ in curvilinear coordinates frame and the element of 4-dimensional integral:** In 4-dimensional Minkowski coordinates frame x_M^μ, we use $\tilde{\epsilon}^{\mu\nu\lambda\rho}$ to denote the completely antisymmetric unit tensor of fourth rank. This is the tensor whose components change sign under interchange of any pair of indices, and whose nonzero components are ± 1. From the antisymmetry it follows that the only nonvanishing components are those for which all four indices are different. We set

$$\tilde{\epsilon}^{\,0123} = 1 \tag{2.158}$$

(hence $\tilde{\epsilon}_{0123} = -1$). Then all the other nonvanishing components $\tilde{\epsilon}^{\,\mu\nu\lambda\rho}$ are equal to+1 or -1, according as the number μ' ν, λ, ρ can be brought to the arrangement 0, 1, 2, 3 by an even or odd number of permutations. The number of such components is $4! = 24$. Thus

$$\tilde{\epsilon}^{\,\mu\nu\lambda\rho}\,\tilde{\epsilon}_{\,\mu\nu\lambda\rho} = -24. \tag{2.159}$$

Now let's consider the transformation from the Minkowski coordinates frame to the curvilinear coordinates frame. Explicitly the components transformations are as follows

$$x_M^\mu \to x^\mu = x^\mu(x_M^\nu), \tag{2.160}$$

and hence

$$\tilde{\epsilon}^{\,\mu\nu\lambda\rho} \to \epsilon^{\mu\nu\lambda\rho} = \frac{\partial x^\mu}{\partial x_M^\alpha}\frac{\partial x^\nu}{\partial x_M^\beta}\frac{\partial x^\lambda}{\partial x_M^\gamma}\frac{\partial x^\rho}{\partial x_M^\delta}\tilde{\epsilon}^{\,\alpha\beta\gamma\delta}, \tag{2.161}$$

or

$$\epsilon^{\mu\nu\lambda\rho} = J\tilde{\epsilon}^{\,\mu\nu\lambda\rho}, \tag{2.162}$$

where J is the determinant formed from the $\partial x^\mu/\partial x_M^\alpha$, i.e., it is just the Jacobian of the transformation from the Minkowski coordinates to the curvilinear coordinates:

$$J = \frac{\partial(x^0,\, x^1,\, x^2,\, x^3)}{\partial(x_M^0,\, x_M^1,\, x_M^2,\, x_M^3)}. \tag{2.163}$$

This Jacobian can be expressed in terms of the determinant of the metric tensor $g_{\mu\nu}$ in the system x^μ. The notation of $g \equiv \det(g_{\mu\nu}) = |g_{\mu\nu}|$ or $g^{-1} \equiv \det(g^{\mu\nu}) = |g^{\mu\nu}|$ will be used below. To do this we write the transformation expression of the metric tensor:

$$g^{\mu\nu} = \frac{\partial x^\mu}{\partial x_M^\alpha}\frac{\partial x^\nu}{\partial x_M^\beta}\eta^{\alpha\beta}, \tag{2.164}$$

and equate the determinants of the two sides of this equation. Noting $|g^{\mu\nu}| = 1/g$, $|\frac{\partial x^\mu}{\partial x_M^\alpha}| \cdot |\frac{\partial x^\nu}{\partial x_M^\beta}| = J^2$, $|\eta^{\alpha\beta}| = -1$, we have $1/g = -J^2$, and so $J = 1/\sqrt{-g}$. Thus, in the curvilinear coordinates the antisymmetric unit tensor of rank four must be define as

$$\epsilon^{\mu\nu\lambda\rho} = \frac{1}{\sqrt{-g}}\tilde{\epsilon}^{\,\mu\nu\lambda\rho}. \tag{2.165}$$

Noting the formula $\tilde{\epsilon}^{\,\mu\nu\lambda\rho}g_{\mu\alpha}g_{\nu\beta}g_{\lambda\gamma}g_{\rho\delta} = -g\tilde{\epsilon}_{\alpha\beta\gamma\delta}$, the covariant components of Eq.(2.165) is :

$$\epsilon_{\mu\nu\lambda\rho} = \sqrt{-g}\,\tilde{\epsilon}_{\,\mu\nu\lambda\rho}. \tag{2.166}$$

In a Minkowski coordinate system x_M^μ the integral of a scalar with respect to $d^4x_M = dx^0\, dx^1\, dx^2\, dx^3$ is also a scalar, i.e., the element d^4x_M behaves like a scalar in the integration. On transforming to curvilinear coordinates x^μ, the element of integration goes over into

$$d^4x_M \to \frac{1}{J}d^4x = \sqrt{-g}d^4x. \qquad (2.167)$$

Thus, in curvilinear coordinates, when integrating over a four-volume the quantity $\sqrt{-g}d^4x$ behaves like an invariant.

(4) **Covariant differentiation:** In curvilinear coordinates the differentials dA_μ of a vector A_μ is not a vector, and $\partial A_\mu/\partial x^\nu$ is not a tensor. It is easy to verify these statements directly. To do this we determine the transformation formulas for the differentials dA_μ in curvilinear coordinates. A covariant vector is transformed according to the formula (see Eq.(2.154))

$$A_\mu = \frac{\partial x'^\nu}{\partial x^\mu}A'_\nu; \qquad (2.168)$$

therefore

$$dA_\mu = \frac{\partial x'^\nu}{\partial x^\mu}dA'_\nu + A_\nu d\frac{\partial x'^\nu}{\partial x^\mu} = \frac{\partial x'^\nu}{\partial x^\mu}A'_\nu + A'_\nu\frac{\partial^2 x'^\nu}{\partial x^\mu \partial x^\lambda}dx^\lambda. \ (2.169)$$

Thus dA_μ does not transform like a vector (the same also applies, of course, to the differential of a contravariant vector). This result is understandable. It is learned from Eq.(2.168) that since the coefficients in the transformation formula (2.168) are functions of the coordinates, and at different points in spacetime vectors transform differently, the transformation of dA_μ which represents the difference of vectors located at different (infinitesimally separated) points of spacetime, of course, cannot behaves like a vector described by Eq.(2.168) at an exact one point of spacetime. Thus, in curvilinear coordinates, in order to obtain a differential of a vector which behaves like a vector, it is necessary that the two vectors to be subtracted from each other be located at the same point in spacetime. In other words, we must somehow "translate" one of the vectors (which are separated infinitesimally from each other) to the point where the second is located, after which we determine the difference of two vectors which now refer to one and the same point in spacetime. By means of this consideration, we define the difference DA_μ between two vectors which are now located at the same point is

$$DA_\mu = dA_\mu - \delta A_\mu, \qquad (2.170)$$

where additional change δA_μ is determined by following considerations: δA_μ comes from an infinitesimal displacement and depends on the values of the A-components themselves, where the dependence must clearly be linear. Thus δA_μ has the form

$$\delta A_\mu = \Gamma^\lambda_{\mu\nu} A_\lambda dx^\nu, \qquad (2.171)$$

where the $\Gamma^\lambda_{\mu\nu}$ are certain functions of the coordinates. Suppose the transformation formula for $\Gamma^\lambda_{\mu\nu}$ under the curvilinear coordinates transformation is:

$$\Gamma^\lambda_{\mu\nu} = \Gamma'^\rho_{\alpha\beta} \frac{\partial x^\lambda}{\partial x'^\rho} \frac{\partial x'^\alpha}{\partial x^\mu} \frac{\partial x'^\beta}{\partial x^\nu} + \frac{\partial^2 x'^\rho}{\partial x^\mu \partial x^\nu} \frac{\partial x^\lambda}{\partial x'^\rho}, \qquad (2.172)$$

then the formula for DA_μ's transformation takes the form

$$DA_\mu = dA_\mu - \delta A_\mu = dA_\mu - \Gamma^\lambda_{\mu\nu} A_\lambda dx^\nu$$

$$= \frac{\partial x'^\nu}{\partial x^\mu} A'_\nu + A'_\nu \frac{\partial^2 x'^\nu}{\partial x^\mu \partial x^\lambda} dx^\lambda$$

$$- \left(\Gamma'^\rho_{\alpha\beta} \frac{\partial x^\lambda}{\partial x'^\rho} \frac{\partial x'^\alpha}{\partial x^\mu} \frac{\partial x'^\beta}{\partial x^\nu} + \frac{\partial^2 x'^\rho}{\partial x^\mu \partial x^\nu} \frac{\partial x^\lambda}{\partial x'^\rho} \right) A_\lambda dx^\nu$$

$$= \frac{\partial x'^\nu}{\partial x^\mu} \left(A'_\nu - \Gamma'^\rho_{\nu\beta} A'_\rho dx'^\beta \right)$$

$$= \frac{\partial x'^\nu}{\partial x^\mu} D' A'_\nu. \qquad (2.173)$$

Comparing Eq.(2.173) with Eq.(2.168), we conclude that when the transformation of $\Gamma^\lambda_{\mu\nu}$ follows formula of (2.172), the corresponding covariant differential DA_μ of vector A_μ behaves like a vector in curvilinear coordinates. This covariant vector can also be written as

$$DA_\mu = dA_\mu - \Gamma^\lambda_{\mu\nu} A_\lambda dx^\nu = \left(\frac{\partial A_\mu}{\partial x^\nu} - \Gamma^\lambda_{\mu\nu} A_\lambda \right) dx^\nu. \qquad (2.174)$$

Similarly, we find for a contravariant vector,

$$DA^\mu = dA^\mu + \Gamma^\mu_{\lambda\nu} A^\lambda dx^\nu = \left(\frac{\partial A^\mu}{\partial x^\nu} + \Gamma^\mu_{\lambda\nu} A^\lambda \right) dx^\nu. \qquad (2.175)$$

The expressions in parentheses in Eqs.(2.174) and (2.175) are tensors, since when multiplied by the vector dx^ν they give a vector. Clearly, these are the tensors which give the desired generalization of the concept of a derivative to curvilinear coordinates. These tensors are called the *covariant derivatives* of the vectors A_μ and A^μ respectively. We shall denote them by $A_{\mu\;;\nu}$ and $A^\mu{}_{;\nu}$. Thus

$$DA_\mu = A_{\mu\;;\nu} dx^\nu, \quad DA^\mu = A^\mu{}_{;\nu}, \qquad (2.176)$$

while the covariant derivatives themselves are:

$$A_{\mu\,;\nu} = \frac{\partial A_\mu}{\partial x^\nu} - \Gamma^\lambda_{\mu\nu} A_\lambda \,, \tag{2.177}$$

$$A^\mu{}_{;\nu} = \frac{\partial A^\mu}{\partial x^\nu} + \Gamma^\mu_{\lambda\nu} A^\lambda \,. \tag{2.178}$$

Symbol $\Gamma^\lambda_{\mu\nu}$ is not a tensor, and is called affine connection of the spacetime with curvilinear coordinates, which transforms according to Eq.(2.172). Generally, $\Gamma^\lambda_{\mu\nu} \neq \Gamma^\lambda_{\nu\mu}$. But when the spacetime is torsionless, we have $\Gamma^\lambda_{\mu\nu} = \Gamma^\lambda_{\nu\mu}$, and such affine connections are called Christoffel symbols at sometimes. In this book we do not consider the spacetime torsion.

Let us show that the covariant derivative of the metric tensor $g_{\mu\nu}$ is zero. To do this we note that the relation

$$DA_\mu = g_{\mu\nu} DA^\nu$$

is valid for the vector DA_μ, as for any vector. On the other hand, $A_\mu = g_{\mu\nu} A^\nu$, so that

$$DA_\mu = D(g_{\mu\nu} A^\nu) = g_{\mu\nu} DA^\nu + A^\nu Dg_{\mu\nu}.$$

Comparing with $DA_\mu = g_{\mu\nu} A^\nu$, and remembering that the vector A^ν is arbitrary,

$$Dg_{\mu\nu} = 0.$$

Therefore the covariant derivative

$$g_{\mu\nu;\,\lambda} = 0. \tag{2.179}$$

Equation (2.179) can be used to express the Christoffel symbols $\Gamma^\lambda_{\mu\nu}$ in terms of the metric tensor $g_{\mu\nu}$. To do this we write in accordance with the general definition (2.177):

$$g_{\mu\nu;\,\lambda} = \frac{\partial g_{\mu\nu}}{\partial x^\lambda} - \Gamma^\rho_{\mu\lambda} g_{\rho\nu} - \Gamma^\rho_{\nu\lambda} g_{\mu\rho} = \frac{\partial g_{\mu\nu}}{\partial x^\lambda} - \Gamma_{\nu,\,\mu\lambda} - \Gamma_{\mu,\,\nu\lambda} = 0,$$

where $\Gamma_{\mu,\,\nu\lambda} = g_{\mu\rho} \Gamma^\rho_{\nu\lambda}$, and etc. Permuting the indices μ, ν, λ, we have

$$\frac{\partial g_{\mu\nu}}{\partial x^\lambda} = \Gamma_{\nu,\,\mu\lambda} + \Gamma_{\mu,\,\nu\lambda} \,,$$

$$\frac{\partial g_{\lambda\mu}}{\partial x^\nu} = \Gamma_{\mu,\,\nu\lambda} + \Gamma_{\nu,\,\mu\lambda} \,,$$

$$-\frac{\partial g_{\nu\lambda}}{\partial x^\mu} = -\Gamma_{\lambda,\,\nu\mu} - \Gamma_{\nu,\,\lambda\mu} \,.$$

Taking half the sum of these equations, we find (noting that $\Gamma_{\mu,\,\nu\lambda} = \Gamma_{\mu,\,\lambda\nu}$)

$$\Gamma_{\mu,\,\nu\lambda} = \frac{1}{2}\left(\frac{\partial g_{\mu\nu}}{\partial x^\lambda} + \frac{\partial g_{\mu\lambda}}{\partial x^\nu} - \frac{\partial g_{\nu\lambda}}{\partial x^\mu}\right). \tag{2.180}$$

From this we have the Christoffel symbols

$$\Gamma^\mu_{\nu\lambda} = \frac{1}{2}g^{\mu\rho}\left(\frac{\partial g_{\rho\nu}}{\partial x^\lambda} + \frac{\partial g_{\rho\lambda}}{\partial x^\nu} - \frac{\partial g_{\nu\lambda}}{\partial x^\rho}\right). \tag{2.181}$$

(5) **The curvature tensor:** If we twice differentiate a vector A_ρ with respect to x^μ and x^ν, then the result generally depends on the order of differentiation, contrary to the situation for ordinary differentiation. It turns out that the difference $A_{\rho\,;\mu\,;\nu} - A_{\rho\,;\nu\,;\mu}$ results a new tensor. The straightforward calculations give following result

$$A_{\rho\,;\mu\,;\nu} - A_{\rho\,;\nu\,;\mu} = A_\lambda \mathcal{R}^\lambda{}_{\rho\mu\nu}, \tag{2.182}$$

where

$$\mathcal{R}^\lambda{}_{\rho\mu\nu} = \frac{\partial \Gamma^\lambda_{\rho\nu}}{\partial x^\mu} - \frac{\partial \Gamma^\lambda_{\rho\mu}}{\partial x^\nu} + \Gamma^\lambda_{\alpha\mu}\Gamma^\alpha_{\rho\nu} - \Gamma^\lambda_{\alpha\nu}\Gamma^\alpha_{\rho\mu}. \tag{2.183}$$

$\mathcal{R}^\lambda{}_{\rho\mu\nu}$ is called the curvature tensor, or Riemann tensor. Similarly, for a contravariant vector,

$$A^\lambda{}_{;\mu\,;\nu} - A^\lambda{}_{;\nu\,;\mu} = -A_\rho \mathcal{R}^\lambda{}_{\rho\mu\nu}. \tag{2.184}$$

The curvature tensor has symmetry properties which can be made completely apparent by changing from mixed components $\mathcal{R}^\alpha{}_{\rho\mu\nu}$ to covariant ones:

$$\mathcal{R}_{\lambda\rho\mu\nu} = g_{\lambda\alpha}\mathcal{R}^\alpha{}_{\rho\mu\nu}.$$

From the explicit expression of $\mathcal{R}_{\lambda\rho\mu\nu}$, one can see the following symmetry properties:

$$\mathcal{R}_{\lambda\rho\mu\nu} = -\mathcal{R}_{\rho\lambda\mu\nu} = -\mathcal{R}_{\lambda\rho\nu\mu} = \mathcal{R}_{\mu\nu\lambda\rho}, \tag{2.185}$$

i.e., the tensor is antisymmetric in each of the index pairs λ, ρ and μ, ν, and is symmetric under the interchange of the two pairs with one another. One can also verify that the cyclic sum of components of $\mathcal{R}_{\lambda\rho\mu\nu}$, formed by permutation of any three indices, is equal to zero; e.g.,

$$\mathcal{R}_{\lambda\rho\mu\nu} + \mathcal{R}_{\lambda\nu\rho\mu} + \mathcal{R}_{\lambda\mu\nu\rho} = 0. \tag{2.186}$$

Now, we prove the *Bianchi identity*:

$$\mathcal{R}^\lambda{}_{\rho\mu\nu\;;\alpha} + \mathcal{R}^\lambda{}_{\rho\alpha\mu\;;\nu} + \mathcal{R}^\lambda{}_{\rho\nu\alpha\;;\mu} = 0. \tag{2.187}$$

It is conveniently verified by using a locally-geodesic coordinate system. Because of its tensor character, the relation (2.187) will then be valid in any other system. Differentiating Eq.(2.183) and then substituting in it $\Gamma^\lambda_{\rho\mu} = 0$, we find for the point under consideration

$$\mathcal{R}^\lambda{}_{\rho\mu\nu\;;\alpha} = \frac{\partial \mathcal{R}^\lambda{}_{\rho\mu\nu}}{\partial x^\alpha} = \frac{\partial^2 \Gamma^\lambda_{\rho\nu}}{\partial x^\mu \partial x^\alpha} - \frac{\partial^2 \Gamma^\lambda_{\rho\mu}}{\partial x^\nu \partial x^\alpha}.$$

With the aid of this expression it is easy to verify that the Bianchi identity (2.187) actually holds.

Finally, from the curvature tensor we can, by contraction, construct a tensor of the second rank. This contraction can be carried out in only one way: contraction of the tensor $\mathcal{R}_{\lambda\rho\mu\nu}$ on the indices λ and ρ or μ and ν gives zero because of the antisymmetry in this indices, while contraction on any other pair always gives the same result, except for sign. We define the tensor $\mathcal{R}_{\rho\nu}$ (the *Ricci tensor*) as

$$\mathcal{R}_{\rho\mu} = g^{\lambda\nu} \mathcal{R}_{\lambda\rho\mu\nu} = \mathcal{R}^\lambda{}_{\rho\mu\lambda}. \tag{2.188}$$

According to Eq.(2.183), we have

$$\mathcal{R}_{\rho\nu} = \frac{\partial \Gamma^\lambda_{\rho\lambda}}{\partial x^\nu} - \frac{\partial \Gamma^\lambda_{\rho\nu}}{\partial x^\lambda} + \Gamma^\lambda_{\alpha\nu}\Gamma^\alpha_{\rho\lambda} - \Gamma^\lambda_{\alpha\lambda}\Gamma^\alpha_{\rho\nu}. \tag{2.189}$$

This tensor is clearly symmetric:

$$\mathcal{R}_{\rho\nu} = \mathcal{R}_{\nu\rho}. \tag{2.190}$$

Furthermore, contracting $\mathcal{R}_{\rho\nu}$, we obtain the invariant

$$\mathcal{R} = g^{\rho\nu} \mathcal{R}_{\rho\nu} = g^{\rho\nu} g^{\lambda\mu} \mathcal{R}_{\lambda\rho\nu\mu}, \tag{2.191}$$

which is called the *scalar curvature* of the spacetime.

2.9 Localization of E-SR spacetime symmetry and Einstein's General Relativity (II)

After introducing the gauge theory, gauge transformation arising from localization of $ISO(3,1)$, and mathematical tool to deal with curvilinear coordinates systems of the spacetime in the previous section, we continue to pursue the theory of localization of E-SR spacetime symmetry.

In Section 2.6, we have proved the energy of particle is one of the Noether charges corresponding to the $ISO(3,1)$ symmetry of E-SR mechanics (see Eq.(2.83)). Due to Newtonian gravity law, the force arising from the particle's mass (or energy in E-SR) is gravitational, and hence we realize that the $ISO(3,1)$-gauge field theory must be a gravitational field theory. So we could call it also *gauge theory of gravity*. Many efforts (e.g., see [Utiyama, R. (1956); Kibble, B.W.T. (1961)]) have been made to construct a *full satisfactory* gauge theory of gravity. Although there is no established formulation for that as yet, it is clear conceptually that the Einstein's General Relativity (E-GR) relates to some gauge theory of localized external symmetry. In this section, in order to show E-SR is the foundation of E-GR, we describe a minimal version (or simplest version) of gauge gravity theory by means of the concepts of localization of E-SR spacetime symmetry. This description should be helpful pedagogically.

We showed in the last section that the internal $U(1)$-gauge transformations of matter fields $\phi(x)$, $\phi^*(x)$ and gauge fields $A_\mu(x)$ all arose from the gauge symmetry in the Hilbert function space, and the functions of $\phi(x)$, $\phi^*(x)$, $A_\mu(x)$ represent "coordinates components" in that abstract function space. However, the situation for the external spacetime gauge theory is very different. The $ISO(3,1)$-gauge transformation $x^\mu \to x'^\mu = f^\mu(x)$ (see Eq.(2.151)) is ordinary coordinates transformation in 4-dim real world spacetime described in curvilinear coordinates system. Therefore, the gauge theory of gravity should be a geometry theory of spacetime, and gravitational fields as gauge fields should be characterized by geometric quantity. In the following, the gravity fields will be described by the spacetime metric $g_{\mu\nu}(x)$ in gauge theory of gravity or GR, because it is basic quantity in Riemann geometry.

Under the localization transformation Eq.(2.148) the global $ISO(3,1)$ symmetry of E-SR becomes local $ISO(3,1)$ symmetry, or gauge $ISO(3,1)$ symmetry Eq.(2.149). According to the principle of gauge theory, the forms of physics laws should be invariant under the gauge transformation (or general coordinates transformations) Eq.(2.149). Namely, under the general coordinates transformations Eqs.(2.149) and (2.151) the transformed expressions of physics laws have to be of the same functional form as before.

Now let us determine the action of gravity fields S_G. Just as for the expression of action of complex scalar field (2.125), the action S_G must be expressed in terms of a scalar integral $\int \mathcal{G}\sqrt{-g}d^4x$ (see Eq.(2.167)). To determine scalar $\mathcal{G}$ we should also consider the fact that the equation of the gravitational field must contain derivatives of the "potentials" (i.e., $g_{\mu\nu}(x)$) no higher than the second order (just as is the case for the electromagnetic field). From Eqs.(2.191), (2.189), (2.183) and (2.181), it is found that only $\mathcal{R}$ and trivial constant $\Lambda \equiv constant$ satisfies all requirements. Therefore, $\mathcal{G} = a(\mathcal{R} - 2\Lambda)$ where a is also a constant. In *Gaussian system of units*, $a = -c^3/(16\pi G)$ where $G = 6.67 \times 10^{-8}cm^3 \cdot gm^{-1} \cdot sec^{-2}$ is the universal gravitational constant. Thus we obtain the action of gauge gravity:

$$S_G = -\frac{c^3}{16\pi G} \int d^4x\sqrt{-g}(\mathcal{R} - 2\Lambda). \tag{2.192}$$

Denoting the matter action in gravity fields by:

$$S_M = \int d^4x\sqrt{-g}\mathcal{L}_{matter}(g_{\mu\nu}, \phi_{matter}(x)),$$

one has the total action:

$$S_{tot} = S_G + S_M = S_G[g] + S_M[g, \phi] \tag{2.193}$$

where $S.[\cdot]$ denotes a functional.

Now, we derive the equations of motion of the gravitational field. These equations are obtained from the principle of least action

$$\delta S_{tot} = \delta(S_G + S_M) = 0. \tag{2.194}$$

We now let the gravitational field, that is, the quantities $g_{\mu\nu}$, to undergo variation.

Calculating the variation δS_G, we have

$$\delta \int \mathcal{R}\sqrt{-g}d^4x = \delta \int g^{\mu\nu}\mathcal{R}_{\mu\nu}\sqrt{-g}d^4x$$

$$= \int \{\mathcal{R}_{\mu\nu}\sqrt{-g}\delta g^{\mu\nu} + \mathcal{R}_{\mu\nu}g^{\mu\nu}\delta\sqrt{-g} + g^{\mu\nu}\sqrt{-g}\delta\mathcal{R}_{\mu\nu}\}d^4x.$$

From formula $dg = gg^{\mu\nu}dg_{\mu\nu} = -gg_{\mu\nu}dg^{\mu\nu}$, we have

$$\delta\sqrt{-g} = -\frac{1}{2\sqrt{-g}}\delta g = -\frac{1}{2}\sqrt{-g}g_{\mu\nu}dg^{\mu\nu}; \tag{2.195}$$

and

$$\delta \int \mathcal{R}\sqrt{-g}d^4x = \int \left(\mathcal{R}_{\mu\nu} - \frac{1}{2}g_{\mu\nu}\mathcal{R}\right)\delta g^{\mu\nu}\sqrt{-g}d^4x + \int g^{\mu\nu}\delta\mathcal{R}_{\mu\nu}\sqrt{-g}d^4x.$$

$$\tag{2.196}$$

For the calculation of $\delta \mathcal{R}_{\mu\nu}$ we note that although the quantities $\Gamma^\lambda_{\mu\nu}$ do not constitute a tensor, their variations $\delta\Gamma^\lambda_{\mu\nu}$ do form a tensor, for $\Gamma^\lambda_{\mu\nu}A_\lambda dx^\nu$ is the change in a vector under parallel displacement (see Eq.(2.171)) from some point P to an infinitesimally separated point P'. Therefore $\delta\Gamma^\lambda_{\mu\nu}A_\lambda dx^\nu$ is the difference between the two vectors, obtained as result of two parallel displacements (one with the unvaried, the other with the varied $\Gamma^\lambda_{\mu\nu}$) from P to one and the same point P'. The difference between two vectors at the same point is a vector, and therefore $\delta\Gamma^\lambda_{\mu\nu}$ is a tensor.

From the Ricci tensor's definition (2.188), we have

$$-\delta\mathcal{R}_{\mu\nu} = \delta\mathcal{R}^\lambda{}_{\mu\lambda\nu}$$
$$= \delta\Gamma^\lambda_{\mu\lambda}{}_{,\nu} - \delta\Gamma^\lambda_{\mu\nu}{}_{,\lambda} + \delta(\Gamma^\sigma_{\mu\lambda}\Gamma^\lambda_{\nu\sigma} - \Gamma^\sigma_{\mu\nu}\Gamma^\lambda_{\lambda\sigma})$$
$$= \left\{(\delta\Gamma^\lambda_{\mu\lambda})_{,\nu} - \Gamma^\sigma_{\nu\mu}\delta\Gamma^\lambda_{\sigma\lambda}\right\} - \left\{(\delta\Gamma^\lambda_{\mu\nu})_{,\lambda} + \Gamma^\lambda_{\lambda\sigma}\delta\Gamma^\sigma_{\mu\nu} - \Gamma^\sigma_{\lambda\mu}\delta\Gamma^\lambda_{\sigma\nu} - \Gamma^\sigma_{\lambda\nu}\delta\Gamma^\lambda_{\sigma\mu}\right\}$$
$$= (\delta\Gamma^\lambda_{\mu\lambda})_{;\nu} - (\delta\Gamma^\lambda_{\mu\nu})_{;\lambda}. \tag{2.197}$$

Thus we find that the integrand of the last term of Eq.(2.196) becomes

$$-\sqrt{-g}g^{\mu\nu}\delta\mathcal{R}_{\mu\nu} = \sqrt{-g}\left\{(g^{\mu\nu}\delta\Gamma^\lambda_{\mu\lambda})_{;\nu} - (g^{\mu\nu}\delta\Gamma^\lambda_{\mu\nu})_{;\lambda}\right\}$$
$$= \partial_\lambda W^\lambda, \tag{2.198}$$

where a formula $T^\lambda{}_{;\lambda} = (\sqrt{-g})^{-1}\partial_\lambda(\sqrt{-g}T^\lambda)$ has been used, and

$$W^\lambda = \sqrt{-g}\left\{g^{\lambda\mu}\delta\Gamma^\sigma_{\mu\sigma} - g^{\mu\nu}\delta\Gamma^\lambda_{\mu\nu}\right\}.$$

Consequently the second integral on the right side of (2.196) is equal to

$$\int g^{\mu\nu}\delta\mathcal{R}_{\mu\nu}\sqrt{-g}d^4x = -\int \frac{\partial W^\lambda}{\partial x^\lambda}d^4x, \tag{2.199}$$

and by Gauss's theorem can be transformed into an integral of W^λ over the hypersurface surrounding the whole four-volume. Since the variation of the field are zero at the integration limits, this term drops out. Thus, the variation of the first term in right side of S_G (see (2.192)) is equal to

$$\delta\left(-\frac{c^3}{16\pi G}\int d^4x\sqrt{-g}\mathcal{R}\right) = -\frac{c^3}{16\pi G}\int\left(\mathcal{R}_{\mu\nu} - \frac{1}{2}g_{\mu\nu}\mathcal{R}\right)\delta g^{\mu\nu}\sqrt{-g}d^4x. \tag{2.200}$$

From Eq.(2.195), the variation of the secod term on the right side of S_G (see (2.192)) is equal to

$$\delta\left(-\frac{c^3}{16\pi G}\int d^4x\sqrt{-g}(-2\Lambda)\right) = -\frac{c^3}{16\pi G}\int\Lambda g_{\mu\nu}\delta g^{\mu\nu}\sqrt{-g}d^4x. \tag{2.201}$$

Combining Eqs.(2.199) and (2.201) and noting Eq.(2.192), we have

$$\delta S_G = -\frac{c^3}{16\pi G} \int \left(\mathcal{R}_{\mu\nu} - \frac{1}{2} g_{\mu\nu} \mathcal{R} + \Lambda g_{\mu\nu} \right) \delta g^{\mu\nu} \sqrt{-g} d^4 x. \quad (2.202)$$

The energy-momentum tensor of matter fields is defined by

$$T_{\mu\nu} = \frac{-2c}{\sqrt{-g}} \frac{\delta}{\delta g^{\mu\nu}} S_M[g, \phi], \quad (2.203)$$

where $\delta/\delta g^{\mu\nu}$ means the functional derivative. Thus, the variation of $S_M[g, \phi]$ due to $\delta g^{\mu\nu}$ is:

$$\delta S_M = \frac{1}{2c} \int T_{\mu\nu} \delta g^{\mu\nu} \sqrt{-g} d^4 x. \quad (2.204)$$

From Eqs.(2.202), (2.204) and $\delta S_{tot} = \delta(S_G + S_M) = 0$ (see (2.194)), we have

$$\begin{aligned}
\delta S_{tot} &= \delta(S_G + S_M) \\
&= -\frac{c^3}{16\pi G} \int \left(\mathcal{R}_{\mu\nu} - \frac{1}{2} g_{\mu\nu} \mathcal{R} + \Lambda g_{\mu\nu} + \frac{8\pi G}{c^4} T_{\mu\nu} \right) \delta g^{\mu\nu} \sqrt{-g} d^4 x \\
&= 0, \quad (2.205)
\end{aligned}$$

from which we have in view of the arbitrariness of the $\delta g^{\mu\nu}$:

$$\mathcal{R}_{\mu\nu} - \frac{1}{2} g_{\mu\nu} \mathcal{R} + \Lambda g_{\mu\nu} = -\frac{8\pi G}{c^4} T_{\mu\nu}. \quad (2.206)$$

This is the required equation of the gravitational field.

E-SR tells us that in empty spacetime the metric must be Lorentz-Minkowski spacetime metric $\eta_{\mu\nu} = \mathrm{diag}\{1, -1, -1, -1\}$. Namely, when $T_{\mu\nu} = 0$, the solution of (2.206) is required to be $g_{\mu\nu} = \eta_{\mu\nu}$. Then the value of constant Λ is determined to be

$$\Lambda = 0, \quad (2.207)$$

and Eq.(2.206) becomes

$$\mathcal{R}_{\mu\nu} - \frac{1}{2} g_{\mu\nu} \mathcal{R} = -\frac{8\pi G}{c^4} T_{\mu\nu}. \quad (2.208)$$

This is the basic equation of the Einstein's General Relativity (E-GR) and also the equation of gauge gravity theory based on the localization of E-SR spacetime symmetry $ISO(3,1)$. It is called *Einstein equation*. Thus, we have achieved the theory of E-GR via localizing the spacetime symmetry of E-SR.

E-GR is based on two principles: (I) General relativity principle, and (II) The Equivalence Principle. Since the general coordinates transformations can well be viewed as the gauge transformations of gauge theory with external localized symmetries, the principle (I) is well consistent with the gauge principle. Now we discuss the principle (II). The Equivalence Principle states that *experiments in a sufficiently small freely falling laboratory, over sufficiently short time, give results that are indistinguishable from those of the same experiments in an inertial frame in empty space.* This principle suggests that the local properties of curved spacetime should be indistinguishable from those of the flat spacetime of E-SR. A concrete expression of this physical idea is the requirement that, given a metric $g_{\mu\nu}$ in one system of coordinates, at each point P of spacetime it is possible to introduce new coordinates x'^{μ} such that

$$g'_{\mu\nu}(x'_P) = \eta_{\mu\nu}, \qquad (2.209)$$

where $\eta_{\mu\nu} = \mathrm{diag}\{1, -1, -1, -1\}$ is Minkowski metric of flat spacetime and x'^{μ}_P are the coordinates locating the point P. It is called Local Inertial Frame (LIF). From Eq.(2.17), we have learnt that the spacetime group to preserve $\eta_{\mu\nu}$ is $ISO(3,1)$ (see Eq.(8.26)). Therefore, the external symmetry group to preserve $\eta_{\mu\nu}$ at the point P with coordinates x'^{μ} is just localized $ISO(3,1)$ at x'^{μ}, or gauge $ISO(3,1)$ (see Eq.(2.149)). In other wards, the requirement of existing LIF in curved spacetime corresponds to existence of the gauge group in the gauge theory of gravity. LIF is an important conception in both GR and the gauge theory of gravity, because it indicates that E-SR is the foundation of E-GR.

Chapter 3

De Sitter Invariant Special Relativity

3.1 Beltrami metrics

We consider a 4-dimensional pseudo-sphere (or hyperboloid) $\mathcal{S}_\Lambda$ embedded in a 5-dimensional Minkowski spacetime with metric $\eta_{AB} = \mathrm{diag}(1, -1, -1, -1, -1)$:

$$\mathcal{S}_\Lambda : \ \eta_{AB}\xi^A\xi^B = -R^2,$$
$$ds^2 = \eta_{AB}d\xi^A d\xi^B, \tag{3.1}$$

where index $A, \ B = \{0, 1, 2, 3, 5\}$, $R^2 := 3\Lambda^{-1}$ and Λ is the cosmological constant. $\mathcal{S}_\Lambda$ is also called de Sitter pseudo-spherical surface with radii R. We define

$$x^\mu := R\frac{\xi^\mu}{\xi^5}, \ \text{ with } \ \xi^5 \neq 0, \text{ and } \mu = \{0, 1, 2, 3\}. \tag{3.2}$$

Treating x^μ are Cartesian-type coordinates of a 4-dimensional spacetime with metric $g_{\mu\nu}(x) \equiv B_{\mu\nu}(x)$, we denote this 4-dimensional spacetime as $\mathcal{B}_\Lambda$ and call it Beltrami spacetime.

Now we derive $B_{\mu\nu}(x)$ by means of the geodesic projection of $\{\mathcal{S}_\Lambda \mapsto \mathcal{B}_\Lambda\}$ (see Fig.1). From the definition (3.1), we have

$$ds^2 = \eta_{AB}d\xi^A d\xi^B|_{\xi^{A,B}\in\mathcal{S}_\Lambda}$$
$$= \eta_{\mu\nu}d\xi^\mu d\xi^\nu - (d\xi^5)^2$$
$$:= B_{\mu\nu}(x)dx^\mu dx^\nu. \tag{3.3}$$

Since $\xi^{A,B} \in \mathcal{S}_\Lambda$, and from Eqs.(3.2) and (3.1), it is easy to obtain:

$$\xi^\mu = \frac{x^\mu}{R}\xi^5, \ \ d\xi^\mu = \frac{1}{R}(\xi^5 dx^\mu + x^\mu d\xi^5), \ \ (\xi^5)^2 = \frac{R^2}{\sigma(x)},$$
$$d\xi^5 = \eta_{\mu\nu}\frac{\xi^\mu}{\xi^5}d\xi^\nu = \frac{1}{R}\eta_{\mu\nu}x^\mu d\xi^\nu = \frac{\eta_{\mu\nu}x^\mu dx^\nu}{\xi^5\sigma(x)^2}, \tag{3.4}$$

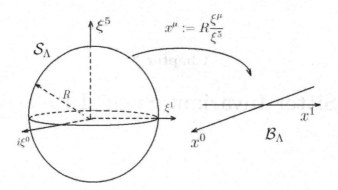

Fig. 3.1   Sketch of the geodesic projection from de Sitter pseudo-spherical surface $\mathcal{S}_\Lambda$ to the Beltrami spacetime $\mathcal{B}_\Lambda$ via Eq.(3.2).

where

$$\sigma(x) = 1 - \frac{\eta_{\mu\nu}x^\mu x^\nu}{R^2}. \tag{3.5}$$

Substituting the above into Eq.(3.3), we have

$$ds^2 = \frac{\eta_{\mu\nu}dx^\mu dx^\nu}{\sigma(x)} + \frac{(\eta_{\mu\nu}x^\mu dx^\nu)^2}{R^2\sigma(x)^2} := B_{\mu\nu}(x)dx^\mu dx^\nu.$$

Then, we obtain the Beltrami metric as follows

$$B_{\mu\nu}(x) = \frac{\eta_{\mu\nu}}{\sigma(x)} + \frac{\eta_{\mu\lambda}x^\lambda \eta_{\nu\rho}x^\rho}{R^2\sigma(x)^2}. \tag{3.6}$$

According to Riemann geometry, the following expressions can be proved by means of straightforward calculations from Eq.(3.6):

$$B^{\mu\nu}(x) = \sigma(x)(\eta^{\mu\nu} - R^{-2}x^\mu x^\nu), \tag{3.7}$$

$$\Gamma^\rho_{\mu\nu}(x) = \frac{1}{R^2\sigma(x)}(\delta^\rho_\mu\eta_{\nu\lambda}x^\lambda + \delta^\rho_\nu\eta_{\mu\lambda}x^\lambda), \tag{3.8}$$

$$\mathcal{R}^\rho{}_{\lambda\mu\nu}(x) = R^{-2}(B_{\lambda\mu}(x)\delta^\rho_\nu - B_{\lambda\nu}(x)\delta^\rho_\mu), \tag{3.9}$$

$$\mathcal{R}_{\mu\nu}(x) = \frac{3}{R^2}B_{\mu\nu}(x), \tag{3.10}$$

$$\mathcal{R} = \frac{12}{R^2} = \text{constant}. \tag{3.11}$$

Exercise

(1) Problem 1, To $(1 + 1)$-dimension case (i.e., $\eta_{\mu\nu} = diag\{1, -1\}$), please write out expressions of $B_{00}(x)$, $B_{01}(x)$, $B_{11}(x)$ in terms of $x^0 \equiv ct$, x^1.

(2) Problem 2, Derive expression of the interval $ds^2_{Bel} = B_{\mu\nu}(x)dx^\mu dx^\nu$ in the polar coordinates of the Beltrami spacetime, where $x^0 = ct$, $x^1 = r\sin\theta\cos\phi$, $x^2 = r\sin\theta\sin\phi$, $x^3 = r\cos\theta$.

3.2 Motion of a particle in the Beltrami spacetime

The motion of a free massive particle is determined in the special theory of relativity from the principle of least action,

$$\delta S = -mc\,\delta \int ds = 0 \qquad (3.12)$$

according to which the particle moves so that its world line is an extremal between a given pair of world points. We have shown in previous chapter (see Eq.(2.20)) that $ds = \sqrt{\eta_{\mu\nu}dx^\mu dx^\nu}$ in usual Einstein's Special Relativity (E-SR), and the inertial motion law holds for free particle. Namely, the world line is a straight line, and the motion is uniform rectilinear.

The motion of a particle in the Beltrami spacetime $\mathcal{B}_\Lambda$ is determined by the principle of least action in the same form (3.12) with Beltrami metric $B_{\mu\nu}$ in place of the Minkowski metric $\eta_{\mu\nu}$ manifested in the world-line interval as follows

$$ds = \sqrt{B_{\mu\nu}(x)dx^\mu dx^\nu}. \qquad (3.13)$$

Thus, in $\mathcal{B}_\Lambda$ with x^μ-dependent metric $B_{\mu\nu}(x)$ the particle moves so that its world point moves along an extremal or, as it is called, a *geodesic line* in the four-space $\{x^0,\ x^1,\ x^2,\ x^3\}$ of $\mathcal{B}_\Lambda$. A geodesic line could be thought as an generalization of usual "straight line" in Euclidean space mathematically.

Now we derive the equation of motion of particles in $\mathcal{B}_\Lambda$ based on the principle of least action (3.12). We have

$$\delta\,ds^2 = 2ds\,\delta ds = \delta(B_{\mu\nu}(x)dx^\mu dx^\nu)$$
$$= dx^\mu dx^\nu \frac{\partial B_{\mu\nu}(x)}{\partial x^\lambda}\delta x^\lambda + 2B_{\mu\nu}(x)dx^\mu d\delta x^\nu. \qquad (3.14)$$

Therefore

$$\delta S = -mc \int \delta ds = mc \int \left\{ \frac{1}{2} \frac{dx^\mu}{ds} \frac{dx^\nu}{ds} \frac{\partial B_{\mu\nu}(x)}{\partial x^\lambda} \delta x^\lambda + B_{\mu\nu}(x) \frac{dx^\mu}{ds} \frac{d\delta x^\nu}{ds} \right\} ds$$

$$= mc \int \left\{ \frac{1}{2} \frac{dx^\mu}{ds} \frac{dx^\nu}{ds} \frac{\partial B_{\mu\nu}(x)}{\partial x^\lambda} \delta x^\lambda - \frac{d}{ds} \left(B_{\mu\nu}(x) \frac{dx^\mu}{ds} \right) \delta x^\nu \right\} ds$$

$$+ mc B_{\mu\nu}(x) \frac{dx^\mu}{ds} \delta x^\nu |_. \, ,$$

where the last term represents value of the function at the limit. Using the fact that $\delta x^\nu |_. = 0$, we have

$$\delta S = mc \int \left\{ \frac{1}{2} \frac{dx^\mu}{ds} \frac{dx^\nu}{ds} \frac{\partial B_{\mu\nu}(x)}{\partial x^\lambda} \delta x^\lambda - \frac{d}{ds} \left(B_{\mu\nu}(x) \frac{dx^\mu}{ds} \right) \delta x^\nu \right\} ds$$

$$= mc \int \left\{ \frac{1}{2} \frac{dx^\mu}{ds} \frac{dx^\nu}{ds} \frac{\partial B_{\mu\nu}(x)}{\partial x^\lambda} - \frac{d}{ds} \left(B_{\mu\lambda}(x) \frac{dx^\mu}{ds} \right) \right\} \delta x^\lambda ds. \quad (3.15)$$

We then find, by equating to zero the coefficient of arbitrary variation δx^λ:

$$\frac{1}{2} \frac{dx^\mu}{ds} \frac{dx^\nu}{ds} \frac{\partial B_{\mu\nu}(x)}{\partial x^\lambda} - \frac{d}{ds} \left(B_{\mu\lambda}(x) \frac{dx^\mu}{ds} \right)$$

$$= \frac{1}{2} \frac{dx^\mu}{ds} \frac{dx^\nu}{ds} \frac{\partial B_{\mu\nu}(x)}{\partial x^\lambda} - B_{\mu\lambda}(x) \frac{d^2 x^\mu}{ds^2} - \frac{dx^\mu}{ds} \frac{dx^\nu}{ds} \frac{\partial B_{\mu\lambda}(x)}{\partial x^\nu} = 0. \quad (3.16)$$

Noting the third term can be written as

$$-\frac{1}{2} \left(\frac{\partial B_{\mu\lambda}(x)}{\partial x^\nu} + \frac{\partial B_{\nu\lambda}(x)}{\partial x^\mu} \right) \frac{dx^\mu}{ds} \frac{dx^\nu}{ds},$$

and introducing the Christoffel symbols $\Gamma_{\lambda,\mu\nu}$ in accordance with Eq.(2.180), we have

$$B_{\mu\lambda}(x) \frac{d^2 x^\mu}{ds^2} + \Gamma_{\lambda,\mu\nu} \frac{dx^\mu}{ds} \frac{dx^\nu}{ds} = 0.$$

Multiplying $B^{\rho\lambda}(x)$, we obtain the geodesic line equation in $\mathcal{B}_\Lambda$ as follows

$$\frac{d^2 x^\rho}{ds^2} + \Gamma^\rho_{\mu\nu} \frac{dx^\mu}{ds} \frac{dx^\nu}{ds} = 0, \quad (3.17)$$

where $\Gamma^\rho_{\mu\nu}$ has been given in Eq.(3.8).

Equation (3.17) serves as an equation of motion of free particle in Beltrami spacetime $\mathcal{B}_\Lambda$. Let us solve it. Substituting Eq.(3.8) into Eq.(3.17), we have

$$\frac{d^2 x^\rho}{ds^2} + \frac{2}{R^2 \sigma(x)} \frac{dx^\rho}{ds} \frac{dx^\mu}{ds} x^\nu \eta_{\mu\nu} = 0. \quad (3.18)$$

From $\sigma(x) = 1 - \eta_{\mu\nu}x^\mu x^\nu / R^2$ (see Eq.(3.5)), we have the following formula

$$\frac{d}{ds}\left(\frac{1}{\sigma(x)}\right) = \frac{2}{R^2\sigma(x)^2}\eta_{\mu\nu}x^\mu\frac{dx^\nu}{ds}, \tag{3.19}$$

and hence

$$\frac{d}{ds}\left(\frac{1}{\sigma(x)}\frac{dx^\rho}{ds}\right) = \frac{1}{\sigma(x)}\left(\frac{d^2x^\rho}{ds^2} + \frac{2}{R^2\sigma(x)}\frac{dx^\rho}{ds}\frac{dx^\mu}{ds}x^\nu\eta_{\mu\nu}\right). \tag{3.20}$$

Substituting the equation of motion (3.18) into (3.20), we obtain

$$\frac{d}{ds}\left(\frac{1}{\sigma(x)}\frac{dx^\rho}{ds}\right) = 0,$$

and hence

$$\frac{1}{\sigma(x)}\frac{dx^\rho}{ds} = A^\rho, \tag{3.21}$$

where A^ρ =consts. are of four integration constants: $\{A^0,\ A^1,\ A^2,\ A^3\}$. Since $\sigma(x)^{-1}dx^i/ds = A^i$ and $\sigma(x)^{-1}dx^0/ds = A^0$, we have

$$\frac{\sigma(x)^{-1}dx^i/ds}{\sigma(x)^{-1}dx^0/ds} = \frac{dx^i}{dx^0} = \frac{A^i}{A^0} = constant,$$

$$\text{or}: \quad \frac{dx^i}{dt} = \dot{x}^i \equiv v^i = c\frac{A^i}{A^0} = constant,$$

$$\text{or}: \quad \frac{d^2x^i}{dt^2} = \ddot{x}^i = 0. \tag{3.22}$$

Equation (3.22) is a highly non-trivial result: $dx^i/dt = \dot{x}^i \equiv v^i = constant$ means the velocity of free particle along the straight (or geodesic) line in Beltrami spacetime $\mathcal{B}_\Lambda$ is uniform, and $\ddot{x}^i = 0$ indicates that there is no inertial force acting to the free particle in that spacetime, i.e., $F^i \equiv m\ddot{x}^i = 0$. Usually, one may instinctively think that the inertial reference frames can only be built in the flat spacetime described by Minkowski metric. Above result (3.22), however, tells us that the inertial frames can also be built in Beltrami spacetime $\mathcal{B}_\Lambda$ which is yet a curved spacetime with constant Riemann curvature $\mathcal{R} = 12/R^2 \neq 0$ (see Eq.(3.11)). Equation (3.22) is a significant discovery [Lu (1970); Lu, Zou and Guo (1974)], which inspires a deep thinking of that a new Special Relativity in $\mathcal{B}_\Lambda$ based on $B_{\mu\nu}$ may exist in parallel with the usual Einstein's Special Relativity (E-SR).

3.3 Lagrangian mechanics of free particles in the Beltrami spacetime

Similar to Eq.(2.20), the free particle in $\mathcal{B}_\Lambda$ is described by the corresponding Landau-Lifshitz (L-L) action:

$$S_{dS} = -mc \int ds = -mc \int \sqrt{B_{\mu\nu}(x)dx^\mu dx^\nu} \equiv \int dt L_{dS}(t, x^i, \dot{x}^i), \quad (3.23)$$

where subscript dS means that $\mathcal{B}_\Lambda \subset \{dS - \text{spacetime}\}$. From (3.23), the L-L Lagrange for free particle in $\mathcal{B}_\Lambda$ is

$$L_{dS} = -mc\frac{ds}{dt} = -mc\frac{\sqrt{B_{\mu\nu}(x)dx^\mu dx^\nu}}{dt} = -mc\sqrt{B_{\mu\nu}(x)\dot{x}^\mu \dot{x}^\nu}, \quad (3.24)$$

where $\dot{x}^\mu = \frac{d}{dt}x^\mu$, $B_{\mu\nu}(x)$ is Beltrami metric (3.6). Setting up the time $t = x^0/c$, $B_{\mu\nu}(x)$ can be rewritten as follows

$$ds^2 = B_{\mu\nu}(x)dx^\mu dx^\nu = \tilde{g}_{00}d(ct)^2 + \tilde{g}_{ij}\left[(dx^i + N^i d(ct))(dx^j + N^j d(ct))\right]$$
$$= c^2(dt)^2\left[\tilde{g}_{00} + \tilde{g}_{ij}(\frac{1}{c}\dot{x}^i + N^i)(\frac{1}{c}\dot{x}^j + N^j)\right], \quad (3.25)$$

where

$$\tilde{g}_{00} = \frac{R^2}{\sigma(x)(R^2 - c^2t^2)}, \quad (3.26)$$

$$\tilde{g}_{ij} = \frac{\eta_{ij}}{\sigma(x)} + \frac{1}{R^2\sigma(x)^2}\eta_{il}\eta_{jm}x^l x^m, \quad (3.27)$$

$$N^i = \frac{ctx^i}{R^2 - c^2t^2}. \quad (3.28)$$

Substituting Eqs.(3.25)–(3.28) into Eq.(3.24), we obtain the Lagrange for free particle in $\mathcal{B}_\Lambda$:

$$L_{dS} = -mc^2\sqrt{\tilde{g}_{00} + \tilde{g}_{ij}(\frac{1}{c}\dot{x}^i + N^i)(\frac{1}{c}\dot{x}^j + N^j)}. \quad (3.29)$$

And from the principle of least action, we have the Euler-Lagrangian equation:

$$\frac{\partial L_{dS}}{\partial x^i} = \frac{d}{dt}\frac{\partial L_{dS}}{\partial \dot{x}^i}. \quad (3.30)$$

For convenience, we introduce Cartesian space-coordinates notations:

$$\mathbf{x} = x^1\mathbf{i} + x^2\mathbf{j} + x^3\mathbf{k}, \qquad \dot{\mathbf{x}} = \dot{x}^1\mathbf{i} + \dot{x}^2\mathbf{j} + \dot{x}^3\mathbf{k},$$
$$\text{and } \mathbf{i}\cdot\mathbf{i} = \mathbf{j}\cdot\mathbf{j} = \mathbf{k}\cdot\mathbf{k} = 1, \qquad \mathbf{i}\cdot\mathbf{j} = \mathbf{i}\cdot\mathbf{k} = \mathbf{j}\cdot\mathbf{k} = 0. \quad (3.31)$$

By these notations and expressions of (3.5), (3.26), (3.27) and (3.28), the explicit expression of Lagrange (3.29) is:

$$
L_{dS} = -mc^2 \left[\frac{R^4}{(R^2 + \mathbf{x}^2 - c^2t^2)(R^2 - c^2t^2)} + \frac{-R^2}{R^2 + \mathbf{x}^2 - c^2t^2} \right.
$$

$$
\times \left(\frac{\dot{\mathbf{x}}^2}{c^2} + \frac{c^2t^2\mathbf{x}^2}{(R^2 - c^2t^2)^2} + \frac{2t(\mathbf{x} \cdot \dot{\mathbf{x}})}{R^2 - c^2t^2} \right)
$$

$$
\left. + \frac{R^2}{(R^2 + \mathbf{x}^2 - c^2t^2)^2} \left(\frac{\dot{\mathbf{x}} \cdot \mathbf{x}}{c} + \frac{ct\mathbf{x}^2}{R^2 - c^2t^2} \right)^2 \right]^{1/2}
$$

$$
= -mc^2R \sqrt{ \frac{R^2(c^2 - \dot{\mathbf{x}}^2) - \mathbf{x}^2\dot{\mathbf{x}}^2 + (\mathbf{x} \cdot \dot{\mathbf{x}})^2 + c^2(\mathbf{x} - \dot{\mathbf{x}}t)^2}{c^2(R^2 + \mathbf{x}^2 - c^2t^2)^2} } \quad (3.32)
$$

where $\mathbf{x}^2 = (\mathbf{x} \cdot \mathbf{x}) = -\eta_{ij}x^ix^j$. The Euler-Lagrangian equation is

$$
\frac{\partial L_{dS}}{\partial \mathbf{x}} = \frac{d}{dt}\frac{\partial L_{dS}}{\partial \dot{\mathbf{x}}} = \frac{\partial}{\partial t}\frac{\partial L_{dS}}{\partial \dot{\mathbf{x}}} + \left(\dot{\mathbf{x}} \cdot \frac{\partial}{\partial \mathbf{x}} \right) \frac{\partial L_{dS}}{\partial \dot{\mathbf{x}}} + \left(\ddot{\mathbf{x}} \cdot \frac{\partial}{\partial \dot{\mathbf{x}}} \right) \frac{\partial L_{dS}}{\partial \dot{\mathbf{x}}},
$$

$$(3.33)$$

where $L_{dS} = L_{dS}(t, \mathbf{x}, \dot{\mathbf{x}})$, $\partial/\partial\mathbf{x} \equiv \nabla := (\partial/\partial x^1)\mathbf{i} + (\partial/\partial x^2)\mathbf{j} + (\partial/\partial x^3)\mathbf{k}$ etc. All of the terms in Eq.(3.33) can be calculated in terms of the explicit expression Eq.(3.32) of L_{dS}. Thus, by means of the straightforward calculations from Eq.(3.32) we can prove following identity:

$$
\frac{\partial L_{dS}}{\partial \mathbf{x}} = \frac{\partial}{\partial t}\frac{\partial L_{dS}}{\partial \dot{\mathbf{x}}} + \left(\dot{\mathbf{x}} \cdot \frac{\partial}{\partial \mathbf{x}} \right) \frac{\partial L_{dS}}{\partial \dot{\mathbf{x}}}. \quad (3.34)
$$

(For a full understanding of these calculations, please refer to the exercise in the end of this section.)

Substituting Eq.(3.34) to Eq.(3.33), we have

$$
\left(\ddot{\mathbf{x}} \cdot \frac{\partial}{\partial \dot{\mathbf{x}}} \right) \frac{\partial L_{dS}}{\partial \dot{\mathbf{x}}} = 0. \quad (3.35)
$$

Since

$$
\| \frac{\partial}{\partial \dot{\mathbf{x}}}\frac{\partial L_{dS}}{\partial \dot{\mathbf{x}}} \| \equiv \det \left(\frac{\partial^2 L_{dS}}{\partial \dot{x}^i \partial \dot{x}^j} \right) \neq 0, \quad (3.36)
$$

(see problem 6 in Exercise of this section) we have

$$
\ddot{\mathbf{x}} = 0, \quad \dot{\mathbf{x}} = \mathbf{v} = \text{Const.}, \quad (3.37)
$$

which indicates that the particle motion described by Lagrangian function (3.32) is inertial, (i.e., the first Newton motion law holds) and the corresponding inertial reference systems can be built. Thus, the result (3.22) from the geodesic calculations in Beltrami spacetime $\mathcal{B}_\Lambda$ has been reproduced in the Lagrangian formulation (3.32).

Some remarks on L_{dS} of (3.32) are as follows:

(1) From Eq.(3.37) and

$$\lim_{R\to\infty} L_{dS} = L_{Einstein} \equiv -mc^2\sqrt{1 + \frac{\eta_{ij}\dot{x}^i\dot{x}^j}{c^2}}, \qquad (3.38)$$

we learned that a new kind of Special Relativity based on L_{dS} (3.32) serving as an extension of the Einstein's Special Relativity (E-SR) should exist.

(2) The precondition of Special Relativity principle is existence of inertial reference frames, in which the equation of motion for free particles is $\ddot{\mathbf{x}} = 0$. This condition leads to the following constraint that the Lagrange of free particles $L(t, \mathbf{x}, \dot{\mathbf{x}})$ must satisfy the equations as follows:

$$\frac{\partial L}{\partial \mathbf{x}} = \frac{\partial}{\partial t}\frac{\partial L}{\partial \dot{\mathbf{x}}} + \left(\dot{\mathbf{x}}\cdot\frac{\partial}{\partial \mathbf{x}}\right)\frac{\partial L}{\partial \dot{\mathbf{x}}}, \qquad (3.39)$$

$$\left\|\frac{\partial}{\partial \dot{\mathbf{x}}}\frac{\partial L}{\partial \dot{\mathbf{x}}}\right\| \equiv \det\left(\frac{\partial^2 L}{\partial \dot{x}^i \partial \dot{x}^j}\right) \neq 0. \qquad (3.40)$$

Obviously, since $L_{Einstein}$ of (A.12) depends only on $\dot{x}^i$, it satisfies Eq.(3.39) trivially. To L_{dS}, function of (3.32) could be considered a non-trivial solution of Eqs.(3.39) and (3.40), therefore the equation of motion from L_{dS} is inertial.

Exercise

(1) Problem 1, Carry out calculations of the partial derivative of the function $L_{dS}(t, \mathbf{x}, \dot{\mathbf{x}})$ with respect to x^1. Note, Eq.(3.32) indicates:

$$L_{dS} = -mc^2 R\sqrt{\frac{R^2(c^2 - \dot{\mathbf{x}}^2) - \mathbf{x}^2\dot{\mathbf{x}}^2 + (\mathbf{x}\cdot\dot{\mathbf{x}})^2 + c^2(\mathbf{x} - \dot{\mathbf{x}}t)^2}{c^2(R^2 + \mathbf{x}^2 - c^2t^2)^2}}.$$

(2) Problem 2, Derive the partial derivative of $L_{dS}(t, \mathbf{x}, \dot{\mathbf{x}})$ with respect to $\dot{x}^1$.

(3) Problem 3, Carry out calculations of the partial derivative of the function $\partial_{\dot{x}^1} L_{dS}(t, \mathbf{x}, \dot{\mathbf{x}})$ with respect to t.

(4) Problem 4, Carry out calculations of the partial derivative of the function $\partial_{\dot{x}^1} L_{dS}(t, \mathbf{x}, \dot{\mathbf{x}})$ with respect to x^i, and then times $\dot{x}^i$ with $i = 1, 2, 3$.

(5) Problem 5: Use above results to verify

$$\frac{\partial L_{dS}}{\partial \mathbf{x}} = \frac{\partial}{\partial t}\frac{\partial L_{dS}}{\partial \dot{\mathbf{x}}} + \left(\dot{\mathbf{x}}\cdot\frac{\partial}{\partial \mathbf{x}}\right)\frac{\partial L_{dS}}{\partial \dot{\mathbf{x}}}.$$

(6) Problem 6: Prove

$$\left\| \frac{\partial}{\partial \mathbf{\dot{x}}} \frac{\partial L_{dS}}{\partial \mathbf{\dot{x}}} \right\| \equiv \det \left(\frac{\partial^2 L_{dS}}{\partial \dot{x}^i \partial \dot{x}^j} \right) \neq 0.$$

3.4 Spacetime transformation preserving Beltrami metric

Recalling the descriptions in Chapter 2 on the ordinary E-SR, we learned from Eqs.(2.15)–(2.17) that the Minkowski spacetime metric $\eta_{\mu\nu}$ (to call it *inertial metric*[1] in SR since inertial motion law holds in Minkowski spacetime) is preserved by inhomogeneous Lorentz transformation (2.4). Namely under spacetime transformation of

$$x^\mu \xrightarrow{\ ISO(1,3)\ } x'^\mu, \tag{3.41}$$

we have

$$\eta_{\mu\nu} \to \eta'_{\mu\nu} = \frac{\partial x^\lambda}{\partial x'^\mu} \frac{\partial x^\rho}{\partial x'^\nu} \eta_{\lambda\rho} = \eta_{\mu\nu}. \tag{3.42}$$

To see $ISO(1,3)$ transformation more explicitly, we showed it in (2.71)–(2.74) for the case without rotations in space as follows:

$$t \to t' = \gamma(t - \beta x^1/c) + a^0/c, \tag{3.43}$$

$$x^1 \to x'^1 = \gamma(x^1 - \beta ct) + a^1, \tag{3.44}$$

$$x^2 \to x'^2 = x^2 + a^2, \tag{3.45}$$

$$x^3 \to x'^3 = x^3 + a^3, \tag{3.46}$$

where β, γ are Lorentz factors, and a^μ are translation parameters in spacetime.

We come back to continue the discussions on $B_{\mu\nu}$. In the previous sections we have proved that the Beltrami metric $B_{\mu\nu}(x)$ is also *inertial metric*. Similar to E-SR, we should derive the spacetime transformation preserving metric $B_{\mu\nu}(x)$. This transformation were firstly discovered by Lu, Zou, Guo (LZG) in 1970's [Lu (1970)][Lu, Zou and Guo (1974)]. We call it LZG-transformation in short hereafter.

---

[1] *"Inertial metric"* is a new phrase in the relativity theory. The definition is that when the motion of a free particle in a spacetime with metric $g_{\mu\nu}$ is *inertial*, we call that metric $g_{\mu\nu}$ *"inertial metric"*. So far there are two inertial metrics in the relativity theory: $\eta_{\mu\nu}$ and $B_{\mu\nu}(x)$.

Under the LZG-transformation, the $B_{\mu\nu}(x)$ were required to transfer as follows

$$B_{\mu\nu}(x) \xrightarrow{LZG} B'_{\mu\nu}(x') = \frac{\partial x^\lambda}{\partial x'^\mu} \frac{\partial x^\rho}{\partial x'^\nu} B_{\lambda\rho}(x) = B_{\mu\nu}(x'). \qquad (3.47)$$

This requirement will lead to determine the LZG spacetime transformation. What follows is to verify this property and to find out the LZG transformation.

An infinitesimal interval squared in ordinary 4-spacetime can always be written as trace of a matrix, for instance:

$$ds^2 \equiv g_{\mu\nu}dx^\mu dx^\nu = \text{tr}\{\hat{I}d\hat{X}d\hat{X}^T\hat{G}\}, \qquad (3.48)$$

where $\hat{I}$ is 1×1-identity matrix, $\hat{X} := (x^0, x^1, x^2, x^3)$ is 1×4-matrix, $\hat{X}^T$ is the transpose of $\hat{X}$ and $\hat{G} := \{g_{\mu\nu}\}$ is 4×4 matrix. Therefore, the symmetry properties of $g_{\mu\nu}$ can be revealed through discussion of the automorphism mapping in some set of $m \times n$-matrices.

Now we set X to be $m \times n$ matrix, and define the set $\mathfrak{D}_\lambda(m,n)$ to be all matrix X such that

$$I - \lambda XJX^T > 0. \qquad (3.49)$$

Here, I is the $m \times m$ identity matrix, $J = \text{diag}[1,-1,\ldots,-1]$ is $n \times n$ matrix, $\lambda = 1/R^2 \neq 0$ is a real number, and X^T is the transpose of matrix X. A real matrix $A > 0$ means that A is positive definite. Let A, B, C, D be $m \times m$, $n \times m$, $m \times n$, $n \times n$ matrices respectively, satisfying

$$\begin{pmatrix} A & C \\ B & D \end{pmatrix}\begin{pmatrix} I & 0 \\ 0 & -\lambda J \end{pmatrix}\begin{pmatrix} A & C \\ B & D \end{pmatrix}^T = \begin{pmatrix} I & 0 \\ 0 & -\lambda J \end{pmatrix}. \qquad (3.50)$$

Writing out the entries we get

$$AA^T - \lambda CJC^T = I, \quad AB^T = \lambda CJD^T, \quad BB^T - \lambda DJD^T = -\lambda J. \quad (3.51)$$

Equation (3.50) is also equivalent to

$$\begin{pmatrix} A & C \\ B & D \end{pmatrix}\begin{pmatrix} I & 0 \\ 0 & -\lambda J \end{pmatrix}\begin{pmatrix} A & C \\ B & D \end{pmatrix}^T\begin{pmatrix} I & 0 \\ 0 & -\lambda^{-1}J \end{pmatrix} = \begin{pmatrix} I & 0 \\ 0 & I \end{pmatrix}$$

$$\Leftrightarrow \begin{pmatrix} I & 0 \\ 0 & -\lambda J \end{pmatrix}\begin{pmatrix} A & C \\ B & D \end{pmatrix}^T\begin{pmatrix} I & 0 \\ 0 & -\lambda^{-1}J \end{pmatrix}\begin{pmatrix} A & C \\ B & D \end{pmatrix} = \begin{pmatrix} I & 0 \\ 0 & I \end{pmatrix}$$

$$\Leftrightarrow \begin{pmatrix} A & C \\ B & D \end{pmatrix}^T\begin{pmatrix} I & 0 \\ 0 & -\lambda^{-1}J \end{pmatrix}\begin{pmatrix} A & C \\ B & D \end{pmatrix} = \begin{pmatrix} I & 0 \\ 0 & -\lambda^{-1}J \end{pmatrix}$$

$$\Leftrightarrow \begin{pmatrix} A & C \\ B & D \end{pmatrix}^T\begin{pmatrix} \lambda I & 0 \\ 0 & -J \end{pmatrix}\begin{pmatrix} A & C \\ B & D \end{pmatrix} = \begin{pmatrix} \lambda I & 0 \\ 0 & -J \end{pmatrix}. \qquad (3.52)$$

Writing out the entries we get

$$\lambda A^T A - B^T J B = \lambda I, \quad \lambda A^T C = B^T J D, \quad \lambda C^T C - D^T J D = -J. \quad (3.53)$$

Therefore, Eqs.(3.51) and (3.53) are equivalent. Observing that Eq.(3.49) can be written as

$$\begin{pmatrix} I & X \end{pmatrix} \begin{pmatrix} I & 0 \\ 0 & -\lambda J \end{pmatrix} \begin{pmatrix} I \\ X^T \end{pmatrix} > 0 \quad (3.54)$$

we use Eq.(3.50) to get

$$\begin{pmatrix} I & X \end{pmatrix} \begin{pmatrix} A & C \\ B & D \end{pmatrix} \begin{pmatrix} I & 0 \\ 0 & -\lambda J \end{pmatrix} \begin{pmatrix} A & C \\ B & D \end{pmatrix}^T \begin{pmatrix} I \\ X^T \end{pmatrix} > 0$$

$$\Leftrightarrow (A + XB) \begin{pmatrix} I & Y \end{pmatrix} \begin{pmatrix} I & 0 \\ 0 & -\lambda J \end{pmatrix} \begin{pmatrix} I \\ Y^T \end{pmatrix} (A + XB)^T > 0$$

$$\Leftrightarrow \begin{pmatrix} I & Y \end{pmatrix} \begin{pmatrix} I & 0 \\ 0 & -\lambda J \end{pmatrix} \begin{pmatrix} I \\ Y^T \end{pmatrix} > 0,$$

here

$$Y = (A + XB)^{-1}(C + XD). \quad (3.55)$$

Therefore the transformation (3.55) maps $\mathfrak{D}_\lambda(m,n)$ to itself and is an automorphism. We can define a metric

$$ds^2 = \mathrm{Tr}\left\{ (I - \lambda X J X^T)^{-1} dX (J - \lambda X^T X)^{-1} dX^T \right\}. \quad (3.56)$$

We claim that this metric is invariant under the transformation (3.55)

Proof: Note that from the above discussion we have

$$(I - \lambda X J X^T) = (A + XB)(I - \lambda Y J Y^T)(A + XB)^T. \quad (3.57)$$

Since $X = (AY - C)(D - BY)^{-1}$, we also have

$$J - \lambda Y^T Y$$

$$= \begin{pmatrix} Y^T & -I \end{pmatrix} \begin{pmatrix} -\lambda I & 0 \\ 0 & J \end{pmatrix} \begin{pmatrix} Y \\ -I \end{pmatrix}$$

$$= \begin{pmatrix} Y^T & -I \end{pmatrix} \begin{pmatrix} A & C \\ B & D \end{pmatrix}^T \begin{pmatrix} -\lambda I & 0 \\ 0 & J \end{pmatrix} \begin{pmatrix} A & C \\ B & D \end{pmatrix} \begin{pmatrix} Y \\ -I \end{pmatrix}$$

$$= (D - BY)^T \begin{pmatrix} X^T & -I \end{pmatrix} \begin{pmatrix} -\lambda I & 0 \\ 0 & J \end{pmatrix} \begin{pmatrix} X \\ -I \end{pmatrix} (D - BY)$$

$$= (D - BY)^T (J - \lambda X^T X)(D - BY)$$

$$dY = d((A + XB)^{-1}(C + XD))$$
$$= [-(A+XB)^{-1}d(A+XB)(A+XB)^{-1}(C+XD)+(A+XB)^{-1}d(C+XD)]$$
$$= [-(A + XB)^{-1}dXBY + (A + XB)^{-1}dXD]$$
$$= (A + XB)^{-1}dX(D - BY)$$
$$dY^T = d((A + XB)^{-1}(C + XD))^T = (D - BY)^T dX^T (A + XB)^{T-1}$$

hence

$$tr\left\{(I - \lambda Y J Y^T)^{-1}dY(J - \lambda Y^T Y)^{-1}dY^T\right\}$$
$$= tr\{(A+XB)^T(I-\lambda XJX^T)^{-1}(A+XB)(A+XB)^{-1}dX(D-BY)$$
$$(D-BY)^{-1}(J - \lambda X^T X)^{-1}(D - BY)'^{-1}(D - BY)^T dX^T(A+XB)^{T-1}\}$$
$$= tr\left\{(I - \lambda XJX^T)^{-1}dX(J - \lambda X^T X)^{-1}dX^T\right\} \tag{3.58}$$

which states that the metric (3.56) is invariant under transformation (3.55).

$$\square$$

If we let $X_0 = -CD^{-1}$ (i.e., $X_0 \equiv X(Y = 0)$), then

$$Y = (A + XB)^{-1}(C + XD) = A^{-1}(I + XBA^{-1})^{-1}(X + CD^{-1})D$$

The conditions in (3.53) are equivalent to the following

$$BA^{-1} = (\lambda CD^{-1}J)^T = \lambda JX_0^T$$
$$(AA^T)^{-1} = A^{T-1}(A^T A - \lambda^{-1}B^T JB)A^{-1} = (I - \lambda X_0 JX_0^T)$$
$$(DJD^T)^{-1} = D^{T-1}JD^{-1} = D^{T-1}(D^T JD - \lambda C^T C)D^{-1} = J - \lambda X_0^T X_0$$

Hence we get the formula

$$Y = A^{-1}(I - \lambda XJX_0)^{-1}(X - X_0)D \tag{3.59}$$

where the matrices A, D satisfy

$$AA^T = (I - \lambda X_0 JX_0^T)^{-1}, \quad DJD^T = (J - \lambda X_0^T X_0)^{-1} \tag{3.60}$$

For the special case $\mathfrak{D}_\lambda(1,4)$ in this book, $X = (x^0, x^1, x^2, x^3)$, condition for $\mathfrak{D}_\lambda(1,4)$ is just

$$1 - \lambda \eta_{\mu\nu}x^\mu x^\nu > 0 \tag{3.61}$$

The metric (3.56) now takes the form

$$ds^2 = \frac{dX(J - \lambda X^T X)^{-1}dX^T}{1 - \lambda XJX^T} = g_{\mu\nu}dx^\mu dx^\nu \tag{3.62}$$

$$g_{\mu\nu} = \frac{\eta_{\mu\nu}}{1 - \lambda \eta_{\lambda\rho}x^\lambda x^\rho} + \frac{\lambda \eta_{\mu\lambda}\eta_{\nu\rho}x^\lambda x^\rho}{(1 - \lambda \eta_{\alpha\beta}x^\alpha x^\beta)^2}$$

Comparing Eq.(3.56) with Eq.(3.6), we see $g_{\mu\nu}$ is just the Beltrami metric, i.e., $g_{\mu\nu} = B_{\mu\nu}(x)$. By our claim, this metric is invariant under the transformation (3.59), which now becomes

$$y^\mu \equiv x'^\mu = \sqrt{1 - \lambda \eta_{\lambda\rho} a^\lambda a^\rho} \frac{(x^\nu - a^\nu) D_\nu^\mu}{1 - \lambda \eta_{\alpha\beta} a^\alpha x^\beta} \tag{3.63}$$

where we denote $X_0 = (a^0, a^1, a^2, a^3)$ and $\{D_\nu^\mu\}$ are constants satisfying

$$\eta_{\lambda\rho} D_\mu^\lambda D_\nu^\rho = \eta_{\mu\nu} + \frac{\lambda \eta_{\mu\lambda} \eta_{\nu\rho} a^\lambda a^\rho}{1 - \lambda \eta_{\alpha\beta} a^\alpha a^\beta}. \tag{3.64}$$

Using the notations: $x'^\mu = y^\mu$, $\sigma(x) = 1 - \lambda \eta_{\alpha\beta} x^\alpha x^\beta$, $\sigma(a, x) = 1 - \lambda \eta_{\alpha\beta} a^\alpha x^\beta$, we rewrite Eqs.(3.63) and (3.64) as follows

$$x'^\mu = \sqrt{\sigma(a)} \frac{(x^\nu - a^\nu) D_\nu^\mu}{\sigma(a, x)} \quad , \tag{3.65}$$

$$\eta_{\lambda\rho} D_\mu^\lambda D_\nu^\rho = \eta_{\mu\nu} + \frac{\lambda \eta_{\mu\lambda} \eta_{\nu\rho} a^\lambda a^\rho}{\sigma(a)}. \tag{3.66}$$

Taking ansatz

$$D_\nu^\mu = \pm(L_\nu^\mu + A\lambda \eta_{\nu\lambda} a^\lambda a^\rho L_\rho^\mu) \tag{3.67}$$

where A is a constant which is determined by the normalization constraint Eq.(3.66):

$$A = \frac{1}{\sigma(a) + \sqrt{\sigma}}. \tag{3.68}$$

and substituting Eqs.(3.66), (3.67) and (3.68) into Eq.(3.65), we finally obtain

$$x'^\mu = \pm\sqrt{\sigma(a)} \sigma(a, x)^{-1} (x^\nu - a^\nu) \left(L_\nu^\mu + R^{-2} \frac{1}{\sigma(a) + \sqrt{\sigma}} \eta_{\nu\rho} a^\rho a^\lambda L_\lambda^\mu \right) \tag{3.69}$$

where $\lambda = R^{-2}$ has been used. Equation (3.69) means that when we transform from one initial Beltrami frame x^μ to another Beltrami frame x'^μ, and when the origin of the new frame is a^μ in the original frame, the transformations between them with 10 parameters is as follows

$$x^\mu \xrightarrow{\text{LZG}} x'^\mu = \pm\sigma(a)^{1/2} \sigma(a, x)^{-1} (x^\nu - a^\nu) D_\nu^\mu, \tag{3.70}$$

$$D_\nu^\mu = L_\nu^\mu + R^{-2} \eta_{\nu\rho} a^\rho a^\lambda (\sigma(a) + \sigma^{1/2}(a))^{-1} L_\lambda^\mu,$$

$$L := (L_\nu^\mu) \in SO(1, 3),$$

$$\sigma(x) = 1 - \frac{1}{R^2} \eta_{\mu\nu} x^\mu x^\nu,$$

$$\sigma(a, x) = 1 - \frac{1}{R^2} \eta_{\mu\nu} a^\mu x^\nu.$$

On the other hand, for $\mathfrak{D}_\lambda(1,4)$, Eq.(3.58) indicates that under LZG transformation of (3.55) $X \to Y = (A + XB)^{-1}(C + XD)$ we have:

$$
\begin{aligned}
ds'^2 &= B_{\mu\nu}(x')dx'^\mu dx'^\nu = \text{Tr}\left\{(I - \lambda YJY^T)^{-1}dY(J - \lambda Y^TY)^{-1}dY^T\right\} \\
&= \text{Tr}\left\{(I - \lambda XJX^T)^{-1}dX(J - \lambda X^TX)^{-1}dX^T\right\} \\
&= ds^2 = B_{\mu\nu}(x)dx^\mu dx^\nu,
\end{aligned} \tag{3.71}
$$

which means

$$
B_{\mu\nu}(x') = \frac{\partial x^\lambda}{\partial x'^\mu}\frac{\partial x^\rho}{\partial x'^\nu}B_{\lambda\rho}(x). \tag{3.72}
$$

Since $B_{\mu\nu}(x)$ is a tensor, under spacetime transformation $x^\mu \to x'^\mu$, it transforms as (see Eq.(2.154))

$$
B_{\mu\nu}(x) \to B'_{\mu\nu}(x') = \frac{\partial x^\lambda}{\partial x'^\mu}\frac{\partial x^\rho}{\partial x'^\nu}B_{\lambda\rho}(x). \tag{3.73}
$$

Combining Eq.(3.72) with Eq.(3.73), we conclude that under LZG transformation of (3.70) the Beltrami metric's transformation is

$$
B_{\mu\nu}(x) \xrightarrow{\text{LZG}} B_{\mu\nu}(x'), \tag{3.74}
$$

which is just Eq.(3.47). From Eqs.(3.23) and (3.74), we have

$$
\begin{aligned}
S_{dS} \xrightarrow{\text{LZG}} S'_{dS} &= -mc\int\sqrt{B_{\mu\nu}(x')dx'^\mu dx'^\nu} \\
&= -mc\int\sqrt{B_{\mu\nu}(x)dx^\mu dx^\nu} = S_{dS}.
\end{aligned} \tag{3.75}
$$

Thus, in parallel to the ordinary E-SR, the de Sitter invariant Special Relativity has been formulated.

Exercise

(1) Problem 1, Derive Eqs.(3.51) and (3.53) from Eq.(3.50).
(2) Problem 2, Basics: let $\mathcal{A}$ be a $m \times n$ matrix whose elements are functions of the scalar parameter α. Then the derivative of the matrix $\mathcal{A}$ with respect to the scalar parameter α is the $m \times n$ matrix of element-by-element derivatives:

$$
\mathcal{A} = \begin{pmatrix} a_{11} & a_{12} & \cdots & a_{1n} \\ \cdots & \cdots & \cdots & \cdots \\ a_{m1} & a_{m2} & \cdots & a_{mn} \end{pmatrix}, \tag{3.76}
$$

$$
\frac{\partial \mathcal{A}}{\partial \alpha} = \begin{pmatrix} \frac{\partial a_{11}}{\partial \alpha} & \frac{\partial a_{12}}{\partial \alpha} & \cdots & \frac{\partial a_{1n}}{\partial \alpha} \\ \cdots & \cdots & \cdots & \cdots \\ \frac{\partial a_{m1}}{\partial \alpha} & \frac{\partial a_{m2}}{\partial \alpha} & \cdots & \frac{\partial a_{mn}}{\partial \alpha} \end{pmatrix}, \tag{3.77}
$$

and

$$dA \equiv \frac{\partial A}{\partial \alpha} d\alpha = \begin{pmatrix} \frac{\partial a_{11}}{\partial \alpha} d\alpha & \frac{\partial a_{12}}{\partial \alpha} d\alpha & \cdots & \frac{\partial a_{1n}}{\partial \alpha} d\alpha \\ \cdots & \cdots & \cdots & \cdots \\ \frac{\partial a_{m1}}{\partial \alpha} d\alpha & \frac{\partial a_{m2}}{\partial \alpha} d\alpha & \cdots & \frac{\partial a_{mn}}{\partial \alpha} d\alpha \end{pmatrix}. \tag{3.78}$$

Now, let A be a nonsingular, $m \times m$ matrix. Please prove:

$$dA^{-1} = -A^{-1} dA A^{-1}. \tag{3.79}$$

3.5 Noether charges of dS-SR mechanics

In Section 2.5, we have generally proved the Noether theorem. This theorem claims that the invariance of Action of mechanics under symmetry transformations will lead to constants of motion, or conserved Noether charges. In the last section, we proved that under the LZG-transformation (3.70) the Action of dS-SR S_{dS} (3.23) is invariant, i.e., Eq.(3.75). Therefore we can derive the Noether charges of dS-SR mechanics. In Section 2.6, the procedure to derive Noether charges of E-SR has been shown. What follows is to do so for dS-SR.

The Action

$$S_{dS} = \int L_{dS} dt, \tag{3.80}$$

where L_{dS} is the Lagrange Eq.(3.32) of dS-SR mechanics, is invariant under the transformation (3.70). When space rotations were neglected temporarily for simplicity, the LZG-transformation that entails a Lorentz-like boost in the x^1 direction (with parameters $\beta = u^1/c$ and a^1 respectively) and a time transition (with parameter a^0) can be explicitly written as follows:

$$t \to t' = \frac{\sqrt{\sigma(a)}}{c\sigma(a,x)} \gamma \left[ct - \beta x^1 - a^0 + \beta a^1 + \frac{a^0 - \beta a^1}{R^2} \frac{a^0 ct - a^1 x^1 - (a^0)^2 + (a^1)^2}{\sigma(a) + \sqrt{\sigma(a)}} \right]$$

$$x^1 \to x'^1 = \frac{\sqrt{\sigma(a)}}{\sigma(a,x)} \gamma \left[x^1 - \beta ct + \beta a^0 - a^1 + \frac{a^1 - \beta a^0}{R^2} \frac{a^0 ct - a^1 x^1 - (a^0)^2 + (a^1)^2}{\sigma(a) + \sqrt{\sigma(a)}} \right]$$

$$x^2 \to x'^2 = \frac{\sqrt{\sigma(a)}}{\sigma(a,x)} x^2 \tag{3.81}$$

$$x^3 \to x'^3 = \frac{\sqrt{\sigma(a)}}{\sigma(a,x)} x^3$$

The Noether theorem condition Eq.(2.46) becomes

$$\left[\frac{\partial}{\partial \epsilon} \{ L_{dS} \left(t'(t, \mathbf{x}, \dot{\mathbf{x}}, \epsilon), \mathbf{x}'(t, \mathbf{x}, \dot{\mathbf{x}}, \epsilon), \dot{\mathbf{x}}'(t, \mathbf{x}, \dot{\mathbf{x}}, \ddot{\mathbf{x}}, \epsilon) \right) \dot{t}'(t, \mathbf{x}, \dot{\mathbf{x}}, \ddot{\mathbf{x}}, \epsilon) \} \right]_{\epsilon=0} = \dot{F}, \tag{3.82}$$

and the corresponding Noether charge (2.47) is:

$$G \equiv -L_{dS}\xi - \sum_i \frac{\partial L_{dS}(t, \mathbf{x}, \dot{\mathbf{x}})}{\partial \dot{x}^i}(\eta^i - \dot{x}^i \xi) + F \qquad (3.83)$$

where

$$\xi = \left[\frac{\partial t'(t, \mathbf{x}, \dot{\mathbf{x}}, \epsilon)}{\partial \epsilon} \right]_{\epsilon=0}, \qquad (3.84)$$

$$\eta^i = \left[\frac{\partial x'^i(t, \mathbf{x}, \dot{\mathbf{x}}, \epsilon)}{\partial \epsilon} \right]_{\epsilon=0}. \qquad (3.85)$$

All the Noether charges for dS-SR are as follows:

(1) Energy: The Lagrangian function (3.32) is time dependent. When let $\beta = a^1 = a^2 = a^3 = 0$, $a^0 = \epsilon c$ in Eq.(3.81), we have

$$t' = t - \left(1 - \frac{c^2 t^2}{R^2} \right)\epsilon, \qquad (3.86)$$

$$x'^i = x^i \left(1 + \frac{c^2 t}{R^2}\epsilon \right), \qquad (3.87)$$

and then

$$\xi = -1 + \frac{c^2 t^2}{R^2}, \quad \eta^i = x^i \frac{c^2 t}{R^2}, \qquad (3.88)$$

$$\dot{t}' = 1 + \frac{2c^2 t}{R^2}\epsilon, \quad \dot{x}'^i = \dot{x}^i \left(1 + \frac{c^2 t\epsilon}{R^2} \right) + x^i \frac{c^2 \epsilon}{R^2}, \qquad (3.89)$$

$$\acute{x}'^i = \dot{x}^i \left(1 - \frac{c^2 t\epsilon}{R^2} \right) + x^i \frac{c^2 \epsilon}{R^2} + \mathcal{O}(\epsilon^2). \qquad (3.90)$$

Firstly we verify the Noether theorem condition (3.82). Substituting Eqs.(3.86), (3.87) and (3.90) into $L_{dS}(t', \mathbf{x}', \acute{\mathbf{x}}'))$, we can get

$$L_{dS}^{(a^0)}(t', \mathbf{x}', \acute{\mathbf{x}}') = L_{dS}(t, \mathbf{x}, \dot{\mathbf{x}}) \left(1 - \frac{2c^2 t}{R^2}\epsilon + \mathcal{O}(\epsilon^2) \right), \qquad (3.91)$$

where the $L_{dS}^{(a^0)}$'s superscript (a^0) indicates that the variables of $t', \mathbf{x}', \acute{\mathbf{x}}'$ in the Lagrange arise from a time transition in LZG-transformation (3.81) (see Eqs.(3.86) and (3.87)). And then, from Eq.(3.89) we have

$$L_{dS}^{(a^0)}(t', \mathbf{x}', \acute{\mathbf{x}}')\dot{t}' = L_{dS}(t, \mathbf{x}, \dot{\mathbf{x}}) + \mathcal{O}(\epsilon^2). \qquad (3.92)$$

This equation leads to

$$\left[\frac{\partial}{\partial \epsilon} \left\{ L_{dS}^{(a^0)}(t', \mathbf{x}', \acute{\mathbf{x}}')\dot{t}' \right\} \right]_{\epsilon=0} = 0, \quad \text{and then}: F = 0. \qquad (3.93)$$

Therefore we conclude that the Noether theorem condition Eq.(3.82) is satisfied and $F = 0$ (where the convention (2.79) were used).

Inserting Eqs.(3.88) and (3.93) into Eq.(3.83) and noting (G.3), we obtain the energy Noether charge of dS-SR:

$$G_{a^0} = -L_{dS}\left(-1 + \frac{c^2 t^2}{R^2}\right) - \sum_{i=1}^{3}\left[x^i \frac{c^2 t}{R^2} - \dot{x}^i\left(-1 + \frac{c^2 t^2}{R^2}\right)\right]\left[\left(\frac{m^2 c^4 R^2}{L_{dS}}\right)\right.$$

$$\left. \times \left(\frac{-R^2 \dot{x}^i - \mathbf{x}^2 \dot{x}^i + (\mathbf{x}\cdot\dot{\mathbf{x}})x^i - c^2 t(x^i - \dot{x}^i t)}{c^2(R^2 + \mathbf{x}^2 - c^2 t^2)^2}\right)\right]. \tag{3.94}$$

Substituting Eq.(3.32) (or (3.32)) into Eq.(3.94), we get the energy of dS-SR mechanics:

$$G_{a^0} \equiv E = \frac{mc^2}{\sqrt{1 - \frac{\dot{\mathbf{x}}^2}{c^2} + \frac{(\mathbf{x}\cdot\dot{\mathbf{x}})^2 - \mathbf{x}^2\dot{\mathbf{x}}^2}{c^2 R^2} + \frac{(\mathbf{x} - \dot{\mathbf{x}}t)^2}{R^2}}}. \tag{3.95}$$

Two remarks are follows:

(i) Introducing a notation as follows

$$\Gamma = \frac{1}{\sqrt{1 - \frac{\dot{\mathbf{x}}^2}{c^2} + \frac{(\mathbf{x}\cdot\dot{\mathbf{x}})^2 - \mathbf{x}^2\dot{\mathbf{x}}^2}{c^2 R^2} + \frac{(\mathbf{x} - \dot{\mathbf{x}}t)^2}{R^2}}} \tag{3.96}$$

$$= \frac{R}{\sqrt{(R^2 - \eta_{ij}x^i x^j)(1 + \frac{\eta_{ij}\dot{x}^i\dot{x}^j}{c^2}) + 2t\eta_{ij}x^i\dot{x}^j - \eta_{ij}\dot{x}^i\dot{x}^j t^2 + \frac{(\eta_{ij}x^i\dot{x}^j)^2}{c^2}}},$$

which is called *dS-SR Lorentz factor*, we can shortly re-write Eq.(3.95) and Eq.(3.32) as

$$E = mc^2\Gamma, \tag{3.97}$$

$$L_{dS} = -mc^2(\sigma\Gamma)^{-1}, \tag{3.98}$$

where σ has been given in Eq.(3.5). When $|R| \to \infty$, we have $\Gamma \to \gamma \equiv (1 - \dot{\mathbf{x}}^2/c^2)^{-1/2}$ that is usual *Lorentz factor*. And then $E = mc^2\Gamma$ approaches to $E = mc^2\gamma$ which is the usual energy formula in E-SR, and L_{dS} approaches to the usual E-SR's Lagrange (2.69).

(ii) From the equation of motion $\ddot{\mathbf{x}} = 0$ (i.e., (3.37)), we can easily check that

$$\dot{E} = mc^2\dot{\Gamma} = 0, \tag{3.99}$$

which indicates that the energy in dS-SR is conserved.

(2) Momentum: We let $\beta = a^0 = a^2 = a^3 = 0,\ a^1 = \epsilon$ in Eq.(3.81). Then we have

$$t' = t\left(1 - \frac{x^1 \epsilon}{R^2}\right), \tag{3.100}$$

$$x'^1 = x^1 - \epsilon\left(1 + \frac{(x^1)^2}{R^2}\right),\quad x'^2 = x^2\left(1 + \frac{x^1 \epsilon}{R^2}\right),\quad x'^3 = x^3\left(1 + \frac{x^1 \epsilon}{R^2}\right), \tag{3.101}$$

and

$$\xi = -\frac{1}{R^2}(x^1 + \dot{x}^1),\ \eta^1 = -\left(1 + \frac{(x^1)^2}{R^2}\right),\ \eta^2 = \frac{x^1 x^2}{R^2},\ \eta^3 = \frac{x^1 x^3}{R^2} \tag{3.102}$$

$$\dot{t}' = 1 - \frac{1}{R^2}(x^1 + t\dot{x}^1), \tag{3.103}$$

$$\dot{x}'^1 = \dot{x}^1\left(1 - \frac{2x^1 \epsilon}{R^2}\right),\ \dot{x}'^2 = \dot{x}^2\left(1 - \frac{x^1 \epsilon}{R^2}\right) - \frac{x^2 \epsilon}{R^2}\dot{x}^1,$$

$$\dot{x}'^3 = \dot{x}^3\left(1 - \frac{x^1 \epsilon}{R^2}\right) - \frac{x^3 \epsilon}{R^2}\dot{x}^1, \tag{3.104}$$

$$\acute{x}'^i \equiv \frac{\dot{x}'^i}{\dot{t}'}. \tag{3.105}$$

Next, we verify the Noether theorem condition (3.82). Substituting Eqs.(3.100), (3.101) and (3.105) into $L_{dS}\left(t', \mathbf{x}', \acute{\mathbf{x}}'\right)$, we get

$$L_{dS}^{(a^1)}\left(t', \mathbf{x}', \acute{\mathbf{x}}'\right) = L_{dS}(t, \mathbf{x}, \dot{\mathbf{x}})\left(1 + \frac{1}{R^2}(x^1 + t\dot{x}^1)\epsilon + \mathcal{O}(\epsilon^2)\right) \tag{3.106}$$

where the $L_{dS}^{(a^1)}$'s superscript (a^1) indicates that the variables of $t', \mathbf{x}', \acute{\mathbf{x}}'$ in the Lagrange arise from a x^1-transition in LZG-transformation (3.81) (see Eqs.(3.100) and (3.101)). And then, from Eqs.(3.106) and (3.103) we have

$$L_{dS}^{(a^1)}\left(t', \mathbf{x}', \acute{\mathbf{x}}'\right)\dot{t}' = L_{dS}(t, \mathbf{x}, \dot{\mathbf{x}}) + \mathcal{O}(\epsilon^2). \tag{3.107}$$

This equation leads to

$$\left[\frac{\partial}{\partial \epsilon}\left\{L_{dS}^{(a^1)}\left(t', \mathbf{x}', \acute{\mathbf{x}}'\right)\dot{t}'\right\}\right]_{\epsilon=0} = 0,\quad \text{and then} : F = 0. \tag{3.108}$$

Therefore we see also that the Noether theorem condition Eq.(3.82) is satisfied and $F = 0$ in the x^1-transition case (the convention (2.79) were used). Inserting Eqs.(3.102) and (3.108) into Eq.(3.83) and noting (G.3), we obtain the energy Noether charge of dS-SR:

$$
\begin{aligned}
G_{a^1} = L_{dS}\frac{x^1 t}{R^2} &- \sum_{i=1}^{3}\left[\frac{-x^1 x^i}{R^2} - \dot{x}^i\frac{tx^1}{R^2}\right]\left[\left(\frac{m^2 c^4 R^2}{L_{dS}}\right)\right.\\
&\times \left.\left(\frac{-R^2\dot{x}^i - \mathbf{x}^2\dot{x}^i + (\mathbf{x}\cdot\dot{\mathbf{x}})x^i - c^2 t(x^i - \dot{x}^i t)}{c^2(R^2 + \mathbf{x}^2 - c^2 t^2)^2}\right)\right]\\
&+ \left[\left(\frac{m^2 c^4 R^2}{L_{dS}}\right)\left(\frac{-R^2\dot{x}^1 - \mathbf{x}^2\dot{x}^1 + (\mathbf{x}\cdot\dot{\mathbf{x}})x^1 - c^2 t(x^1 - \dot{x}^1 t)}{c^2(R^2 + \mathbf{x}^2 - c^2 t^2)^2}\right)\right].
\end{aligned}
$$
(3.109)

Substituting Eq.(3.32) (or (3.32)) into Eq.(3.94), we get the momentum p^1 of dS-SR mechanics:

$$
G_{a^1} \equiv p^1 = \frac{m\dot{x}^1}{\sqrt{1 - \frac{\dot{x}^2}{c^2} + \frac{(\mathbf{x}\cdot\dot{\mathbf{x}})^2 - \mathbf{x}^2\dot{\mathbf{x}}^2}{c^2 R^2} + \frac{(\mathbf{x} - \dot{\mathbf{x}}t)^2}{R^2}}} = \text{constant.} \quad (3.110)
$$

Similarly, $G_{a^i} \equiv p^i = $ constant to $i = 2,\ 3$. Therefore, we have

$$
\begin{aligned}
p^i &= \frac{m\dot{x}^i}{\sqrt{1 - \frac{\dot{x}^2}{c^2} + \frac{(\mathbf{x}\cdot\dot{\mathbf{x}})^2 - \mathbf{x}^2\dot{\mathbf{x}}^2}{c^2 R^2} + \frac{(\mathbf{x} - \dot{\mathbf{x}}t)^2}{R^2}}}\\
&= m\dot{x}^i\Gamma.
\end{aligned}
\quad (3.111)
$$

Under $\ddot{\mathbf{x}}^i = 0$, we have

$$
\dot{p}^i = 0, \quad \text{or} \quad \dot{\mathbf{p}} = 0. \quad (3.112)
$$

(3) Lorentz boost: Taking $\beta = \epsilon$, $a^0 = a^1 = a^2 = a^3 = 0$ in Eq.(3.81), then we have

$$
\begin{aligned}
t' &= t - \frac{\epsilon x^1}{c}, \quad (\text{note}: \ \gamma \equiv 1/\sqrt{1 - \beta^2} \simeq 1 + \mathcal{O}(\epsilon^2))\\
x'^1 &= x^1 - \epsilon ct, \quad x'^2 = x^2, \quad x'^3 = x^3,
\end{aligned}
\quad (3.113)
$$

As always, we firstly check the condition Eq.(3.82) and determine the function F. From Eq.(3.113) we have

$$
\dot{x}'^i \equiv \frac{d\dot{x}'^i/dt}{dt'/dt} = \frac{\dot{x}^i - c\epsilon\delta_1^i}{1 - \frac{\epsilon}{c}\dot{x}^1},
$$

and $L_{dS}^{(\beta)}(t', \mathbf{x}', \dot{\mathbf{x}}') = L_{dS}(t, \mathbf{x}, \dot{\mathbf{x}})\left(1 + \frac{\dot{x}^1}{c}\epsilon + \mathcal{O}(\epsilon^2)\right).$ (3.114)

Noting $\dot{t}' = 1 - \frac{\dot{x}^1}{c}\epsilon$, we have

$$L_{dS}^{(\beta)}(t', \mathbf{x}', \dot{\mathbf{x}}')\dot{t}' = L_{dS}(t, \mathbf{x}, \dot{\mathbf{x}})\left(1 + \frac{\dot{x}^1}{c}\epsilon\right)\left(1 - \frac{\dot{x}^1}{c}\epsilon\right) + \mathcal{O}(\epsilon^2)$$

$$= L_{dS}(t, \mathbf{x}, \dot{\mathbf{x}}) + \mathcal{O}(\epsilon^2),$$

and then $\left[\dfrac{\partial}{\partial\epsilon}\left\{L_{dS}^{(\beta)}(t', \mathbf{x}', \dot{\mathbf{x}}')\dot{t}'\right\}\right]_{\epsilon=0} = \left[\dfrac{\partial}{\partial\epsilon}\{\mathcal{O}(\epsilon^2)\}\right]_{\epsilon=0} = 0.$ (3.115)

Hence, when $F = 0$ (noting the convention (2.79)), the condition of Eq.(3.82) is satisfied. Next, from the choice of Eq.(3.113), we have

$$\xi = \frac{\partial t'}{\partial\epsilon} = -\frac{x^1}{c}, \quad \eta^1 = \frac{\partial x'^1}{\partial\epsilon} = -ct, \quad \eta^2 = \eta^3 = \dot{x}^2 = \dot{x}^3 = 0.$$

(3.116)

And from Eqs.(3.83) and (3.32), we obtain the boost charge:

$$G_{\beta^1} \equiv K^1 = -L_{dS}\xi - \sum_i \frac{\partial L_{dS}(t, \mathbf{x}, \dot{\mathbf{x}})}{\partial \dot{x}^i}(\eta^i - \dot{x}^i\xi) + F$$

$$= -\frac{mc}{\sqrt{1 - \frac{\dot{\mathbf{x}}^2}{c^2} + \frac{(\mathbf{x}\cdot\dot{\mathbf{x}})^2 - \mathbf{x}^2\dot{\mathbf{x}}^2}{c^2 R^2} + \frac{(\mathbf{x}-\dot{\mathbf{x}}t)^2}{R^2}}}(x^1 - t\dot{x}^1). \text{ (3.117)}$$

Similarly, one can repeat above calculations for the cases of $\{\beta^{(2)} \neq 0, \beta^{(1)} = \beta^{(3)} = 0\}$ (or $\{\beta^{(3)} \neq 0, \beta^{(1)} = \beta^{(2)} = 0\}$), and obtains the generic expression for the Lorentz boost Noether charge:

$$K^i = -\frac{mc}{\sqrt{1 - \frac{\dot{\mathbf{x}}^2}{c^2} + \frac{(\mathbf{x}\cdot\dot{\mathbf{x}})^2 - \mathbf{x}^2\dot{\mathbf{x}}^2}{c^2 R^2} + \frac{(\mathbf{x}-\dot{\mathbf{x}}t)^2}{R^2}}}(x^i - t\dot{x}^i)$$

$$= -mc\Gamma(x^i - t\dot{x}^i), \quad \text{with } i = 1, 2, 3. \tag{3.118}$$

It is easy to check

$$\dot{K}^i = 0, \quad \text{or } \dot{\mathbf{K}} = 0. \tag{3.119}$$

(4) Angular momentum: Letting $a^\mu = 0$, we can see that the LZG- transformation (3.70) becomes

$$x^\mu \to x'^\mu = M^\mu{}_\nu x^\nu, \tag{3.120}$$

which is just the usual Lorentz transformation (see: Eq.(2.16)). Therefore the Lagrangian of L_{dS} (3.32) is invariant under space rotations. The Noether charges corresponding to such invariance of Lagrange L of E-SR (2.69) have been derived in Section 2.6 in detail (see:

Eqs.(2.96)–(2.104)). Such derivations and discussions are valid for dS-SR except that L of (2.69) E-SR should be replaced by L_{dS} of Eq.(3.32). After such a replacement we obtain the angular momentums of dS-SR mechanics L^i which are the Noether charges arises from the space rotation invariance:

$$L^i = m\Gamma\epsilon^i{}_{jk}x^j\dot{x}^k, \quad \text{with } i = 1,\ 2,\ 3,\ \text{and } \epsilon^{123} = +1, \quad (3.121)$$

where Γ has been given in Eq.(3.96). It is also easy to check that

$$\dot{L}^i = 0. \tag{3.122}$$

Summary: In above we have derived all 10 conserved Noether chargers (or motion integrals) in the de Sitter invariant special relativistic mechanics. Here we listed them as follows:

$$
\begin{aligned}
E &= mc^2\Gamma, \\
p^i &= m\Gamma\dot{x}^i, \\
K^i &= mc\Gamma(x^i - t\dot{x}^i) = mc\Gamma x^i - tp^i, \\
L^i &= -m\Gamma\epsilon^i{}_{jk}x^j\dot{x}^k = -\epsilon^i{}_{jk}x^j p^k.
\end{aligned}
\tag{3.123}
$$

Here $E, \mathbf{p}, \mathbf{L}, \mathbf{K}$ are conserved physical energy, momentum, angular-momentum and boost charges respectively, and Γ is the Lorentz factor of de Sitter invariant special relativity (see Eq.(3.96)).

3.6 Hamilton and canonical momenta

As an essential advantage in the Lagrangian-Hamiltonian formulation over others, both canonical momentum π_i conjugate to the Beltrami -coordinate x^i and canonical energy H_{cR} (or Hamilton) conjugate to the Beltrami -time t for free particles in the inertial reference frame can be determined rationally by the mechanical principle. In the Section 2.4, the canonical formulation of E-SR mechanics was described. This Section is devoted to the description of the canonical formulation of E-SR mechanics.

From Eq.(3.29) (or (3.32)), the canonical momentum and the canonical energy (or Hamiltonian) can be derived:

$$\pi_i = \frac{\partial L_{dS}}{\partial \dot{x}^i} = -m\sigma(x)\Gamma B_{i\mu}\dot{x}^\mu \tag{3.124}$$

$$H_{dS} = \sum_{i=1}^{3} \frac{\partial L_{dS}}{\partial \dot{x}^i}\dot{x}^i - L_{dS} = mc\sigma(x)\Gamma B_{0\mu}\dot{x}^\mu. \tag{3.125}$$

where Γ has been given in Eq.(3.96). Generally, canonical momenta π_i and H_{cR} are not the conserved physical momenta and the energy of the particle respectively. The conserved physical momentum and the energy of the particles have been determined in Eqs.(3.111) and (3.97) in the last Section from the Noether theorem.

When $R \to \infty$ limit, π_i and H_{dS} go back to the standard E-SR's expressions:

$$\pi_i|_{R\to\infty} = \frac{mv_i}{\sqrt{1 - \frac{v^2}{c^2}}}, \quad H_{dS}|_{R\to\infty} = \frac{mc^2}{\sqrt{1 - \frac{v^2}{c^2}}}. \tag{3.126}$$

where $v_i = \eta_{ij}\dot{x}^j$. Furthermore, at the original point of space-time coordinates $t = x^i = 0$, but $R =$finite, we have also expressions like Eq.(3.126):

$$\pi_i|_{t=x^i=0} = \frac{mv_i}{\sqrt{1 - \frac{v^2}{c^2}}}, \quad H_{dS}|_{t=x^i=0} = \frac{mc^2}{\sqrt{1 - \frac{v^2}{c^2}}}. \tag{3.127}$$

In the Table 3.1, we listed some results of Lagrange formalism both in the ordinary special relativity E-SR and in the de Sitter invariant special relativity dS-SR. Comparing the results in dS-SR with those in the well known E-SR, we learned that as an extension theory, dS-SR can simply be formulated by a variable alternating in E-SR: 1) $\eta_{\mu\nu} \Rightarrow B_{\mu\nu}$; 2) $\gamma \Rightarrow \sigma\Gamma$. This is a natural and nice feature for the Lagrangian formulism of dS-SR.

Table 3.1 Metric, Lagrangian, equation of motions, canonical momenta, and Hamiltonian in the special relativity(E-SR), and in the de Sitter special relativity(dS-SR). The quantities $\gamma^{-1} = \sqrt{1 + \frac{\eta_{ij}\dot{x}^i\dot{x}^j}{c^2}}$ and $\Gamma^{-1} = \sqrt{1 - \frac{\dot{x}^2}{c^2} + \frac{(\mathbf{x}\cdot\dot{\mathbf{x}})^2 - \mathbf{x}^2\dot{\mathbf{x}}^2}{c^2 R^2} + \frac{(\mathbf{x} - \dot{\mathbf{x}}t)^2}{R^2}}$ (see Eq.(3.96)).

	E-SR	dS-SR
space-time metric	$\eta_{\mu\nu}$	$B_{\mu\nu}(x)$, (Eq.(3.6))
Lagrangian	$L = -mc^2\gamma^{-1}$	$L_{dS} = -mc^2\sigma^{-1}\Gamma^{-1}$
equation of motion	$v^i = \dot{x}^i$=constant, (or $\dot{\gamma} = 0$)	$v^i = \dot{x}^i$=constant, (or $\dot{\Gamma} = 0$)
canonical momenta	$\pi_i = -m\gamma\eta_{i\mu}\dot{x}^\mu$	$\pi_i = -m\sigma\Gamma B_{i\mu}\dot{x}^\mu$
Hamiltonian	$H = mc\gamma\eta_{0\mu}\dot{x}^\mu$	$H_{dS} = mc\sigma\Gamma B_{0\mu}\dot{x}^\mu$

Combining Eq.(3.124) with Eq.(3.125), the covariant 4-momentum in $\mathcal{B}_\Lambda$ is:

$$\pi_\mu \equiv (\pi_0, \ \pi_i) = (-\frac{H_{dS}}{c}, \pi_i) = -m\sigma\Gamma B_{\mu\nu}\dot{x}^\nu = -mcB_{\mu\nu}\frac{dx^\nu}{ds}, \quad (3.128)$$

and the dispersion relation is:

$$B^{\mu\nu}\pi_\mu\pi_\nu = m^2c^2. \quad (3.129)$$

From this relation, we have

$$B^{00}(\pi_0)^2 + 2B^{0i}\pi_0\pi_i + B^{ij}\pi_i\pi_j - m^2c^2 = 0,$$

and then

$$H_{dS} \equiv -c\pi_0 = \frac{1}{2B^{00}}\left\{2cB^{0i}\pi_i \pm c\sqrt{4(B^{0i}\pi_i)^2 - 4B^{00}(B^{ij}\pi_i\pi_j - m^2c^2)}\right\}, \quad (3.130)$$

where the metric $B^{\mu\nu}$ has been given in Eq.(3.7) and hence

$$B^{00} = \sigma(x)\left(1 - \frac{c^2t^2}{R^2}\right), \quad B^{0i} = -\sigma(x)\frac{ctx^i}{R^2},$$

$$B^{ij} = \sigma(x)\left(\eta^{ij} - \frac{x^ix^j}{R^2}\right), \quad (3.131)$$

where $\sigma(x)$ is shown in Eq.(3.5). From Eqs.(3.125) and (3.124), we have

$$dH_{dS}(x^i, \pi_i) = \sum_{i=1}^3 \left(\dot{x}^i d\pi_i + \pi_i d\dot{x}^i\right) - dL_{dS}(t, x^i, \dot{x}^i). \quad (3.132)$$

According to

$$dL_{dS}(t, x^i, \dot{x}^i) = \frac{\partial L_{dS}}{\partial t}dt + \sum_{i=1}^{3}\left(\frac{\partial L_{dS}}{\partial x^i}dx^i + \frac{\partial L_{dS}}{\partial \dot{x}^i}d\dot{x}^i\right)$$

$$= \frac{\partial L_{dS}}{\partial t}dt + \sum_{i=1}^{3}\left(\frac{d}{dt}\left(\frac{\partial L_{dS}}{\partial \dot{x}^i}\right)dx^i + \frac{\partial L_{dS}}{\partial \dot{x}^i}d\dot{x}^i\right)$$

$$= \frac{\partial L_{dS}}{\partial t}dt + \sum_{i=1}^{3}\left(\dot{\pi}_i dx^i + \pi_i d\dot{x}^i\right), \qquad (3.133)$$

where the equation of motion (3.30) and π_i's definition (3.124) have been used, and substituting Eq.(3.133) into Eq.(3.132) we have:

$$dH_{dS}(x^i, \pi_i, t) = \sum_{i=1}^{3}\left(\dot{x}^i d\pi_i - \dot{\pi}_i dx^i\right) - \frac{\partial L_{dS}}{\partial t}dt. \qquad (3.134)$$

Therefore we obtain the following canonical equations

$$\dot{x}^i = \frac{\partial H_{dS}}{\partial \pi_i}, \qquad (3.135)$$

$$\dot{\pi}_i = -\frac{\partial H_{dS}}{\partial x^i}, \qquad (3.136)$$

and

$$\frac{\partial H_{dS}}{\partial t} = -\frac{\partial L_{dS}}{\partial t}. \qquad (3.137)$$

For functions $f(\pi_i, x^i, t)$ and $g(\pi_i, x^i, t)$, we define the Poisson Bracket

$$\{f, g\}_{PB} \equiv \sum_{i=1}^{3}\left(\frac{\partial f}{\partial x^i}\frac{\partial g}{\partial \pi_i} - \frac{\partial f}{\partial \pi_i}\frac{\partial g}{\partial x^i}\right). \qquad (3.138)$$

By means of this definition, the canonical equations (3.135) and (3.136) take the following form:

$$\dot{x}^i = \{x^i, H_{dS}\}_{PB}, \qquad (3.139)$$

$$\dot{\pi}_i = \{\pi_i, H_{dS}\}_{PB}. \qquad (3.140)$$

It is easy to prove that:

$$\{x^i, \pi_j\}_{PB} = \delta_j^i, \quad \{x^i, x^j\}_{PB} = 0, \quad \{\pi_i, \pi_j\}_{PB} = 0. \qquad (3.141)$$

Two remarks are in order:

(i) It is also straightforward to check $\dot{x}^i =$ const. by means of Eqs.(3.135) and (3.136). From explicit expression of H_{dS} (3.130) and a straightforward calculation, we can prove:

$$\ddot{x}^i = \frac{\partial}{\partial t}\left(\frac{\partial H_{dS}}{\partial \pi_i}\right) + \sum_{j=1}^{3}\left(\frac{\partial H_{dS}}{\partial \pi_j}\frac{\partial}{\partial x^j} - \frac{\partial H_{dS}}{\partial x^j}\frac{\partial}{\partial \pi_j}\right)\left(\frac{\partial H_{dS}}{\partial \pi_i}\right) = 0.$$

(3.142)

Then, we conclude that $\dot{x}^i =$consts., which is the same of (3.37).

(ii) Generally, for a function $F = F(t, x^i, \pi_i)$, we have

$$\dot{F} \equiv \frac{dF}{dt} = \frac{\partial F}{\partial t} + \sum_{i=1}^{3}\left(\frac{\partial F}{\partial x^i}\frac{\partial H_{dS}}{\partial \pi_i} - \frac{\partial F}{\partial \pi_i}\frac{\partial H_{dS}}{\partial x^i}\right)$$

$$= \frac{\partial F}{\partial t} + \{F, H_{dS}\}_{PB},$$

(3.143)

where Eqs.(3.135) and (3.136) have been used.

Finally, we present the relations between the canonical energy-momentums π_μ and the conserved energy-momentums $p^\mu \equiv \{p^0, p^i\}$. Combining Eq.(3.97) with Eq.(3.111) gives conserved four-momentum in the de Sitter special relativity:

$$p^\mu \equiv \{p^0, p^i\} = m\Gamma\dot{x}^\mu.$$

(3.144)

From Eqs.(3.24) and (3.98), we have:

$$dt = \sigma\Gamma ds, \quad \text{and} \quad \dot{x}^i \equiv \frac{dx^i}{dt} = \frac{1}{\sigma\Gamma}\frac{dx^i}{ds}.$$

(3.145)

Noting

$$\dot{x}^0 = c,$$

(3.146)

and substituting Eqs.(3.145) and (3.146) into Eq.(3.144) we have

$$p^\mu = \frac{mc}{\sigma}\frac{dx^\mu}{ds} = -\frac{1}{\sigma}B^{\mu\nu}\pi_\nu,$$

(3.147)

where Eq.(3.128) has been used.

Exercise

(1) Problem 1, Derive equation (3.141).

(2) Problem 2, Derive equation (3.143).

(3) Problem 3, Suppose $f = f(x, y)$ where f is a function with two independent variables x and y. Derive the corresponding function $g(u, y)$ through the Legendre transformation, whose independent variables are $u \equiv \partial f(x, y)/\partial x$ and y.

3.7 Dispersion relation

We derive the dispersion relation[2] in terms of the Noether charges of dS-SR.

We rewrite angular momentum L^i of (3.121) as the following form:

$$L^{jk} = \epsilon^{jk}{}_i L^i = m\Gamma\epsilon^{jk}{}_i \epsilon^i{}_{lm} x^l \dot{x}^m = m\Gamma(x^j \dot{x}^k - x^k \dot{x}^j), \qquad (3.148)$$

where $\epsilon^{jk}{}_i\epsilon^i{}_{lm} = \delta^j_l \delta^k_m - \delta^j_m \delta^k_l$ is used. From (3.24) and (3.98), we have:

$$dt = \sigma\Gamma ds, \quad \text{and} \quad \dot{x}^i \equiv \frac{dx^i}{dt} = \frac{1}{\sigma\Gamma}\frac{dx^i}{ds}. \qquad (3.149)$$

From (3.2), we have

$$x^j = R\frac{\xi^j}{\xi^5}, \quad \text{and} \quad dx^j = R\frac{d\xi^j}{\xi^5} - R\frac{\xi^j d\xi^5}{(\xi^5)^2}. \qquad (3.150)$$

Substituting Eqs.(3.149) and (3.150) into Eq.(3.148), and noting $(\xi^5)^2 = R^2/\sigma$ (see Eq.(3.4)), we obtain the L^{jk}-expression in ξ-coordinates of 5-dimensional Minkowski spacetime with metric η_{AB} as follows

$$L^{jk} = m\frac{R^2}{(\xi^5)^2\sigma}\left(\xi^j\frac{d\xi^k}{ds} - \xi^k\frac{d\xi^j}{ds}\right) = m\left(\xi^j\frac{d\xi^k}{ds} - \xi^k\frac{d\xi^j}{ds}\right). \qquad (3.151)$$

Here, the elements of antisymmetric contravariant spatial tensor L^{jk} are of the spatial components of a following 5×5-tensor $\mathcal{L}^{AB}$ in 5d Minkowski spacetime:

$$\mathcal{L}^{AB} = m\left(\xi^A\frac{d\xi^B}{ds} - \xi^B\frac{d\xi^A}{ds}\right), \quad \text{with} \quad A, B = 0, 1, 2, 3, 5. \qquad (3.152)$$

We have shown in (3.148) and (3.151) that $L^{jk} = \mathcal{L}^{jk}$ represents the angular momentum Noether charges corresponding to rotation in the $\{j, k\}$-spatial plane of 5d Minkowski spacetime. We should further reveal the physical meanings of $\mathcal{L}^{0i}$, $\mathcal{L}^{05}$ and $\mathcal{L}^{5i}$. Let's determine them by means of the Noether charges:

(i) $\mathcal{L}^{0i}$: From Eq.(3.152), we have

$$\mathcal{L}^{0i} = m\left(\xi^0\frac{d\xi^i}{ds} - \xi^i\frac{d\xi^0}{ds}\right), \qquad (3.153)$$

---

[2]Usually, *dispersion relation* is a term in optics. This term, however, has also been widely received in SR theory to describe the energy-momentum relation such as $E^2 - c^2 p^2 = m^2 c^4$.

From Eq.(3.4),

$$\xi^0 = \frac{x^0}{R}\xi^5, \quad \xi^i = \frac{x^i}{R}\xi^5, \quad (\xi^5)^2 = \frac{R^2}{\sigma}. \tag{3.154}$$

we obtain

$$
\begin{aligned}
\mathcal{L}^{0i} &= m\left\{\frac{x^0}{R}\xi^5\left(\frac{\sigma\Gamma}{R}\xi^5\dot{x}^i + \frac{\xi^i}{R}\frac{d\xi^5}{ds}\right) - \frac{x^i}{R}\xi^5\left(\frac{c\sigma\Gamma}{R}\xi^5 + \frac{\xi^0}{R}\frac{d\xi^5}{ds}\right)\right\} \\
&= \frac{\sigma}{R^2}(\xi^5)^2 mc\Gamma(t\dot{x}^i - x^i) \\
&= K^i,
\end{aligned} \tag{3.155}
$$

where formula of K^i (3.118) is used.

(ii) $\mathcal{L}^{5i}$: Similarly,

$$
\begin{aligned}
\mathcal{L}^{5i} &= m\left(\xi^5\frac{d\xi^i}{ds} - \xi^i\frac{d\xi^5}{ds}\right) \\
&= m\left\{\frac{(\xi^5)^2}{R}\sigma\Gamma\dot{x}^i + \frac{\xi^5 x^i}{R}\frac{d\xi^5}{ds} - \frac{\xi^5 x^i}{R}\frac{d\xi^5}{ds}\right\} \\
&= \frac{\sigma}{R}(\xi^5)^2 p^i \\
&= Rp^i,
\end{aligned} \tag{3.156}
$$

where $p^i = m\Gamma\dot{x}^i$ (see: Eq.(3.111)) is used.

(iii) $\mathcal{L}^{50}$: Finally,

$$
\begin{aligned}
\mathcal{L}^{50} &= m\left(\xi^5\frac{d\xi^0}{ds} - \xi^0\frac{d\xi^5}{ds}\right) \\
&= m(\xi^5)^2\frac{c\sigma\Gamma}{R} \\
&= R\frac{E}{c},
\end{aligned} \tag{3.157}
$$

where $E = mc^2\Gamma$ (see: Eq.(3.97)) is used. Therefore, by using Eqs.(3.148), (3.155), (3.156) and (3.157), we have formula:

$$
\begin{aligned}
-\frac{c^2}{2R^2}\mathcal{L}^{AB}\mathcal{L}_{AB} &= -\frac{c^2}{2R^2}(2\mathcal{L}^{50}\mathcal{L}_{50} + 2\mathcal{L}^{5i}\mathcal{L}_{5i} + 2\mathcal{L}^{0i}\mathcal{L}_{0i} + L^{ij}L_{ij}) \\
&= E^2 - c^2\mathbf{p}^2 + \frac{c^2}{R^2}(\mathbf{K}^2 - \mathbf{L}^2),
\end{aligned} \tag{3.158}
$$

where $\mathcal{L}_{AB} = \eta_{AC}\eta_{BD}\mathcal{L}^{CD}$, $L_{ij} = \eta_{ik}\eta_{jl}L^{kl}$, $\{\eta_{ij}\} = \text{diag}\{+,-,-,-\}$, $\{\eta_{AB}\} = \text{diag}\{+,-,-,-,-\}$, and

$$L^{ij}L_{ij} = \epsilon^{ij}{}_k L^k \epsilon_{ij}{}^l L_l = -2\delta_k^l L_l L^k = 2\mathbf{L}^2.$$

On other hand, the Left Hand Side (LHS) of Eq.(3.158) can be straight-forwardly calculated since $\{\xi\} \in \{\mathcal{S}_\Lambda\}$ in 5d Minkowski spacetime. From Eq.(3.1) we have

$$\eta_{AB}\xi^A\xi^B = \xi^A\xi_A = -R^2, \quad \text{and then} \quad \xi^A\frac{d\xi_A}{ds} = 0. \qquad (3.159)$$

The LHS of Eq.(3.158) then reads

$$
\begin{aligned}
-\frac{c^2}{2R^2}\mathcal{L}^{AB}\mathcal{L}_{AB} &= -\frac{c^4m^2}{2R^2}\left(\xi^A\frac{d\xi^B}{ds} - \xi^B\frac{d\xi^A}{ds}\right)\left(\xi_A\frac{d\xi_B}{ds} - \xi_B\frac{d\xi_A}{ds}\right) \\
&= -\frac{m^2c^4}{2R^2}\left\{2\xi^A\xi_A\frac{d\xi^B}{ds}\frac{d\xi_B}{ds} - 2\xi^A\frac{d\xi_A}{ds}\xi^B\frac{d\xi_B}{ds}\right\} \\
&= -\frac{m^2c^4}{2R^2}(-2R^2)\frac{(ds)^2}{(ds)^2} \\
&= m^2c^4, \qquad\qquad\qquad\qquad\qquad\qquad\qquad\qquad (3.160)
\end{aligned}
$$

where Eq.(3.159) is used. Substituting Eq.(3.160) into Eq.(3.158), we obtain the dispersion relation of dS-SR for free particles:

$$E^2 - c^2\mathbf{p}^2 + \frac{c^2}{R^2}(\mathbf{K}^2 - \mathbf{L}^2) = m^2c^4, \qquad (3.161)$$

which is a generalized Einstein's famous formula of $E^2 - c^2\mathbf{p}^2 = m^2c^4$.

3.8   Minkowski point in Beltrami spacetime

Different from the Minkowski metric $\eta_{\mu\nu}$, the Beltrami metric $B_{\mu\nu}(x)$ (3.6) is spacetime dependent. We may ask a question: is there a spacetime point (or a region) where the metric equals to $\eta_{\mu\nu}$? Namely, can we solve equation:

$$B_{\mu\nu}(x) \equiv \frac{\eta_{\mu\nu}}{\sigma(x)} + \frac{\eta_{\mu\lambda}x^\lambda\eta_{\nu\rho}x^\rho}{R^2\sigma(x)^2} = \eta_{\mu\nu}. \qquad (3.162)$$

The solution of x^μ will be called Minkowski point in Beltrami spacetime $\mathcal{B}$. From Eq.(3.162), when R is finite (i.e., $|R| \nrightarrow \infty$), such Minkowski point is just the origin of coordinates:

$$x^\mu = 0. \qquad (3.163)$$

We shall use constants M^μ to denote the position of the Minkowski point and sometimes use notations:

$$B_{\mu\nu}^{(0)}(x) \equiv B_{\mu\nu}^{(M=0)}(x) = B_{\mu\nu}(x), \qquad (3.164)$$

for the Beltrami metric with $M^\mu = 0$ case. A variable shift of $x^\mu \longrightarrow x^\mu - M^\mu$ with constants $M^\mu \neq 0$ gives

$$B^{(0)}_{\mu\nu}(x) \longrightarrow B^{(M)}_{\mu\nu}(x) \equiv B_{\mu\nu}(x - M)$$
$$= \frac{\eta_{\mu\nu}}{\sigma(x - M)} + \frac{\eta_{\mu\lambda}\eta_{\nu\rho}(x^\lambda - M^\lambda)(x^\rho - M^\rho)}{R^2\sigma(x - M)^2},$$

$$(3.165)$$

where

$$\sigma(x - M) = 1 - \frac{\eta_{\mu\nu}(x^\mu - M^\mu)(x^\nu - M^\nu)}{R^2}. \qquad (3.166)$$

Equation (3.165) is the general expression of Beltrami metric with $M^\mu \neq 0$, and once $x^\mu \longrightarrow M^\mu$ the $B^{(M)}_{\mu\nu}(x)$ will tend to $\eta_{\mu\nu}$. Since M^μ are all constants, it is easy to be sure that $B^{(M)}_{\mu\nu}(x)$ is still the *inertial metric* and preserved by de Sitter transformations like $B_{\mu\nu}(x)$.

Two remarks are as follows:

(1) The various aspects of predictions of the E-SR to the experiments at the Earth laboratories have been well verified with high accuracy [Zhang (1997)]. Since the basic metric of E-SR is $\eta_{\mu\nu}$, this fact indicates that the metric of the spacetime around the region of the Earth must be very close to Lorentz metric $\eta_{\mu\nu}$. Namely the Earth nears the Minkowski point in $\mathcal{B}$:

$$|x^\mu - M^\mu| << |R|, \qquad (3.167)$$

where x^μ and M^μ are the coordinates of the Earth and the Minkowski position in $\mathcal{B}$ respectively. As long as $|R|$ is cosmologically large enough, it is fully possible to avoid violating any experiments verified usual special relativity within the error bands reported in literatures (e.g., in [Zhang (1997)]).

(2) $c \equiv 299\ 792\ 458\ m\ s^{-1}$ is a universal parameter in the relativity theories, which can be used to convert the time dimensionality to the space's. Thus the universality of c makes one dimension time plus three dimension space to form a physical 4-dimensional spacetime. But the photon velocity c_{photon} is not always equal to c in the de Sitter invariant special relativity. To see this point, let's consider the photon velocity at original point of $\mathcal{B}$ with $M = \{M^0 \neq 0,\ M^i = 0\}$. Suppose our Earth laboratory locates about

the origin of the $\mathcal{B}$-universe where $x^\mu \simeq 0$. According to Eq.(3.123), the Noether charges for the mechanics generated from $B^{(M)}_{\mu\nu}(x)$ are:

$$E = mc^2\Gamma,$$

$$p^i = m\Gamma\dot{x}^i,$$

$$K^i = mc\Gamma(x^i - (t - M^0/c)\dot{x}^i) = mc\Gamma x^i - (ct - M^0)p^i, \quad (3.168)$$

$$L^i = -m\Gamma\epsilon^i{}_{jk}x^j\dot{x}^k = -\epsilon^i{}_{jk}x^j p^k.$$

As usual, the photon is treated as a massless particle. So the photon velocity that we want to calculate is:

$$c_{photon} = \dot{x}|_{\{m=0,x^\mu=0\}} = \left.\frac{c^2 p}{E}\right|_{\{m=0,x^\mu=0\}}, \quad (3.169)$$

where $x \equiv |\mathbf{x}|$, $p \equiv |\mathbf{p}|$. Substituting Eq.(3.157) into Eq.(3.169) gives

$$c_{photon} = \left.\frac{cp}{\sqrt{p^2 - \frac{1}{R^2}(K^2 - L^2)}}\right|_{\{m=0,x^\mu=0\}}. \quad (3.170)$$

From Eq.(3.168) we have

$$L|_{x^i=0} = 0, \quad and \quad K|_{\{t=0,\ x^i=0\}} = M^0 p.$$

Therefore we obtain:

$$c_{photon} = \frac{c}{\sqrt{1 - \frac{(M^0)^2}{R^2}}} \neq c. \quad (3.171)$$

This is an equation beyond the usual special relativity. Only when $|R| \to \infty$ or $M^0 = 0$, the usual special relativistic identity $c_{photon} = c$ can hold. For $SO(4,1)$ de Sitter invariant special relativity, $R^2 > 0$, and hence $c_{photon} > c$. For example, suppose $R \sim 10^4 Gly$, $M^0 \simeq 13.7Gly$ (i.e., the so-called "age" of the Universe), then $c_{photon} - c \simeq 9.4 \times 10^{-7}c$. Maybe, the OPERA-type experiments[3] could verify such sort of results which relate to the principles of special relativity [Yan, Xiao, Huang, Hu (2012)] [Hu, Li, Yan (2012)].

---

[3] T. Adam, *et al.*, *"Measurement of the neutrino velocity with OPERA detector in the CNGS beam"*, arXev:1109.4897 [hep-ex], arXiv:1212.1276 [hep-ex].

Chapter 4

De Sitter Invariant General Relativity

In the previous two chapters, we elaborated systematically the formulations of Einstein's Special Relativity (E-SR) and de Sitter invariant Special Relativity (dS-SR). Several points were addressed: (i) Parallel to the well-known E-SR, the dS-SR is also self-consistent. When the universal parameter R of dS-SR tends to infinity, this relativity goes back to the Einstein's; (ii) The basic spacetime metric of E-SR is Minkowski metric $\eta_{\mu\nu}$, and the basic metric of dS-SR is Beltrami metric $B_{\mu\nu}(x)$; (iii) The spacetime transformation preserving Minkowski metric in E-SR is the inhomogeneous Lorentz transformation (or Poincaré transformation), and the transformation preserving the Beltrami metric in de Sitter special relativity is Lu-Zou-Guo (LZG) transformation; (iv) The spacetime symmetry for E-SR is $ISO(3,1)$-group, and that for the de Sitter invariant special relativity is de Sitter group $SO(4,1)$ (or $SO(3,2)$); (v) According to the gauge principle, the ordinary general relativity (or Einstein's general relativity) can be constructed by means of localizing the spacetime symmetry group $ISO(3,1)$ of the E-SR. This mechanism provides a bridge connecting the general relativity with the special relativity. It must work also for the de Sitter invariant special relativity which has been described in detail in the previous chapter. Now we are ready to employ this mechanism to construct a general relativity. Namely we will achieve the de Sitter invariant general relativity from localizing the de Sitter invariant special relativity in this chapter.

4.1 Localization of Lu-Zou-Guo transformation and action of de Sitter invariant General Relativity

The basic spacetime metric of the de Sitter invariant special relativity is the Beltrami metric (3.6) and the Lu-Zou-Guo (LZG) transformation (3.70) preserves Beltrami metric. The transformations are the elements of de Sitter group: i.e., $\{LZG\text{-}trans.\} \in \{SO(4,1)\}$. There are 10 parameters in LZG-transformation: spacetime translation parameters a^μ, space rotation's α^i (Euler angles) and Lorentz boost's β^i, all of which are spacetime independent constants. Hence the corresponding $SO(4,1)$ represents a global symmetry. Localizing Lu-Zou-Guo transformations is equivalent to localizing $SO(4,1)$. We showed the localization of $ISO(3,1)$ In Section 2.8. Based on the same procedures, the global symmetry $SO(4,1)$ will be localized via following substitutions

$$a^\mu \to a^\mu(x), \ \beta^i \to \beta^i(x), \ \alpha^i \to \alpha^i(x) \qquad (4.1)$$

where $a^\mu(x)$, $\beta^i(x)$, $\alpha^i(x)$ are arbitrary functions of variable x. Under (4.1), the global transformation (3.70) becomes

$$x^\mu \to x'^\mu = f(x), \qquad (4.2)$$

where $f^\mu(x)$ are four arbitrary functions of x. Hence, Eq.(4.2) represents a general spacetime coordinates transformation, or curvilinear coordinates transformation. Assuming the spacetime is torsion-free just like Einstein did in ordinary general relativity, the affine connection here is also Christoffel symbol: $\Gamma^\lambda_{\mu\nu} = \Gamma^\lambda_{\nu\mu}$, which has been shown in Eq.(2.181).

Now let us determine the action of gravity fields $S_G \equiv \int d^4x \sqrt{-g} \, \mathcal{G}(x)$ in empty spacetime, where $\mathcal{G}(x)$ is a scalar. To determine $\mathcal{G}(x)$ we should also consider the fact that the equation of the gravitational field must contain derivatives of the "potentials" (i.e., $g_{\mu\nu}(x)$) no higher than the second order (just as is the case for the electromagnetic field). From the Riemann geometry, it is found that only $\mathcal{R}$ and trivial constant $\Lambda \equiv constant$ satisfies all requirements. Therefore, $\mathcal{G}(x) = a(\mathcal{R} - 2\Lambda)$ where a is also a constant. In *Gaussian system of units*, $a = -c^3/(16\pi G)$ where $G = 6.67 \times 10^{-8} \text{cm}^3 \cdot \text{gm}^{-1} \cdot \text{sec}^{-2}$ is the universal gravitational constant. Thus we obtain the action of gauge gravity in empty spacetime:

$$S_G = -\frac{c^3}{16\pi G} \int d^4x \sqrt{-g}(\mathcal{R} - 2\Lambda). \qquad (4.3)$$

From $\delta S_G = 0$, we obtain (see Eqs.(2.194)−(2.206))

$$\mathcal{R}_{\mu\nu} - \frac{1}{2}g_{\mu\nu}\mathcal{R} + \Lambda g_{\mu\nu} = 0. \tag{4.4}$$

The de Sitter invariant special relativity tells us that to empty spacetime the metric must be Beltrami metric (3.6). Namely, one solution of Eq.(4.4) is required to be $g_{\mu\nu} = B_{\mu\nu}$. Then the value of constant Λ is determined to be

$$\Lambda = \frac{3}{R^2}, \tag{4.5}$$

and Eq.(4.4) becomes

$$\mathcal{R}_{\mu\nu} - \frac{1}{2}g_{\mu\nu}\mathcal{R} + \frac{3}{R^2}g_{\mu\nu} = 0, \tag{4.6}$$

$$\text{or} \quad \mathcal{R}_{\mu\nu} = \frac{3}{R^2}g_{\mu\nu}. \tag{4.7}$$

This is the basic equation of the *de Sitter invariant general relativity* in empty spacetime, which is different from the usual general relativity's (see Eqs.(2.2) and (2.209)).

Denoting the matter action in gravity fields by:

$$S_M = \int d^4x \sqrt{-g}\mathcal{L}_{matter}(g_{\mu\nu}, \phi_{matter}(x)),$$

one has the total action:

$$S_{tot} = S_G + S_M = S_G[g] + S_M[g, \phi] \tag{4.8}$$

where $S.[\cdot]$ denotes a functional. Similar to the discussions in Section 2.9, from the principle of least action

$$\delta S_{tot} = \delta(S_G + S_M) = 0, \tag{4.9}$$

and the definition of energy-momentum tensor of matter fields (see Eq.(2.203)):

$$T_{\mu\nu} = \frac{-2c}{\sqrt{-g}}\frac{\delta}{\delta g^{\mu\nu}}S_M[g, \phi], \tag{4.10}$$

we obtain the gravitational field equation of *de Sitter invariant general relativity*:

$$\mathcal{R}_{\mu\nu} - \frac{1}{2}g_{\mu\nu}\mathcal{R} + \Lambda g_{\mu\nu} = -\frac{8\pi G}{c^4}T_{\mu\nu}, \tag{4.11}$$

where $\Lambda = 3/R^2$. It is easy to see that when $\Lambda \to 0$ (or $|R| \to \infty$) the de Sitter general relativity goes back to the Einstein's general relativity.

4.2   Static spherically symmetric solution in empty spacetime

In this section, we seek the metric tensor field representing the static spherically symmetric gravitational field in empty spacetime surrounding some massive spherical object like the sun by means of solving Eq.(4.7) in the de Sitter invariant general relativity. The guiding assumptions are: (a) the metric is static, (b) the metric is spherically symmetric, (c) the spacetime is empty, (d) the dynamics guiding the gravitational fields is the de Sitter invariant general relativity. Taking spherical coordinates $\{x^0, x^1, x^2, x^3\} = \{ct, r, \theta, \phi\}$ (with $x^1 = r \sin\theta \cos\phi$, $x^2 = r \sin\theta \sin\phi$, $x^3 = r \cos\theta$), then we can set

$$ds^2 = e^{\nu(r)}c^2 dt^2 - e^{\mu(r)}dr^2 - r^2 d\theta^2 - r^2 \sin^2\theta d\phi^2, \qquad (4.12)$$

where $\nu = \nu(r)$, $\mu = \mu(r)$ are the unknown functions of r, which are expected to be determined by Eq.(4.7). From Eq.(4.12) the non-zero metric elements are:

$$g_{00} = e^\nu,$$
$$g_{11} = -e^\mu,$$
$$g_{22} = -r^2, \qquad (4.13)$$
$$g_{33} = -r^2 \sin^2\theta,$$

and all off-diagonal elements of $g_{\mu\nu}$ vanish. From the definition of (2.156), the non-zero elements of $g^{\mu\nu}$ are:

$$g^{00} = e^{-\nu},$$
$$g^{11} = -e^{-\mu},$$
$$g^{22} = -\frac{1}{r^2}, \qquad (4.14)$$
$$g^{33} = -\frac{1}{r^2 \sin^2\theta},$$

and all off-diagonal elements of the $g^{\mu\nu}$ are also equal to zero. From Eq.(2.181), the non-zero components of the corresponding Christoffel symbols are:

$$\Gamma^0_{10} = \frac{1}{2}\nu', \quad \Gamma^1_{11} = \frac{1}{2}\mu', \quad \Gamma^1_{00} = \frac{1}{2}e^{\nu-\mu}\nu',$$

$$\Gamma^1_{22} = -re^\mu, \quad \Gamma^1_{33} = -re^{-\mu}\sin^2\theta, \quad \Gamma^1_{12} = \frac{1}{r}, \qquad (4.15)$$

$$\Gamma^2_{33} = -\sin\theta\cos\theta, \quad \Gamma^3_{13} = \frac{1}{r}, \quad \Gamma^3_{23} = \frac{\cos\theta}{\sin\theta},$$

where $\mu' = d\mu/dr$, and $\nu' = d\nu/dr$. Substituting all of Christoffel symbols of Eq.(4.15) into Eq.(2.189), we get all non-zero components of the Ricci tensor:

$$\mathcal{R}_{00} = e^{\nu-\mu}\left\{-\frac{\nu''}{2} - \frac{\nu'}{r} + \frac{\nu'}{4}(\mu' - \nu')\right\},$$

$$\mathcal{R}_{11} = \frac{\nu''}{2} - \frac{\mu'}{r} - \frac{\nu'}{4}(\mu' - \nu'), \tag{4.16}$$

$$\mathcal{R}_{22} = e^{-\mu}\left\{1 - e^{\mu} - \frac{r}{2}(\mu' - \nu')\right\},$$

$$\mathcal{R}_{33} = -\mathcal{R}_{22}\sin^2\theta.$$

Substituting Eqs.(4.12) and (4.16) into the gravitational field equation in empty spacetime (4.7), i.e., $\mathcal{R}_{\mu\nu} = \frac{3}{R^2}g_{\mu\nu}$, we have

$$\frac{\nu''}{2} + \frac{\nu'}{r} - \frac{\nu'}{4}(\mu' - \nu') = -\frac{3}{R^2}e^{\mu}, \tag{4.17}$$

$$-\frac{\nu''}{2} + \frac{\mu'}{r} + \frac{\nu'}{4}(\mu' - \nu') = \frac{3}{R^2}e^{\mu}, \tag{4.18}$$

$$-1 + e^{\mu} + \frac{r}{2}(\mu' - \nu') = \frac{3}{R^2}r^2 e^{\mu}. \tag{4.19}$$

Only two of these equations are independent among these three equations since the Riemannian curvature must satisfy the Bianchi identity (2.186).

It is easy to solve these equations. From Eq.(4.17) plus Eq.(4.18), we have

$$\mu' + \nu' = 0, \tag{4.20}$$

and then

$$\mu + \nu = \lambda, \qquad \lambda = constant. \tag{4.21}$$

Inserting Eq.(4.20) into Eq.(4.19) and eliminating ν', we have

$$1 - r\mu' - e^{\mu}\left(1 - \frac{3}{R^2}r^2\right) = 0,$$

and then

$$(re^{-\mu})' = 1 - \frac{3}{R^2}r^2. \tag{4.22}$$

The solution of Eq.(4.22) is:

$$e^{-\mu} = 1 - \frac{1}{R^2}r^2 - \frac{2\tilde{k}}{r}, \qquad \tilde{k} = constant. \tag{4.23}$$

Combining Eqs.(4.21) and (4.23), we get:

$$e^{\nu} = e^{\lambda - \mu} = e^{\lambda}\left(1 - \frac{1}{R^2}r^2 - \frac{2\tilde{k}}{r}\right). \tag{4.24}$$

From Eq.(4.13) and setting constant $\lambda = 0$, we finally obtain

$$g_{00} = e^{\nu} = 1 - \frac{2\tilde{k}}{r} - \frac{1}{R^2}r^2, \tag{4.25}$$

$$g_{11} = -e^{\mu} = \left(1 - \frac{2\tilde{k}}{r} - \frac{1}{R^2}r^2\right)^{-1}. \tag{4.26}$$

In order to determine the constant $\tilde{k}$, we discuss the limit of $|R| \to \infty$. In the last section, we showed that under that limit the de Sitter invariant general relativity goes back to the ordinary Einstein's general relativity. Therefore, the static spherically symmetric solution in empty spacetime discussed in this section must go back to the Schwarzschild metric under the limit. Namely we have

$$g_{00}|_{|R|\to\infty} = 1 - \frac{2\tilde{k}}{r} = g_{00}(\text{Schwarz}) = 1 - \frac{2GM}{c^2 r}, \tag{4.27}$$

and hence

$$\tilde{k} = \frac{GM}{c^2}. \tag{4.28}$$

The final result is:

$$ds^2 = \left(1 - \frac{2GM}{c^2 r} - \frac{r^2}{R^2}\right)c^2 dt^2 - \left(1 - \frac{2GM}{c^2 r} - \frac{r^2}{R^2}\right)^{-1} dr^2 - r^2(d\theta^2 + \sin^2\theta d\phi^2), \tag{4.29}$$

where $G = 6.67 \times 10^{-8} \text{cm}^3 \cdot \text{gm}^{-1} \cdot \text{sec}^{-2}$ is the universal gravitational constant, and M is *"solar mass"*.

This solution was firstly obtained by Kottler, Weyl and Trefftz during 1918-1922 [Kottler (1918)] [Weyl (1919)] [Trefftz (1922)]. We call it KWT solution below. According to Eq.(4.29) the metric matrix can be written as follows:

$$g_{\mu\nu}^{(M)}(x_K) = \begin{pmatrix} 1 - \frac{2GM}{c^2 r_K} - \frac{r_K^2}{R^2} & 0 & 0 & 0 \\ 0 & -\frac{1}{1 - \frac{2GM}{c^2 r_K} - \frac{r_K^2}{R^2}} & 0 & 0 \\ 0 & 0 & -r_K^2 & 0 \\ 0 & 0 & 0 & -r_K^2\sin^2\theta_K \end{pmatrix}, \tag{4.30}$$

where subscript K means KWT solution.

4.3 Beltrami–de Sitter–Schwarzschild Solution

When $|R| \to \infty$, the KWT metric (4.29) tends to the well-known Schwarzschild metric $^S g_{\mu\nu}^{(M)}(x)$ of Einstein's general relativity:

$$ds^2 \equiv {}^S g_{\mu\nu}^{(M)}(x) dx^\mu dx^\nu$$

$$= \left(1 - \frac{2GM}{c^2 r}\right) c^2 dt^2 - \left(1 - \frac{2GM}{c^2 r}\right)^{-1} dr^2 - r^2 (d\theta^2 + \sin^2\theta d\phi^2) \quad (4.31)$$

where the left superscript S of $g_{\mu\nu}^{(M)}(x)$ indicates that metric is the usual Schwarzschild metric, and M is the *"solar" mass* serving as source of gravity. It is easy to see that when $M \to 0$ (i.e., gravity disappears) we have

$$ds^2 \equiv {}^S g_{\mu\nu}^{(M)}(x) dx^\mu dx^\nu$$
$$\to ds^2_{Mink} = {}^S g_{\mu\nu}^{(0)}(x) dx^\mu dx^\nu$$
$$= c^2 dt^2 - dr^2 - r^2 (d\theta^2 + \sin^2\theta d\phi^2)$$
$$= \eta_{\mu\nu} dx^\mu dx^\nu. \quad (4.32)$$

Therefore, the Schwarzschild metric $^S g_{\mu\nu}^{(M)}(x)$ of Einstein's general relativity can be thought as a solution of following equations:

$$\mathcal{R}_{\mu\nu}(x) = 0, \quad (4.33)$$

$$g_{\mu\nu}^{(M)}(x)\Big|_{M\to 0} = \eta_{\mu\nu}. \quad (4.34)$$

As is well known and addressed in above that $\eta_{\mu\nu}$, the metric of Minkowski spacetime, is *inertial metric*. It is easy to check this point by means of Landau-Lifshitz action for free particle [Landau, Lifshitz (1987)]:

$$A = -mc \int ds = -mc \int \sqrt{\eta_{\mu\nu} dx^\mu dx^\nu} = -mc^2 \int dt \sqrt{1 - \frac{\dot{\mathbf{x}}^2}{c^2}} =: \int dt L(\dot{\mathbf{x}}).$$

Then, from $\delta A = 0$, we have

$$\frac{d}{dt} \frac{\partial L(\dot{\mathbf{x}})}{\partial \dot{\mathbf{x}}} = \frac{\partial L(\dot{\mathbf{x}})}{\partial \mathbf{x}} = 0, \Rightarrow \ddot{\mathbf{x}} = 0. \quad (4.35)$$

Equation (4.35) means $\dot{\mathbf{x}} = constant$, i.e., the inertial motion law holds in Minkowski spacetime with metric $\eta_{\mu\nu}$ for free particle. Therefore we call $\eta_{\mu\nu}$ *inertial metric* (see the footnote in Section 3.4 for the definition of this phrase).

Let's come back to the de Sitter invariant general relativity and pursue an answer to the question whether when gravity vanishes the KWT metric goes to a inertial metric or not. From Eq.(4.30) and setting $M = 0$, we have

$$g^{(0)}_{\mu\nu}(x_K) = \begin{pmatrix} 1 - \frac{r_K^2}{R^2} & 0 & 0 & 0 \\ 0 & -\frac{1}{1-\frac{r_K^2}{R^2}} & 0 & 0 \\ 0 & 0 & -r_K^2 & 0 \\ 0 & 0 & 0 & -r_K^2 \sin^2 \theta_K \end{pmatrix}. \quad (4.36)$$

In the Cartesian space coordinates x_K^i (with $x_K^1 = r_K \cos\phi_K \sin\theta_K$, $x_K^2 = r_K \sin\phi_K \sin\theta_K$, $x_K^3 = r_K \cos\theta_K$), the matrix form of $g^{(0)}_{\mu\nu}(x_K)$ can be written as component's form:

$$ds^2 = g^{(0)}_{\mu\nu}(x_K) dx_K^\mu dx_K^\nu$$

$$= \left(1 + \frac{\eta_{ij} x_K^i x_K^j}{R^2} \right) c^2 dt_K^2 + \eta_{ij} dx_K^i dx_K^j$$

$$+ \left[\left(1 + \frac{\eta_{ij} x_K^i x_K^j}{R^2} \right)^{-1} - 1 \right] \frac{\eta_{lk} \eta_{mn} x_K^l x_K^m dx_K^k dx_K^n}{\eta_{ij} x_K^i x_K^j}, \quad (4.37)$$

where $\eta_{ij} = \mathrm{diag}\{-1,-1,-1\}$. Comparing Eq.(4.37) with Eqs.(2.2) and (3.6), we find:

$$g^{(0)}_{\mu\nu}(x_K) \neq \eta_{\mu\nu}, \quad \text{and} \quad g^{(0)}_{\mu\nu}(x_K) \neq B_{\mu\nu}(x_K). \quad (4.38)$$

This fact indicates that $g^{(0)}_{\mu\nu}(x_K)$ is generally not a inertial metric. To be more concrete, let's see the motion of free particle in KWT-spacetime with $g^{(0)}_{\mu\nu}(x_K)$. The Landau-Lifshitz Lagrangian $L_K(x_K^i, \dot{x}_K^i)$ for a free particle in that spacetime is

$$L_K(x_K^i, \dot{x}_K^i) = -m_0 c \frac{ds}{dt_K} = -m_0 c \frac{\sqrt{g^{(0)}_{\mu\nu}(x_K) dx_K^\mu dx_K^\nu}}{dt_K}$$

$$= -m_0 c \sqrt{ \left(1 - \frac{\mathbf{x}_K^2}{R^2} \right) - \left[\frac{1}{1-\frac{\mathbf{x}_K^2}{R^2}} - 1 \right] \frac{(\mathbf{x}_K \cdot \dot{\mathbf{x}}_K)^2}{c^2 \mathbf{x}_K^2} - \frac{\dot{\mathbf{x}}^2}{c^2} }, \quad (4.39)$$

where $\mathbf{x}_K = x_K^1 \mathbf{i} + x_K^2 \mathbf{j} + x_K^3 \mathbf{k}$ and $\dot{\mathbf{x}}_K \equiv d\mathbf{x}_K/dt_K = \dot{x}_K^1 \mathbf{i} + \dot{x}_K^2 \mathbf{j} + \dot{x}_K^3 \mathbf{k}$. From $\delta S = -m_0 c\, \delta \int ds = \delta \int dt_K L(x_K^i, \dot{x}_K^i) = 0$, we have

$$\frac{\partial L_K(x_K^i, \dot{x}_K^i)}{\partial x_K^i} = \frac{d}{dt_K} \frac{\partial L_K(x_K^i, \dot{x}_K^i)}{\partial \dot{x}_K^i}. \quad (4.40)$$

Substituting Eq.(8.105) into Eq.(8.106), we can easily obtain the equation of motion $f(\ddot{x}_K^i, \dot{x}_K^i, x_K^i) = 0$. For our purpose, an explicit consideration of one-dimensional motion of the particle is enough. Namely, setting $x_K^2 = x_K^3 = 0$ and ignoring the motion of **j**, **k** directions, the equation of motion $f(\ddot{x}_K^i, \dot{x}_K^i, x_K^i) = 0$ becomes:

$$\ddot{x}_K^1 = \frac{x_K^1}{R^2}\left(c^2\left(1 - \frac{(x_K^1)^2}{R^2}\right) - \frac{3(\dot{x}_K^1)^2}{1 - \frac{(x_K^1)^2}{R^2}}\right) \neq 0. \qquad (4.41)$$

This equation explicitly indicates that there exist inertial forces in de Sitter spacetime with metric $g_{\mu\nu}^{(0)}$, which makes the particle's acceleration in direction **i** to be non-zero, i.e., $\ddot{x}_K^1 \neq 0$. Consequently, we conclude that the empty de Sitter spacetime metric $g_{\mu\nu}^{(0)}(x_K)$ is not a inertial metric. This fact indicates that the KWT metric does not possesses an essential property of the Schwarzschild solution in Einstein's general relativity, which is when gravity approaches to zero that solution tends to Minkowski metric that is inertial metric. It is interesting, therefore, to find the Schwarzschild-like metric of the de Sitter invariant general relativity. The criterion for it is that when gravity disappears the metric of empty spacetime is required to be inertial metric of de Sitter invariant general relativity. Namely, we should solve the equation as follows

$$R_{\mu\nu}(x) = \frac{3}{R^2}\,{}^{\mathcal{B}}g_{\mu\nu}^{(M)}(x), \qquad (4.42)$$

$$\left.{}^{\mathcal{B}}g_{\mu\nu}^{(M)}(x)\right|_{M\to 0} = B_{\mu\nu}(x). \qquad (4.43)$$

Here, we use ${}^{\mathcal{B}}g_{\mu\nu}^{(M)}(x)$ to denote new metric in de Sitter invariant general relativity with non-zero "solar mass" M, which is called Beltrami–de Sitter–Schwarzschild metric hereafter, or Beltrami–dSS metric in short. The equation of (4.43) means that ${}^{\mathcal{B}}g_{\mu\nu}^{(M)}(x)$ must satisfy the requirement that the empty spacetime metric (i.e., the metric ${}^{\mathcal{B}}g_{\mu\nu}^{(0)}(x)$) is inertial. This condition ensures ${}^{\mathcal{B}}g_{\mu\nu}^{(M)}(x)$ to be desired new Schwarzschild-like metric in de Sitter invariant general relativity.

Since all of $g_{\mu\nu}^{(M)}(x_K)$, $g_{\mu\nu}^{(0)}(x_K)$ and ${}^{\mathcal{B}}g_{\mu\nu}^{(0)}(x) \equiv B_{\mu\nu}(x)$ have already been known, the reference system transformation T between spacetimes $\{x_K^\mu\} \equiv \{ct_K,\ r_K,\ \theta_K,\ \phi_K\}$ and $\{x^\mu\} \equiv \{ct,\ r,\ \theta,\ \phi\}$ can be derived from the tensor-transformation properties of $g_{\mu\nu}^{(0)}(x_K)$ and $B_{\mu\nu}(x)$ (see (4.48) below). Then the desired result of ${}^{\mathcal{B}}g_{\mu\nu}^{(M)}(x)$ will be determined by means of $^\mathfrak{B}g^{(M)} = T'g^{(M)}T$. We sketch the logic in following diagram (4.44).

$$g_{\mu\nu}^{(M)}(x_K) \xrightarrow[M\to 0]{(4.36)} g_{\mu\nu}^{(0)}(x_K)$$

$$T\bigg\downarrow (4.66) \qquad\qquad T\bigg\downarrow (4.48) \qquad\qquad (4.44)$$

$$^\mathfrak{B}g_{\mu\nu}^{(M)}(x) \xrightarrow[M\to 0]{(4.43)} B_{\mu\nu}(x).$$

Now let us derive the reference system transformation T between spacetimes $\{x_K^\mu\}$ and $\{x^\mu\}$. In spherical coordinates, and from Eqs.(4.36) and (3.6) we have

$$g_{\mu\nu}^{(0)}(x_K) = \begin{pmatrix} 1 - \frac{r_K^2}{R^2} & 0 & 0 & 0 \\ 0 & -\frac{1}{1-\frac{r_K^2}{R^2}} & 0 & 0 \\ 0 & 0 & -r_N^2 & 0 \\ 0 & 0 & 0 & -r_K^2 \sin^2\theta_K \end{pmatrix}, \qquad (4.45)$$

$$B_{\mu\nu}(x) = \begin{pmatrix} \frac{R^2+r^2}{R^2\sigma^2} & -\frac{rct}{R^2\sigma^2} & 0 & 0 \\ -\frac{rct}{R^2\sigma^2} & -\frac{R^2-c^2t^2}{R^2\sigma^2} & 0 & 0 \\ 0 & 0 & -\frac{r^2}{\sigma} & 0 \\ 0 & 0 & 0 & -\frac{r^2\sin^2\theta}{\sigma} \end{pmatrix}, \qquad (4.46)$$

where $\sigma = \frac{R^2-c^2t^2+r^2}{R^2}$, and the following spherical coordinates expression of Beltrami metric has been used:

$$\begin{aligned} ds_{\text{Bel}}^2 &= B_{\mu\nu}(x)dx^\mu dx^\nu \\ &= \frac{R^2(R^2+r^2)}{(R^2+r^2-c^2t^2)^2}c^2dt^2 - \frac{2rR^2ct}{(R^2+r^2-c^2t^2)^2}cdtdr - \frac{R^2(R^2-c^2t^2)}{(R^2+r^2-c^2t^2)^2}dr^2 \\ &\quad - \frac{r^2R^2}{R^2+r^2-c^2t^2}(d\theta^2+\sin^2\theta d\phi^2). \end{aligned} \qquad (4.47)$$

where subindex Bel means Beltrami. Under reference system transformation between $\{x_N^\mu\}$ and $\{x^\mu\}$ the transformation from $g_{\alpha\beta}^{(0)}(x_K)$ to $B_{\mu\nu}(x)$ reads

$$g_{\alpha\beta}^{(0)}(x_K) \to B_{\mu\nu}(x) = \frac{\partial x_K^\alpha}{\partial x^\mu}\frac{\partial x_K^\beta}{\partial x^\nu}g_{\alpha\beta}^{(0)}(x_K), \qquad (4.48)$$

which can be rewritten in matrix form,

$$\mathcal{B} = T'g^{(0)}T, \qquad (4.49)$$

where matrices $\mathcal{B} \equiv \{B_{\mu\nu}(x)\}$, $g^{(0)} \equiv \{g_{\alpha\beta}^{(0)}(x_K)\}$, $T \equiv \{\frac{\partial x_K^\beta}{\partial x^\nu}\}$, and T' is the transpose of the matrix T.

To simplify the problem, we assume T has the form of

$$T = \begin{pmatrix} \frac{\partial t_K}{\partial t} & \frac{\partial t_K}{\partial r} & 0 & 0 \\ \frac{\partial r_K}{\partial t} & \frac{\partial r_K}{\partial r} & 0 & 0 \\ 0 & 0 & \frac{\partial \theta_K}{\partial \theta} & 0 \\ 0 & 0 & 0 & \frac{\partial \phi_K}{\partial \phi} \end{pmatrix}. \tag{4.50}$$

Then from Eq.(4.49) we have

$$\begin{cases} (1 - \frac{r_N^2}{R^2})(\frac{\partial t_K}{\partial t})^2 - \frac{(\frac{\partial r_K}{\partial t})^2}{1 - \frac{r_K^2}{R^2}} = \frac{R^2 + r^2}{R^2\sigma^2} \\ (1 - \frac{r_K^2}{R^2})\frac{\partial t_K}{\partial t}\frac{\partial t_K}{\partial r} - \frac{\frac{\partial r_K}{\partial t}\frac{\partial r_K}{\partial r}}{1 - \frac{r_K^2}{R^2}} = -\frac{rct}{R^2\sigma^2} \\ (1 - \frac{r_K^2}{R^2})(\frac{\partial t_K}{\partial r})^2 - \frac{(\frac{\partial r_K}{\partial r})^2}{1 - \frac{r_K^2}{R^2}} = -\frac{R^2 - c^2 t^2}{R^2\sigma^2} \\ r_K^2(\frac{\partial \theta_K}{\partial \theta})^2 = \frac{r^2}{\sigma} \qquad \Rightarrow (\frac{\partial \theta_K}{\partial \theta})^2 = \frac{r^2}{\sigma r_K^2} \\ r_K^2 \sin^2 \theta_K (\frac{\partial \phi_K}{\partial \phi})^2 = \frac{r^2 \sin^2 \theta}{\sigma} \quad \Rightarrow (\frac{\partial \phi_K}{\partial \phi})^2 = \frac{r^2 \sin^2 \theta}{\sigma r_K^2 \sin^2 \theta_K}. \end{cases} \tag{4.51}$$

We assume $\frac{\partial t_K}{\partial r} = 0$ to solve $\frac{\partial t_K}{\partial t}$, $\frac{\partial r_K}{\partial t}$ and $\frac{\partial r_K}{\partial r}$. Let

$$A = \frac{R^2 + r^2}{R^2\sigma^2}, \quad B = -\frac{R^2 - c^2 t^2}{R^2\sigma^2}, \quad C = -\frac{rct}{R^2\sigma^2}, \quad F = 1 - \frac{r_K^2}{R^2}, \tag{4.52}$$

the first three equations of (4.51) can be reduced to

$$F(\frac{\partial t_K}{\partial t})^2 - \frac{(\frac{\partial r_K}{\partial t})^2}{F} = A, \tag{4.53}$$

$$-\frac{\frac{\partial r_K}{\partial t}\frac{\partial r_K}{\partial r}}{F} = C, \tag{4.54}$$

$$-\frac{(\frac{\partial r_K}{\partial r})^2}{F} = B. \tag{4.55}$$

The solution is

$$(\frac{\partial r_K}{\partial r})^2 = -BF = \frac{1}{\sigma^2}\left(1 - \frac{c^2 t^2}{R^2}\right)\left(1 - \frac{r_K^2}{R^2}\right), \tag{4.56}$$

$$\frac{\partial r_K}{\partial t} = \frac{-CF}{\sqrt{-BF}} = \frac{rct}{R^2\sigma^{3/2}}\sqrt{\frac{1 - \frac{r_K^2}{R^2}}{1 - \frac{c^2 t^2}{R^2}}}, \tag{4.57}$$

$$(\frac{\partial t_K}{\partial t})^2 = \frac{A}{F} - \frac{C^2}{BF} = \frac{1}{\sigma\left(1 - \frac{c^2 t^2}{R^2}\right)\left(1 - \frac{r_K^2}{R^2}\right)}. \tag{4.58}$$

From (4.56) (note $1 - \frac{c^2 t^2}{R^2} = \sigma - \frac{r^2}{R^2}$), and $r'_K \equiv \frac{\partial r_N}{\partial r}$,

$$(\sqrt{\sigma} r'_K)^2 = (1 - \frac{r^2}{R^2\sigma})(1 - \frac{r_K^2}{R^2}), \tag{4.59}$$

we get a special solution of the coordinate transformation between non-inertial and inertial systems

$$r_K = \frac{r}{\sqrt{\sigma}}. \tag{4.60}$$

From Eq.(4.58),

$$\frac{\partial t_K}{\partial t} = \sqrt{\frac{1}{\left(1 - \frac{c^2 t^2}{R^2}\right)\left(\sigma - \frac{(\sqrt{\sigma} r_N)^2}{R^2}\right)}} = \frac{1}{1 - \frac{c^2 t^2}{R^2}}. \tag{4.61}$$

Finally, we get a coordinate transformation to inertial spacetime frame

$$r_K = \frac{r}{\sqrt{1 + \frac{r^2 - c^2 t^2}{R^2}}}, \tag{4.62}$$

$$t_K = \int \frac{dt}{1 - \frac{c^2 t^2}{R^2}} = \frac{R}{c}\arctan\frac{ct}{R}, \tag{4.63}$$

$$\theta_K = \theta, \tag{4.64}$$

$$\phi_K = \phi. \tag{4.65}$$

It's consistent with [Lu, Zou and Guo (1974)].

Beltrami–de Sitter–Schwarzschild metric: [Sun, Yan, Deng, Huang, Hu (2013)]

Under the transformation T of Eqs.(4.62)–(4.65), the KWT metric (4.30) transforms to Beltrami–dSS metric:

$$g_{\mu\nu}^{(M)}(x_K) \to {}^{\mathfrak{B}}g_{\mu\nu}^{(M)}(x) = \frac{\partial x_K^\alpha}{\partial x^\mu}\frac{\partial x_K^\beta}{\partial x^\nu}g_{\alpha\beta}^{(M)}(x_K). \tag{4.66}$$

Substituting Eqs.(4.62)–(4.65) into Eq.(4.66), we finally obtain the desired Beltrami–dSS metric as follows

$$ds^2 = {}^{\mathfrak{B}}g_{\mu\nu}^{(M)}(x)dx^\mu dx^\nu$$

$$= \left(\frac{1 - \frac{r^2}{R^2\sigma} - \frac{2GM\sqrt{\sigma}}{r}}{\left(1 - \frac{c^2 t^2}{R^2}\right)^2} - \frac{\frac{r^2 c^2 t^2}{R^4}}{\left(1 - \frac{r^2}{R^2\sigma} - \frac{2GM\sqrt{\sigma}}{r}\right)\sigma^3}\right)c^2 dt^2$$

$$-2\frac{\frac{rct}{R^2}\left(1 - \frac{c^2 t^2}{R^2}\right)}{\left(1 - \frac{r^2}{R^2\sigma} - \frac{2GM\sqrt{\sigma}}{r}\right)\sigma^3}cdtdr$$

$$-\frac{\left(1 - \frac{c^2 t^2}{R^2}\right)^2}{\left(1 - \frac{r^2}{R^2\sigma} - \frac{2GM\sqrt{\sigma}}{r}\right)\sigma^3}dr^2 - \frac{r^2}{\sigma}(d\theta^2 + \sin^2\theta d\phi^2). \tag{4.67}$$

This is a new metric of de Sitter invariant general relativity, and serves as main result of this section. It is straightforward to check that it satisfies the Einstein field equation of de Sitter invariant general relativity in empty spacetime, (4.42). It is also easy to see that when $R \to \infty$, ${}^{\mathfrak{B}}g_{\mu\nu}^{(M)}(x)$ of (4.67) coincides with Schwarzschild metric of (2.2); when $M \to 0$, it coincides with Beltrami metric of (G.1); and when $R \to \infty$ and $M \to 0$, it goes back to Minkowski metric of (2.2).

Chapter 5

Dynamics of Expansion of the Universe in General Relativity

In this chapter, we adopt the unit with $c = 1$ for convenience.

Just as other branches of physics, modern cosmology is based on reliable observational facts and theoretical analysis. On the other hand, some fundamental quantities such distance in cosmology are not directly measurable since one can look at distant objects but cannot touch them. Yet, most quantities in other branches of physics can be directly or indirectly measurable in laboratories. As always in physics, cosmology also starts from some basic assumptions. As a first step, we have the following assumptions.

Cosmological Principle:

(1)*Isotropy:* The visible universe seems the same in all directions around us, at least if we look out to distances larger than about 300 million light years. The Isotropy is much more precise in the cosmic microwave background, which has been traveling to us for about 14 billion years, supporting the conclusion that the universe at sufficiently large distances is nearly the same in all directions. Namely, the universe should appear isotropic to observers throughout the universe.

(2) *Homogeneity:* "There is no place which is the center of the Universe". This is actually a natural extension of Copernicus principle on planet's orbits in the solar system. In cosmology, one can rightly claim that the space of the Universe is homogenous.

Obviously, both Isotropy and Homogeneity make statistical sense in cosmology since we on earth see only one sun and one moon in a particular direction at a time during the day. The stress energy tensor $T_{\mu\nu}$ that appear on the right hand side of Einstein equation makes sense statistically.

That is, it should be understood as $\bar{T}_{\mu\nu}$ where $\bar{T}_{\mu\nu}$ means taking cosmological average of $T_{\mu\nu}$. (In contrast, the so-called *dark energy* , which is represented by cosmological constant Λ, is a *de facto* homogeneious distribution. It applies to any region at any length scale). Roughly, Homogeneity is valid at the length scale of about 10 Mpc (1Mpc $= 3.3 \times 10^6$ light years)[Peebles: RevModPhys]. So cosmology does not apply to small (even if very large compared with daily life length scale) areas in the universe. The conventional cosmological metrics such as R-W metric does not apply to the detailed structure of any small area of the universe, just like thermodynamics does not apply to a few molecules (there are no concepts like pressure for a few molecules). Therefore in principle, local physics such the energy spectrum of any atom at any specific location in the universe should depend on the local spacetime structure instead of the cosmological spacetime .

It should be emphasized that the two assumptions cannot be true to all observers. They can be reasonably valid only to a special class of observers since observational results depend on the motion state of the observer, as has been told in physics and our daily life.

5.1   Robertson–Walker metric

Accepting the two assumptions above, differential geometry then tells us that the universe is a 4-dimensional spacetime with a 3-dimensional maximally symmetric space, which has $N(N+1)/2$ Killing vectors, here $N=3$. We adopt the standard approach to construction of this spacetime. [Weinberg:G & C] (In this chapter, we adopt the unit with $c=1$ for convenience.)
Hypersphere:
Maximally symmetric spaces can be obtained by any construction due the uniqueness. The simplest way is to consider a hypersphere embedded in a larger space. Let's consider the 4-dimensional Euclidean space

$$d\sigma^2 = (dy)^2 + \delta_{ij}dx^i dx^j, \qquad (5.1)$$

in which a 3-dimensional hypersphere is embed:

$$y^2 + \delta_{ij}x^i x^j = \frac{1}{K}, \qquad (5.2)$$

After differentiating both Left Hand Side (LHS) and RHS of Eq.(5.2), we have

$$dy = -\frac{\delta_{ij}x^i dx^j}{y} \tag{5.3}$$

Hence

$$(dy)^2 = \frac{(\delta_{ij}x^i dx^j)^2}{\frac{1}{K} - \delta_{ij}x^i x^j} = \frac{K(\delta_{ij}x^i dx^j)^2}{1 - K\delta_{ij}x^i x^j}, \tag{5.4}$$

Therefore the line element squared on the hypersphere is

$$d\sigma^2 = \frac{K(\delta_{ij}x^i dx^j)^2}{1 - K\delta_{ij}x^i x^j} + \delta_{ij}dx^i dx^j$$

$$= \frac{dr^2}{1 - Kr^2} + r^2(d\theta^2 + \sin^2\theta d\phi^2), \tag{5.5}$$

where $x^1 = r\sin\theta\cos\phi$, $x^2 = r\sin\theta\sin\phi$, $x^3 = r\cos\theta$ and $r^2 = -\eta_{ij}x^i x^j$ have been used. Equation (5.5) is the expression of line element of 3-dimensional hypersphere in sphere coordinates frame. When $K > 0$, it is an ordinary hypersphere; when $K = 0$, it's a flat plane; and when $K < 0$, it's a 3-dimensional pseudo hypersphere.

Robertson–Walker metric:

The general form of 4-dimensional metric is:

$$ds^2 = g_{00}(dt)^2 + 2g_{0i}(t,x)dtdx^i + g_{ij}(t,x)dx^i dx^j. \tag{5.6}$$

According to the Cosmological Principle, the evolution of the universe must be homogenous. Namely the picture of evolution should be same everywhere in the universe. Therefore, an uniform *"world time"* t must exist, which characters the evolution itself. Thus, letting t to be the proper time of co-moving clock located the galaxies, we have

$$g_{00} = 1, \quad g_{0i} = 0. \tag{5.7}$$

Furthermore, the Cosmological Principle also requires that the curvature of 3-dimensional space K has to be the same everywhere in the universe though it could be time-dependent. Therefore, the general form of 4-dimensional metric of the universe is as follows

$$ds^2 = dt^2 - a(t)^2 \left[\frac{dr^2}{1 - Kr^2} + r^2(d\theta^2 + \sin^2\theta d\phi^2) \right], \tag{5.8}$$

where $a(t)$ is scale (or expansion) factor, which is independent of the choice of galaxies. Metric of (5.8) is called Robertson–Walker metric, or Ricci

Friedmann–Robertson–Walker metric. Since the space coordinates adopted here are of co-moving coordinates, for convenience of discussions, we can always adjust r to make $K > 0$ becoming $K = +1$, and $K < 0$ becoming $K = -1$. As mentioned previously, the universe is isotropic and homogeneous only to a special class of observers. This class of observers are related by the transformations [Weinberg: G & C] which leave the Robertson–Walker metric invariant.

Red-shift in cosmology:

Suppose a galaxy emitted a light signal at time t_e, and it was received by an observer on the Earth at t_0. Later, the galaxy emitted another signal at $t_e + \Delta t_e$, and was received by the observer at later time $t_0 + \Delta t_0$. We are interested in the relationship between Δt_e and Δt_0. Without loss of generality, suppose the light propagates along radial direction, and the Earth location is $r = 0$ and the galaxy is at $r = r_e$. From Eq.(5.8), the light equation $ds = 0$ reads

$$dt^2 - a(t)^2 \frac{dr^2}{1 - Kr^2} = 0. \tag{5.9}$$

To the first light signal, we have

$$\int_{t_e}^{t_0} \frac{dt}{a(t)} = \int_{r_e}^{0} \frac{dr}{\sqrt{1 - Kr^2}}. \tag{5.10}$$

Similarly, to the second signal, the corresponding equation is

$$\int_{t_e + \Delta t_e}^{t_0 + \Delta t_0} \frac{dt}{a(t)} = \int_{r_e}^{0} \frac{dr}{c\sqrt{1 - Kr^2}}, \tag{5.11}$$

and hence

$$\int_{t_e + \Delta t_e}^{t_0 + \Delta t_0} \frac{dt}{a(t)} = \int_{t_e}^{t_0} \frac{dt}{a(t)}. \tag{5.12}$$

Noting the LHS of Eq.(5.11) is

$$\int_{t_e + \Delta t_e}^{t_0 + \Delta t_0} \frac{dt}{a(t)} = \left(\int_{t_e + \Delta t_e}^{t_e} + \int_{t_e}^{t_0} + \int_{t_0}^{t_0 + \Delta t_0} \right) \frac{dt}{a(t)},$$

Eq.(5.11) becomes

$$\int_{t_e + \Delta t_e}^{t_e} \frac{dt}{a(t)} + \int_{t_0}^{t_0 + \Delta t_0} \frac{dt}{a(t)} = 0. \tag{5.13}$$

When Δt_0 and Δt_e are very small, Eq.(5.13) indicates that:

$$\frac{\Delta t_0}{a(t_0)} = \frac{\Delta t_e}{a(t_e)}. \tag{5.14}$$

The Δt_e and Δt_0 in above can be thought as the period of emitted light $\tau_e \equiv 1/\nu_e$ and the period of received light $\tau_0 \equiv 1/\nu_0$ respectively. Then from Eq.(5.14) we have

$$\frac{\nu_e}{\nu_0} = \frac{\lambda_0}{\lambda_e} = \frac{\tau_0}{\tau_e} = \frac{a(t_0)}{a(t_e)}, \tag{5.15}$$

where $\nu = 1/\tau$ is the frequency, and $\lambda = c\tau$ is the wave length of the light. This relation can also be obtained via the general discussion on observables in curved spacetime [Feng]. The definition of frequency-shift z is as follows

$$z = \frac{\lambda_0 - \lambda_e}{\lambda_e} = \frac{a(t_0)}{a(t_e)} - 1. \tag{5.16}$$

Or

$$\frac{\lambda_0}{\lambda_e} = \frac{a(t_0)}{a(t_e)}. \tag{5.17}$$

That is, the wave length increases in tandem with the increase of the scale factor of the universe. To the expansionary universe, the present scale factor $a(t_0) \equiv a_0$ is always larger than the past scale factor $a(t_e) \equiv a(t)$, then

$$z = \frac{a(t_0)}{a(t_e)} - 1 \equiv \frac{a_0}{a(t)} - 1 > 0. \tag{5.18}$$

To this case, z is called *red-shift* in in cosmology. Equation (5.18) can also be rewritten in

$$a(t) = \frac{a_0}{1 + z}. \tag{5.19}$$

5.2   Friedmann equation

In the last section, we discussed the general kinematics of cosmology. The results do not depend on assumptions about the dynamics of the cosmological expansion. To go further we now need to apply the gravitational field equations of Einstein in de Sitter general relativity, with various tentative assumptions about the cosmic energy density and pressure. The aim of this section is to determine the scale factor in the Robertson–Walker metric.

Stress tensor of perfect fluid of the universe:

General relativity is a satisfactory theory of gravitation, correctly predicting the motions of particles and photons in curved spacetime. But in order to apply it to the universe, we must make some simplifying assumptions. We shall grossly idealize the universe, and model it by a simple macroscopic fluid, devoid of shear-viscous, bulk-viscous, and heat-conductive properties. Its stress tensor (or energy-momentum tensor) $T_{\mu\nu}$ is then that of a perfect fluid, so

$$T_{\mu\nu} = (\rho + p)u_\mu u_\nu - pg_{\mu\nu}, \tag{5.20}$$

where ρ is its proper density, p is its pressure, u_μ is the covariant world velocity of the fluid particles (stars etc.), i.e., $u_\mu = g_{\mu\nu}u^\nu$ and $u^\nu = \{1,0,0,0\}$. Substituting Eq.(5.8) into Eq.(5.20), we have

$$T_{00} = \rho, \tag{5.21}$$

$$T_{11} = \frac{pa^2}{1 - Kr^2}, \quad T_{22} = pa^2r^2, \quad T_{33} = pa^2r^2\sin^2\theta. \tag{5.22}$$

Friedmann equation:

We firstly consider $\Lambda = 0$ case, and re-write the Einstein equation (4.11) as follows

$$\mathcal{R}_{\mu\nu} = -8\pi G\left(T_{\mu\nu} - \frac{1}{2}Tg_{\mu\nu}\right), \tag{5.23}$$

where $T = T^\mu{}_\mu = c^2(\rho - 3p)$. If we label our coordinates according to $ct \equiv x^0$, $r \equiv x^1$, $\theta \equiv x^2$, $\phi \equiv x^3$, then the nonzero connection coefficients arisen from Robertson–Walker metric (5.8) are

$$\Gamma^0_{11} = a\dot{a}/(1 - Kr^2), \quad \Gamma^0_{22} = a\dot{a}r^2, \quad \Gamma^0_{33} = a\dot{a}r^2\sin^2\theta,$$
$$\Gamma^1_{01} = \dot{a}/a, \quad \Gamma^1_{11} = Kr/(1 - Kr^2), \quad \Gamma^1_{22} = -r(1 - Kr^2),$$
$$\Gamma^1_{33} = -r(1 - Kr^2)\sin^2\theta, \tag{5.24}$$
$$\Gamma^2_{02} = \dot{a}/a, \quad \Gamma^2_{12} = 1/r, \quad \Gamma^2_{33} = -\sin\theta\cos\theta,$$
$$\Gamma^3_{03} = \dot{a}/a, \quad \Gamma^3_{13} = 1/r, \quad \Gamma^3_{23} = \cot\theta$$

Feeding the connections into

$$\mathcal{R}_{\mu\nu} \equiv \Gamma^\sigma_{\mu\sigma,\nu} - \Gamma^\sigma_{\mu\nu,\sigma} + \Gamma^\rho_{\mu\sigma}\Gamma^\sigma_{\rho\nu} - \Gamma^\rho_{\mu\nu}\Gamma^\sigma_{\rho\sigma}$$

(Ricci tensor $\mathcal{R}_{\mu\nu}$ has been defined in Eq.(2.189)) gives

$$\mathcal{R}_{00} = 3\ddot{a}/a$$
$$\mathcal{R}_{11} = -(a\ddot{a} + 2\dot{a}^2 + 2K)/(1 - Kr^2)$$
$$\mathcal{R}_{22} = -r^2(a\ddot{a} + 2\dot{a}^2 + 2K) \qquad (5.25)$$
$$\mathcal{R}_{33} = -r^2 \sin^2 \theta(a\ddot{a} + 2\dot{a}^2 + 2K)$$
$$\mathcal{R}_{\mu\nu} = 0, \quad for \ \mu \neq \nu.$$

(Note, like common conventions, notation $\dot{a}$ means $\frac{da}{dt}$, simple notation " , " means derivative and " ; " represents covariant derivative, etc.) Extracting $g_{\mu\nu}$ from the line element (5.8), and using Eq.(5.20), we see that

$$T_{00} - \frac{1}{2}Tg_{00} = \frac{1}{2}(\rho + 3p)$$
$$T_{11} - \frac{1}{2}Tg_{11} = \frac{1}{2}(\rho - p)a^2/(1 - Kr^2)$$
$$T_{22} - \frac{1}{2}Tg_{22} = \frac{1}{2}(\rho - p)a^2r^2$$
$$T_{33} - \frac{1}{2}Tg_{33} = \frac{1}{2}(\rho - p)a^2r^2 \sin^2 \theta$$
$$T_{\mu\nu} - \frac{1}{2}Tg_{\mu\nu} = 0, \quad for \ \mu \neq \nu.$$

So the field equations in the form Eq.(5.23) yield just two equations

$$\frac{\ddot{a}}{a} = -\frac{4\pi G}{3}(\rho + 3p), \qquad (5.26)$$
$$a\ddot{a} + 2\dot{a}^2 + 2K = 4\pi G(\rho - p)a^2. \qquad (5.27)$$

The fact that the three (nontrivial) spatial equations are equivalent is essentially due to the Homogeneity and Isotropy of the Robertson–Walker metric.

Eliminating $\ddot{a}$ from Eqs.(5.26) and (5.27) gives

$$\dot{a}^2 = \frac{8\pi G}{3}\rho a^2 - K. \qquad (5.28)$$

We shall refer to this equation as *Friedmann equation* which does not depend on p. It is completely cancelled out of this equation.

Furthermore, from Einstein equation (4.11), we have

$$G^{\mu\nu} + \Lambda g^{\mu\nu} = -8\pi G T^{\mu\nu}, \qquad (5.29)$$

where $G^{\mu\nu} \equiv \mathcal{R}^{\mu\nu} - \frac{1}{2}g^{\mu\nu}\mathcal{R}$ is the *Einstein tensor*. Since $G^{\mu\nu}{}_{;\mu} = 0$ and $g^{\mu\nu}{}_{;\mu} = 0$, Eq.(5.29) means

$$T^{\mu\nu}{}_{;\mu} = 0. \tag{5.30}$$

Equation (5.20) can be re-written in

$$T^{\mu\nu} = (\rho + p)u^\mu u^\nu - pg^{\mu\nu}. \tag{5.31}$$

Thus Eq.(5.30) indicates

$$(\rho u^\mu)_{;\mu}u^\nu + \rho u^\mu u^\nu{}_{;\mu} + pu^\mu{}_{;\mu}u^\nu + pu^\mu u^\nu{}_{;\mu} + p_{,\mu}u^\mu u^\nu - p_{,\mu}g^{\mu\nu} = 0. \tag{5.32}$$

Since $u^\nu u_\nu = 1$, and the differentiation gives

$$u^\nu{}_{,\mu}u_\nu + u^\nu u_{\nu;\mu} = 0, \tag{5.33}$$

which implies that $u^\nu{}_{,\mu}u_\nu = 0$. So contracting Eq.(5.32) with u_ν

$$(\rho u^\mu)_{;\mu} + pu^\mu{}_{;\mu} = 0. \tag{5.34}$$

Equation (5.32) therefore simplifies to

$$(\rho + p)u^\nu{}_{;\mu}u^\mu = (g^{\mu\nu} - u^\mu u^\nu)p_{,\mu}. \tag{5.35}$$

The continuity equation (5.34) may be written as

$$\rho_{,\mu}u^\mu + (\rho + p)(u^\mu{}_{,\mu} + \Gamma^\mu_{\nu\mu}u^\nu) = 0, \tag{5.36}$$

and with $u^\mu = \delta^\mu_0$ this reduces to

$$\dot{\rho} + (\rho + p)\frac{3\dot{a}}{a} = 0, \tag{5.37}$$

which does contain the pressure. As for Eq.(5.35), both sides turn out to be identically zero, and it is automatically satisfied. This means that the fluid particle (galaxies) follow geodesics, which is as expected since with p a function of t alone, there is no pressure gradient (i.e., no 3-gradient ∇p) to push them off geodesics.

System (5.26) and (5.27) is equivalent to system (5.28) and (5.37) since we can take derivative of (5.28) and obtain (5.26) by using (5.37). We shall make use of Eqs.(5.34) and (5.37) in the next sections where we discuss the models of the universe.

Exercise

(1) Problem 1, The Robertson–Walker metric with $K = 0$ is: $ds^2 = g_{\mu\nu}dx^\mu dx^\nu = dt^2 - a(t)^2 \left[dr^2 + r^2(d\theta^2 + \sin^2\theta d\phi^2) \right]$. Labeling the coordinates according to $t \equiv x^0$, $r \equiv x^1$, $\theta \equiv x^2$, $\phi \equiv x^3$, please derive the all nonzero connection coefficients arisen from that Robertson–Walker metric.

(2) Problem 2, By means of results in Problem 1, derive Ricci tensor $\mathcal{R}_{\mu\nu}$ and scalar curvature $\mathcal{R}_{\mu\nu}$.

5.3   The Friedmann models

If the universe is matter-dominated, and then the pressure is negligible when compared with the density. The standard Friedmann models arose from setting $p = 0$. With this as an input, and substituting $p = 0$ into Eq.(5.37), we obtain an integral of this differential equation:

$$\rho a^3 = \text{Constant.} \tag{5.38}$$

As we shall see, this leads to three possible models, each of which has $a(t) = 0$ at some point in time, and it is natural to take this point as the origin of t, so that $a(0) = 0$, and t is then the age of the universe. Let us use a subscript zero to denote present-day values of quantities, so that t_0 is the present age of the universe, and $a_0 \equiv a(t_0)$ and $\rho_0 \equiv \rho(t_0)$ are the present-day values of a and ρ. We may then write Eq.(5.38) as

$$\rho a^3 = \rho_0 a_0^3. \tag{5.39}$$

The Friedmann equation (5.28) then becomes

$$\dot{a}^2 + K = \frac{A^2}{a}, \tag{5.40}$$

where $A^2 \equiv 8\pi G\rho_0 a_0^3/3$ $(A > 0)$. *Hubble's "constant"* $H(t)$ is defined by

$$H(t) \equiv \frac{\dot{a}(t)}{a(t)}, \tag{5.41}$$

and we denote its present-day value by $H_0 \equiv H(t_0)$, which is determined by the astronomic observation data. To take account of the remaining uncertainty in that Hubble's constant, it is conventional to write

$$H_0 = 100\,h\,\text{kms}^{-1}\text{Mpc}^{-1}, \tag{5.42}$$

with the dimensionless parameter h assumed to be in the neighborhood of 0.7. This corresponds to a Hubble time

$$1/H_0 = 9.778 \times 10^9 \, h^{-1} \text{ years.} \tag{5.43}$$

Equation (5.28) gives

$$\frac{K}{a_0^2} = \frac{8\pi G \rho_0}{3} - H_0^2 = \frac{8\pi G}{3} \left(\rho_0 - \frac{3H_0^2}{8\pi G} \right). \tag{5.44}$$

Hence $K > 0$, $K = 0$, or $K < 0$ as $\rho_0 > \rho_c$, $\rho_0 = \rho_c$, or $\rho_0 < \rho_c$ respectively, where ρ_c is a *critic density* given by

$$\rho_c \equiv \frac{3H_0^2}{8\pi G} = 1.878 \times 10^{-29} \, h^2 \text{ g/cm}^3. \tag{5.45}$$

The *deceleration parameter* q_0 is defined to be the present-day value of $-a\ddot{a}/\dot{a}^2$. Using Eq.(5.26) (with $p = 0$) and (5.27) gives

$$q_0 = \frac{4\pi G \rho_0}{3H_0^2} = \frac{\rho_0}{2\rho_c}. \tag{5.46}$$

The three Friedmann models arise from integrating Eq.(5.40) for the possible values of K: $K = 0$, ± 1.

(i) *Flat model.* $K = 0$: hence $\rho_0 = \rho_c$, $q_0 = 1/2$.
Equation (5.40) gives

$$\frac{da(t)}{dt} = \frac{A}{\sqrt{a(t)}},$$

and integrating gives

$$a(t) = (3A/2)^{2/3} t^{2/3}. \tag{5.47}$$

This model is also known as the Einstein–de Sitter model, since it was firstly suggested by Einstein and de Sitter in 1932. Note that $\dot{a}(t) \to 0$ as $t \to \infty$.

(ii) *Closed model.* $K = 1$: hence $\rho_0 > \rho_c$, $q_0 > 1/2$.
Equation (5.40) gives

$$\frac{da(t)}{dt} = \sqrt{\frac{A^2 - a(t)}{a(t)}},$$

so

$$t = \int_0^{a(t)} \sqrt{\frac{a}{A^2 - a}} \, da \tag{5.48}$$

Putting $a \equiv A^2 \sin^2(\psi/2)$ gives

$$t = A^2 \int_0^\psi \sin^2(\psi/2)d\psi = \frac{1}{2}A^2 \int_0^\psi (1 - \cos\psi)d\psi = \frac{1}{2}A^2(\psi - \sin\psi).$$

So

$$a(t) = \frac{1}{2}A^2(1 - \cos\psi), \qquad t = \frac{1}{2}A^2(\psi - \sin\psi), \qquad (5.49)$$

and these two equations give $a(t)$ via parameter ψ.

(iii) *Open model.* $K = -1$: hence $\rho_0 < \rho_c$, $q_0 < 1/2$.
Equation (5.40) gives

$$\frac{da(t)}{dt} = \sqrt{\frac{A^2 + a(t)}{a(t)}},$$

so

$$t = \int_0^{a(t)} \sqrt{\frac{a}{A^2 + a}}da$$

Putting $a \equiv A^2 \sinh^2(\psi/2)$ gives

$$t = A^2 \int_0^\psi \sinh^2(\psi/2)d\psi = \frac{1}{2}A^2 \int_0^\psi (\cosh\psi - 1)d\psi = \frac{1}{2}A^2(\sinh\psi - \psi).$$

So

$$a(t) = \frac{1}{2}A^2(\cosh\psi - 1), \qquad t = \frac{1}{2}A^2(\sinh\psi - 1), \qquad (5.50)$$

and these two equations give $a(t)$ via parameter ψ. Note that $\dot{a}(t) \to 1$ as ψ (and hence t)$\to \infty$.

5.4   ΛCDM model

In last section, the standard Friedmann universe models for non-relativistic matter arise from setting $p = 0$ which leads to

$$\rho a^3 = constant. \qquad (5.51)$$

These discussions can be extended to relativistic matter models. For the relativistic-matter in the radiation dominating universe, p–ρ relation should be[1]

$$p = \rho/3. \qquad (5.52)$$

---

[1]see, e.g., L.D. Landau and E.M. Lifshitz, *The Classical Theory of Fields*, (Translated from Russian by M. Hamermesh), Pergamon Press, Oxford (1987), *pp.87*, Eq.(35.8) in Section 35. [Landau, Lifshitz (1987)].

Similar to the derivation of Eq.(5.38), we substitute Eq.(5.52) into Eq.(5.37) and obtain an integral of this differential equation:

$$\rho a^4 = constant, \tag{5.53}$$

which leads the quantity ρa^2 in RHS of Friedmann equation (5.28) to grow as a^{-2} as $a \to 0$.

There is a peculiar aspect to these results. The contribution of non-relativistic and relativistic matter to the quantity ρa^2 in Eq.(5.28) grows as a^{-1} and a^{-2}, respectively, as $a \to 0$, so at sufficiently early times in the expansion we may certainly neglect the constant K, and Eq.(5.28) gives

$$\frac{\dot{a}^2}{a^2} \to \frac{8\pi G\rho}{3}. \tag{5.54}$$

That is, at these early times the density becomes essentially equal to the critical density $3H^2/8\pi G$, where $H = \dot{a}/a$ is the value of the Hubble "constant" at those times. On the other hand, we will see later that the total energy density of present universe is still a fair fraction of the critical density. This is sometimes called the *flatness problem*.

The simplest solution to the flatness problem is just that we are in a spatially flat universe, in which $K = 0$ and ρ may be is always precisely equal to ρ_c. A more popular solution is provided by the so called "*inflationary theories*"[2]. In these theories K may not vanish, and ρ may not start out close to ρ_c, but there is an early period of rapid growth in which ρ/ρ_c rapidly approaches unity. Inflationary theories do not require that ρ should now be very close to ρ_c hence K can be different from zero.

For $K = 0$ we get very simple solutions to Eq.(5.28) in the three special cases listed as follows:

Non-relativistic matter: Here $\rho = \rho_0(a/a_0)^{-3}$. The solution of Eq.(5.28) has already been shown in Eq.(5.47), which is

$$a(t) \propto t^{2/3}. \tag{5.55}$$

This gives $q_0 \equiv -a\ddot{a}/\dot{a}^2 = 1/2.$ and simple relation between the age of the universe and the Hubble constant

$$t_0 = \frac{2}{3H_0} = 6.52 \times 10^9 \ h^{-1} \ yr. \tag{5.56}$$

---

[2]see, e.g. S. Weinberg, *Cosmology*, (2008), Chapter 4. [Weinberg (1980)]

Equations (5.55) and (5.26) show that for $K = 0$, the energy density at time t is $\rho = 1/(6\pi Gt^2)$. This is known as the *Einstein–de Sitter model*. It was for many years the most popular cosmological model, though the age (5.56) is uncomfortably short compared with the ages of certain stars.

Relativistic matter: From Eq.(5.53), we have $\rho = \rho_0(a/a_0)^{-4}$. And the solution of Eq.(5.28) with $K = 0$ is

$$a(t) \propto \sqrt{t}. \tag{5.57}$$

This gives $q_0 = +1$, while the age of the universe and the Hubble constant are related by

$$t_0 = \frac{1}{2H_0}. \tag{5.58}$$

The energy density at time t is $\rho = 3/(32\pi Gt^2)$.

Vacuum energy (or dark energy): In the E-SR, the Lorentz invariance requires that in locally inertial coordinate systems the energy-momentum tensor $T_{(v)}^{\mu\nu}$ of the *vacuum* must be proportional to the Minkowski metric $\eta^{\mu\nu}$, and so in general coordinate systems $T_{(v)}^{\mu\nu}$ must be proportional to $g^{\mu\nu}$. Comparing this with Eq.(5.31) shows that the vacuum has $p_{(v)} = -\rho_{(v)}$, so that $T_{(v)}^{\mu\nu} = \rho_{(v)}g^{\mu\nu}$. Usually, the vacuum energy density $\rho_{(v)}$ is also called the *dark energy density*. In the absence of any other form of energy this would satisfy the conservation law $0 = T_{(v);\mu}^{\mu\nu} = g^{\mu\nu}\partial\rho_{(v)}/\partial x^\mu$, so that $\rho_{(v)}$ would be a constant of spacetime position. Equation (5.28) for $K = 0$ requires that $\rho_{(v)} > 0$, and has the solutions

$$a(t) \propto \exp(Ht), \tag{5.59}$$

where H is Hubble constant, now really a constant, given by

$$H = \sqrt{\frac{8\pi G\rho_{(v)}}{3}}. \tag{5.60}$$

Here $q_0 = -1$, and the age of the universe in this case is infinite. This is known as the *de Sitter model* [de Sitter (1917)]. Of course, there is some matter in the universe, so even if the energy density of the universe is now dominated by the constant vacuum energy density, there was a time in the past when matter and/or radiation were more important, and so the expansion has a finite age, although greater than it would be without a vacuum energy.

More generally, ρ is a mixture of vacuum energy and relativistic and non-relativistic matter et al., i.e.

$$\rho = \sum_i \rho_i = \rho_\Lambda + \rho_M + \rho_R. \tag{5.61}$$

Then Eq.(5.28) can be expressed as

$$H^2(a) = H_0^2 \sum_i \frac{\rho_i(a)}{\rho_c} - \frac{K}{a^2}. \tag{5.62}$$

Letting

$$\Omega_i = \frac{\rho_i(a_0)}{\rho_c}, \quad \Omega_K = -\frac{K}{a_0^2 H_0^2}, \tag{5.63}$$

then

$$H^2(a) = H_0^2 \Big[\sum_i \Omega_i \frac{\rho_i(a)}{\rho_i(a_0)} + \Omega_K \frac{a_0^2}{a^2} \Big]. \tag{5.64}$$

Using the local energy conservation law

$$\dot{\rho}_i = -3 \frac{\dot{a}}{a} (\rho_i + p_i), \tag{5.65}$$

and the equation of state of matter ($w_M = 0, w_R = 1/3, w_\Lambda = -1$),

$$p_i = w_i \rho_i, \tag{5.66}$$

we have

$$\rho_i = C_i a^{-3(1+w_i)}. \tag{5.67}$$

Hence

$$\frac{\rho_i(a)}{\rho_i(a_0)} = \left(\frac{a_0}{a}\right)^{3(1+w_i)}. \tag{5.68}$$

Thus

$$H^2(a) = H_0^2 \Big[\sum_i \Omega_i \left(\frac{a_0}{a}\right)^{3(1+w_i)} + \Omega_K \frac{a_0^2}{a^2} \Big], \tag{5.69}$$

and

$$\sum_i \Omega_i + \Omega_K = 1. \tag{5.70}$$

Explicitly, we have

$$\rho = \frac{3H_0^2}{8\pi G} \left[\Omega_\Lambda + \Omega_M \left(\frac{a_0}{a}\right)^3 + \Omega_R \left(\frac{a_0}{a}\right)^4 \right], \tag{5.71}$$

In terms of Ω's, we can express the present energy densities for the vacuum, non-relativistic matter, and relativistic matter (i.e., radiation) as

$$\rho_\Lambda = \frac{3H_0^2\Omega_\Lambda}{8\pi G}, \quad \rho_M = \frac{3H_0^2\Omega_M}{8\pi G}, \quad \rho_R = \frac{3H_0^2\Omega_R}{8\pi G}, \tag{5.72}$$

respectively.

We can establish a relationship between redshift and time elapse of photon propagation. Since

$$1 + z = \frac{a(t_0)}{a(t)}, \tag{5.73}$$

we have

$$H^2(a) = H_0^2 E^2(z), \tag{5.74}$$

where

$$E^2(z) = \sum_i \Omega_i(1+z)^{3(1+w_i)} + \Omega_K(1+z)^2. \tag{5.75}$$

Using the relation

$$dt = \frac{da}{\dot{a}}, \tag{5.76}$$

the time elapse is then

$$t = \int \frac{da}{\dot{a}} = \int \frac{a\,da}{\dot{a}\,a} = H_0^{-1}\int \frac{1}{E(z)}\frac{da}{a} = H_0^{-1}\int \frac{-dz}{E(z)(1+z)}. \tag{5.77}$$

Defining $x \equiv a/a_0 = 1/(1+z)$, we have then

$$dt = \frac{dx}{H_0 x\sqrt{\Omega_\Lambda + \Omega_K x^{-2} + \Omega_M x^{-3} + \Omega_R x^{-4}}}$$

$$= \frac{-dz}{H_0(1+z)\sqrt{\Omega_\Lambda + \Omega_K(1+z)^2 + \Omega_M(1+z)^3 + \Omega_R(1+z)^4}}. \tag{5.78}$$

Therefore, if we define the zero of time as corresponding to an infinite redshift, then the time at which light was emitted that reaches us with redshift z is given by

$$t(z) = \frac{1}{H_0}\int_0^{1/(1+z)} \frac{dx}{x\sqrt{\Omega_\Lambda + \Omega_K x^{-2} + \Omega_M x^{-3} + \Omega_R x^{-4}}}$$

$$= \frac{1}{H_0}\int_z^\infty \frac{dz'}{(1+z')\sqrt{\Omega_\Lambda + \Omega_K(1+z')^2 + \Omega_M(1+z')^3 + \Omega_R(1+z')^4}}. \tag{5.79}$$

In particular, by setting $z = 0$, we find the present age of the universe:

$$t_0 = \frac{1}{H_0} \int_0^1 \frac{dx}{x\sqrt{\Omega_\Lambda + \Omega_K x^{-2} + \Omega_M x^{-3} + \Omega_R x^{-4}}}. \qquad (5.80)$$

It is of some interest to express the deceleration parameter q_0 in terms of Ωs. Equation (5.72) gives the present pressure as

$$p_0 = \frac{3H_0^2}{8\pi G}\left(-\Omega_\Lambda + \frac{1}{3}\Omega_R\right). \qquad (5.81)$$

Then, Eqs.(5.26), (5.41) with p_0 in Eq.(5.81) give the deceleration

$$q_0 \equiv -a_0\ddot{a}_0/\dot{a}_0^2 = \frac{4\pi G(\rho_0 + 3p_0)}{3H_0^2} = \frac{1}{2}(\Omega_M - 2\Omega_\Lambda + 2\Omega_R). \qquad (5.82)$$

Finally, we show the formula of the luminosity distance $d_L(z)$ of a source observed with redshift z without derivations. For $K = 0$, and hence $\Omega_K = 0$, the formula is

$$d_L(z) = \frac{1+z}{H_0} \int_{1/(1+z)}^1 \frac{dx}{x^2\sqrt{\Omega_\Lambda + \Omega_M x^{-3} + \Omega_R x^{-4}}}. \qquad (5.83)$$

One of the interesting questions is whether the ongoing expanding of our universe will ever stop. In this case, we have Eq.(5.28)

$$\frac{8\pi G}{3}\rho a^2 - K = 0. \qquad (5.84)$$

Obviously, we should have $K = 1$. Using Eq.(5.69), we have

$$\sum_i \Omega_i\left(\frac{a_0}{a}\right)^{3(1+w_i)} + \Omega_K\frac{a_0^2}{a^2} = 0, \qquad (5.85)$$

or

$$\Omega_\Lambda + \Omega_M\left(\frac{a_0}{a}\right)^3 + \Omega_R\left(\frac{a_0}{a}\right)^4 + \Omega_K\left(\frac{a_0}{a}\right)^2 = 0. \qquad (5.86)$$

We thus can claim that if the equation

$$\Omega_\Lambda + \Omega_M x^3 + \Omega_R x^4 + \Omega_K x^2 = 0, \qquad (5.87)$$

has solution in (0,1), the expanding will stop sometime in the future. Due to Eq.(5.70), the left hand side of Eq.(5.87) is 1 for $x = 1$. If $\Omega_\Lambda < 0$, for small enough x, the expression will become negative. Hence there must exist some $x \in (0,1)$ such that Eq.(5.87) holds. Even if $\Omega > 0$, for sufficiently negative Ω_K, Eq.(5.87) can still admit a solution in (0,1) since the l.h.s. expression $\approx \Omega_\Lambda + \Omega_K x^2$ for small x.

5.5 Accelerated expansion

We now turn to the measurement of distances as a function of redshift. Considering now redshifts $z > 0.1$, which are large enough so that we can ignore the peculiar motions of the light sources, and also large enough so that we need to take into account the effects of cosmological expansion on distance determination.

For quite many years, the brightest galaxies in rich clusters were used as the "standard candles" in cosmological distance determination. Due to several issues of this method, Type Ia supernovae are used (for review, see [Perlmutter,Schmidt (2003)] [Ruiz-Lapuente (2004)] [Filippenko (2004)] [Panagia (2005)]). Using Type Ia supernovae as distance indicator has several advantages over brightest galaxies. First, they are very bright. Second,the absolute luminosity of Type Ia supernovae can be determined more reliably. Observations of Type Ia supernovae have been compared with theoretical prediction (equivalent to Eq.(5.83)) for luminosity distance as a function of redshift at about the same time by two groups: The Supernova Cosmology Project [Perlmutter, S (1999)] and the high-z Supernova Search Team [Riess, *et al.* (1998)] [Schmidt (1998)]. The Supernova Cosmology Project analyzed the relation between apparent luminosity and reshift for 42 Type Ia supernovae, with redshifts z ranging from 0.18 to 0.83, together with a set of closer supernovae from another supernova survey, at redshift below 0.1. Their original results are shown in Fig.5.1.

With a confidence level of 99%, the data confirm that $\Omega_\Lambda > 0$. For a flat cosmology with $\Omega_K = \Omega_R = 0$, so that $\Omega_\Lambda + \Omega_{NR} = 1$, the data indicate a value

$$\Omega_M = 0.28^{+0.09}_{-0.08} \ (1\sigma \text{ statistical})^{+0.05}_{-0.04} \text{ (identified systematics)}. \quad (5.88)$$

And then we have

$$\Omega_\Lambda = 1 - \Omega_M \approx 0.72^{+0.12}_{-0.14}. \quad (5.89)$$

(These results are independent of the Hubble constant or the absolute calibration of the relation between supernova absolute luminosity and time scale, though they do depend on the shape of this relation.) This gives the age (5.80) as

$$t_0 = 13.4^{+1.3}_{-1.0} \times 10^9 \left(\frac{70 \text{ km s}^{-1} \text{ Mpc}^{-1}}{H_0} \right) \text{ yr}. \quad (5.90)$$

Substituting Eqs.(5.88) and (5.89) into Eq.(5.82) gives the deceleration factor

$$q_0 \equiv -a_0\ddot{a}_0/\dot{a}_0^2 \simeq \frac{1}{2}(\Omega_M - 2\Omega_\Lambda) \approx -0.64^{+0.14}_{-0.12} < 0, \qquad (5.91)$$

which indicates that the present universe is in accelerated expansion.

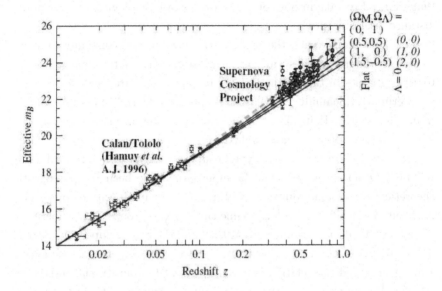

Fig. 5.1 Evidence for dark energy , found in 1998 by the Supernova Cosmology Project, from S. Perlmutter et al., Astrophys. J. 517, 565 (1999) [astro-ph/9812133]. Here the effective blue apparent magnitude (corrected for variations in absolute magnitude, as indicated by supernova light curves) are plotted versus redshift for 42 high redshift Type Ia supernovae observed by the Supernova Cosmology Project, along with 18 lower redshift Type Ia supernovae from the CalxA8xA2nxA8CTololo Supernovae Survey. Horizontal bars indicate the uncertainty in cosmological redshift due to an assumed peculiar velocity uncertainty of 300 km sec.1. Dashed and solid curves give the theoretical effective apparent luminosities for cosmological models with $\Omega_K = 0$ or $\Omega_\Lambda = 0$, respectively, and various possible values of Ω_M.

5.6 Cosmological constant

Once upon a time, Einstein considered his introduction of cosmological constant Λ into his equation his biggest mistake. Yet, the constant seems a necessity based on cosmological observations as of today. There was not even quantum mechanics at the time Einstein proposed his theory of general relativity. We now know that each type of quanta contributes to vacuum from the point of view of quantum field theory. Take an oscillator for instance, due to Heisenberg's uncertainty principle, momentum and displacement cannot be both zero at the same time. Hence the energy of an oscillator cannot be zero. Because of requirement of Lorentz invariance, the stress-energy tensor must be of the form const$\cdot \eta_{\mu\nu}$. Therefore, if we express the total energy momentum tensor $T_{\mu\nu}$ as the sum of a possible vacuum term $-\rho_{\text{vac}}g_{\mu\nu}$ and a term $T_{\mu\nu}^{\text{M}}$ arising from matter (including radiation), then the Einstein equations of the de Sitter invariant general relativity (4.11) take the form

$$\mathcal{R}_{\mu\nu} - \frac{1}{2}g_{\mu\nu}\mathcal{R} + \Lambda g_{\mu\nu} = -8\pi G T_{\mu\nu}^{\text{M}} - 8\pi G \rho_{\text{vac}}g_{\mu\nu}, \tag{5.92}$$

where Λ is originally introduced by Einstein in 1917, and serves as a universal constant in physics. We call it the geometrical cosmological constant (or Einstein cosmological constant) hereafter. Equation (8.21) can be rewritten in form as follows

$$\mathcal{R}_{\mu\nu} - \frac{1}{2}g_{\mu\nu}\mathcal{R} = -8\pi G T_{\mu\nu}^{\text{M}} - 8\pi G \rho_{\text{eff}}g_{\mu\nu}, \tag{5.93}$$

in which

$$\rho_{\text{eff}} = \rho_{\text{vac}} + \frac{\Lambda}{8\pi G} \equiv \frac{\Lambda_{\text{eff}}}{8\pi G},$$
$$\text{with } \Lambda_{\text{eff}} = \Lambda + 8\pi G \rho_{\text{vac}} = \Lambda + \Lambda_{\text{dark energy}}, \tag{5.94}$$

where the fact that ρ_{vac} is the dark energy density has been used, so $\Lambda_{\text{dark energy}} = 8\pi G \rho_{\text{vac}}$. The effective cosmological constant Λ_{eff} instead of $\Lambda_{\text{dark energy}}$ (or ρ_{vac}) is the quantity that one can observe and estimate in testing accelerated expansion of the Universe. The cosmological constant problem which has perplexed physics community for decades is that the limit of Λ_{eff} deduced from observation data is much smaller (less than 10^{-120}) than the contributions calculated from various quantum field theories. This puzzle is open until now.

As discussed previously, cosmological observations have confirmed that $\Lambda_{\text{eff}} > 0$. Recent WMAP data [Jarosik, *et al.* (2011)] show that the result is:

$$\Lambda_{\text{eff}} \simeq 1.26 \times 10^{-56} \text{ cm}^{-2}. \tag{5.95}$$

Therefore, for the dark energy, Eq.(5.94) gives [Yan, Hu, Huang (2012)]

$$\Lambda_{\text{dark energy}} = \Lambda_{\text{eff}} - \Lambda = \Lambda_{\text{eff}} \left(1 - \frac{3}{R^2 \Lambda_{\text{eff}}} \right) < \Lambda_{\text{eff}}, \qquad (5.96)$$

where the universal parameter $\Lambda = 3/R^2$ in the de Sitter Invariant Special Relativity (dS-SR) were used. The parameter R could in principle be determined in the framework of dS-SR, for instance, in a reference [Yan, Xiao, Huang, Hu (2012)] R was estimated to be $\sim 10^{30}$ cm, then (5.96) becomes

$$\Lambda_{\text{dark energy}} = \Lambda_{\text{eff}} (1 - \mathcal{O}(10^{-4})). \qquad (5.97)$$

which implies that the dark-energy-cosmological constant is much more important than the Einstein Λ.

It is essential that the observation result of $\Lambda_{\text{eff}} > 0$ (Eq.(5.95)) does not indicates Einstein Λ (or geometric cosmological constant) must be positive. To Anti de Sitter symmetry case, we have $\Lambda < 0$. Now, Eq.(5.96) becomes

$$\Lambda_{\text{dark energy}} = \Lambda_{\text{eff}} + |\Lambda| = \Lambda_{\text{eff}} \left(1 + \frac{3}{|R^2| \Lambda_{\text{eff}}} \right) > \Lambda_{\text{eff}}, \qquad (5.98)$$

where $R^2 < 0$ is the universal parameter in the Anti de Sitter Invariant Special Relativity (AdS-SR). Both dS-SR and AdS-SR are ambiguity-free at the classical and quantum mechanics level.

In terms of the method described in Chapter 4, the AdS-SR (or AdS-symmetry) could be localized, and then the Anti de Sitter General Relativity (AdS-GR) can be constructed. Thus, in principle, the AdS Quantum Gravity could be discussed. It has been argued that AdS Quantum Gravity seems to be more consistent than dS Quantum Gravity [Witten (2001)].

Chapter 6

Relativistic Quantum Mechanics for de Sitter Invariant Special Relativity

In this chapter, we turn to discussion the quantization of the de Sitter invariant special relativistic mechanics described in Chapter 3. The unit system used below is the same as that chapter, i.e., $c \equiv 299\ 792\ 458$ m s^{-1}.

6.1 Quantization of the de Sitter invariant special relativistic mechanics

Lagrangian-Hamiltonian formulation of mechanics is the foundation of quantization. When the classical Poisson Brackets of canonical coordinates and canonical momenta become operator's commutators, i.e., $\{x, \pi\}_{PB} \Rightarrow \frac{1}{i\hbar}[x, \hat{\pi}]$, the classical mechanics will be quantized. In this way, for instance, the ordinary relativistic one-particle quantum equations have been derived (see Section 2.7). For particles with spin-0, that is just the well-known Klein-Gordon equation. In Section 3.6, the Hamilton and canonical momenta for de Sitter invariant special relativistic mechanics have been achieved. By means of the canonical quantization method described in Section 2.4, we shall accomplish the quantization of the de Sitter invariant special relativistic mechanics in this section.

In the canonical quantization formalism for special relativistic mechanics, the canonical variable operators are x^i, $\hat{\pi}_i$ with $i = 1, 2, 3$, and the basic commutators are:

$$[x^i,\ \hat{\pi}_j] = i\hbar \delta^i_j, \qquad [\hat{\pi}_i,\ \hat{\pi}_j] = 0, \qquad [x_i,\ x_j] = 0. \qquad (6.1)$$

(see Eqs.(2.32), (2.106), and (3.141)). In 4-dimensional spacetime, the Hamiltonian operator $\hat{H}_{cR} \equiv -c\hat{\pi}_0$ represents the generator of time evolution, i.e.,

$$[t, \hat{H}_{cR}] = -i\hbar, \quad \text{or} \quad [x^0, \hat{\pi}_0] = i\hbar. \tag{6.2}$$

As we did in Section 2.7, the commutators (6.1) can be written in a 4-vector form: (see Eq.(2.111))

$$[x^\mu, \hat{\pi}_\nu] = i\hbar\delta^\mu_\nu, \quad [x^\mu, x^\nu] = 0, \quad [\hat{\pi}_\mu, \hat{\pi}_\nu] = 0. \tag{6.3}$$

These 4-dimensional canonical commutation relations could be used to quantize both the mechanics of E-SR and the mechanics of dS-SR. Equation (6.3) is called *covariant Heisenberg algebra* and serves as a basis of covariant quantization for special relativistic mechanics.

Distinguished from the Hamiltonian operator of E-SR (2.29), the Hamiltonian operator $\hat{H}_{dS}(\hat{\pi}, \mathbf{x})$ is $\hat{\pi}$-x ordering dependent (see Eq.(3.130)). Since $\hat{\pi}$-$\mathbf{x}$ is non-commutative, the form of operator $\hat{H}_{dS}(\hat{\pi}, \mathbf{x})$ cannot be determined uniquely from the $H_{dS}(\pi, \mathbf{x})$-function (3.130). For instance, two quantum mechanical Hamiltonian can be proposed based on the same classical functional expression Eq.(3.130)

$$^{(1)}H_{dS}(\pi, \mathbf{x})$$
$$\equiv \frac{1}{2B^{00}(x)}\left\{2cB^{0i}(x)\pi_i \pm c\sqrt{4(B^{0i}(x)\pi_i)^2 - 4B^{00}(x)(B^{ij}(x)\pi_i\pi_j - m^2c^2)}\right\}$$

or

$$^{(2)}H_{dS}(\pi, \mathbf{x})$$
$$\equiv \frac{1}{2B^{00}(x)}\left\{2c\pi_iB^{0i}(x) \pm c\sqrt{4(\pi_iB^{0i}(x))^2 - 4B^{00}(x)(\pi_iB^{ij}(x)\pi_j - m^2c^2)}\right\}.$$

The operators $^{(1)}\hat{H}_{dS}(\hat{\pi}, \mathbf{x})$ and $^{(2)}\hat{H}_{dS}(\hat{\pi}, \mathbf{x})$ are generally different, i.e., $^{(1)}\hat{H}_{dS}(\hat{\pi}, \mathbf{x}) \neq {}^{(2)}\hat{H}_{dS}(\hat{\pi}, \mathbf{x})$, since the orderings of $\hat{\pi}$-$\mathbf{x}$ in $^{(1)}\hat{H}_{dS}(\hat{\pi}, \mathbf{x})$ and in $^{(2)}\hat{H}_{dS}(\hat{\pi}, \mathbf{x})$ are different and $[\hat{\pi}_i, x^j] \neq 0$. This π-$\mathbf{x}$ ordering dependent property can always be used to construct an appropriate $\hat{H}_{dS}(\hat{\pi}, \mathbf{x})$ based on (3.130) so that $[\hat{H}_{dS}(\hat{\pi}, \mathbf{x}), \hat{\pi}_i] = 0$, or equivalently

$$[\hat{\pi}_0, \hat{\pi}_i] = 0, \tag{6.4}$$

where $\hat{H}_{dS}(\hat{\pi}, \mathbf{x}) = -c\hat{\pi}_0$ was used. Equation (8.133) is just required by the last equation of (6.3).

The general solution to Eq.(6.3) is

$$\pi_\mu = -i\hbar\partial_\mu + (\partial_\mu G(t, \mathbf{x})), \tag{6.5}$$

where $G(t, \mathbf{x})$ is a function of t and x^i, and hereafter the hat notations for operators are removed for simplicity. Generally, the most symmetrical ordering (i.e., Weyl ordering) is favored for realistic quantization scheme. For dS-SR, we prefer the quantization scheme that protects the de Sitter symmetry $SO(1, 4)$. This requirement will help fix the function $G(t, \mathbf{x})$ in Eq.(6.5). By this consideration, we take (see, e.g., *pp.29-44* in [Kleinert (1990)])

$$\pi_\mu = -i\hbar \hat{D}_\mu = -i\hbar(\partial_\mu + \frac{\Gamma_\mu}{2}) = -i\hbar(-B)^{-\frac{1}{4}}\partial_\mu(-B)^{\frac{1}{4}}, \qquad (6.6)$$

where $B = \det(B_{\mu\nu})$ and $\Gamma_\mu = \Gamma^\nu_{\mu\nu}$. Equation (6.6) indicates $G(t, \mathbf{x}) = -i\hbar \log(-B)^{\frac{1}{4}}$. In contrast to the ordinary quantization theories in curved space only[1], our treatment presented here is suitable for the theories in generic curved space-time, in which the 4-dimensional metric is time- and space-dependent. The classical dispersion relation (3.129) can be rewritten in symmetric version $(-B)^{-\frac{1}{4}}\pi_\mu(-B)^{\frac{1}{4}}B^{\mu\nu}(-B)^{\frac{1}{4}}\pi_\nu(-B)^{-\frac{1}{4}} = m^2c^2$. Then the one particle de Sitter invariant special relativistic wave equation reads

$$(-B)^{-\frac{1}{4}}\pi_\mu(-B)^{\frac{1}{4}}B^{\mu\nu}(-B)^{\frac{1}{4}}\pi_\nu(-B)^{-\frac{1}{4}}\phi(\mathbf{x}, t) = m^2c^2\phi(\mathbf{x}, t) , \quad (6.7)$$

where $\phi(\mathbf{x}, t)$ is the particle's wave function. Substituting Eq.(6.6) into Eq.(6.7), we have

$$\frac{1}{\sqrt{-B}}\partial_\mu(B^{\mu\nu}\sqrt{-B}\partial_\nu)\phi + \frac{m^2c^2}{\hbar^2}\phi = 0, \qquad (6.8)$$

which is just the Klein-Gordon equation in curved space-time with Beltrami metric $B_{\mu\nu}$. Its explicit form is

$$(\eta^{\mu\nu} - \frac{x^\mu x^\nu}{R^2})\partial_\mu\partial_\nu\phi - 2\frac{x^\mu}{R^2}\partial_\mu\phi + \frac{m^2c^2}{\hbar^2\sigma(x)}\phi = 0, \qquad (6.9)$$

which is the desired de Sitter invariant special relativistic quantum mechanical equation for free particle.

We should note the following:

(i) Substituting Eq.(6.6) into Eq.(3.147), we obtain the physical momentum and energy operators

$$p^\mu = i\hbar[(\eta^{\mu\nu} - \frac{x^\mu x^\nu}{R^2})\partial_\nu + \frac{5x^\mu}{2R^2}]. \qquad (6.10)$$

p^μ together with operator $L^{\mu\nu} = (x^\mu p^\nu - x^\nu p^\mu)/i\hbar$ form a algebra as follows

$$[p^\mu, p^\nu] = \frac{\hbar^2}{R^2} L^{\mu\nu}$$
$$[L^{\mu\nu}, p^\rho] = \eta^{\nu\rho} p^\mu - \eta^{\mu\rho} p^\nu \tag{6.11}$$
$$[L^{\mu\nu}, L^{\rho\sigma}] = \eta^{\nu\rho} L^{\mu\sigma} - \eta^{\nu\sigma} L^{\mu\rho} + \eta^{\mu\sigma} L^{\nu\rho} - \eta^{\mu\rho} L^{\nu\sigma}$$

which is just the de-Sitter algebra SO(1,4). This fact means that the quantization scheme presented in above preserves the external space-time symmetry of dS-SR.

(ii) The wave function $\phi(x)$ of the wave equation (6.7) could be thought as a scalar field $\phi(x)$, and then Eq.(6.7) could be obtained by means of the least action principle with a action A:

$$A = \int d^4 x \sqrt{-B} B^{\mu\nu} \partial_\mu \phi \partial_\nu \phi. \tag{6.12}$$

Namely, Eq.(6.7) can be derived from $\delta A = 0$. This method for deriving wave equations can be called *field-action method*. In the following section we shall derive the Dirac equation in Beltrami spacetime by way of the field-action method.

6.2   Dirac equation in Beltrami spacetime

6.2.1   *Introduction*

In the last section we proved that the wave equation for spin-zero particles in the de Sitter invariant special relativistic quantum mechanics is the standard Klein-Gordon equation in the Beltrami spacetime $\mathcal{B}$, i.e., the curved spacetime with Beltrami metric $B_{\mu\nu}$. Hence, the wave equation for spin-1/2 particle is expected to be Dirac equation in the curved Beltrami spacetime $\mathcal{B}$.

Historically (see, e.g., [Nieh, Yan (1982)]), the generalization of Dirac's equation in curved space (see, e.g., [Brill, Brill (1957)]) took two different approaches. One might call them the Dirac approach and the Cartan approach. As we recall, the Dirac equation (in flat space) was discovered by Dirac [Dirac (1928)] on basis of linearity and relativistic covariance requirements, without emphasizing the geometrical nature of the object involved. It was only later on, through the work of van der Waeden [van der Waerden

(1929)], related to Cartan's spinor (see, e.g., [Cartan (1966)]) with well-defined geometrical meaning. For generalizing to the case of curved space, Schrödinger [Schrödinger (1932)], Bargmann [Bargmann (1932)] and others adopted the approach of a formal generalization of the Dirac equation imposing the requirement of general covariance. This is the spirit of Dirac's original derivation of his equation, and we label it as the Dirac approach. In the author's point of view, Weyl [Weyl (1929)] [Duan (1958)] [Kibble, B.W.T. (1961)]-Fock [Fock (1929)] [Utiyama, R. (1956)] approach is more fundamental. It is especially remarkable that Weyl's work was based on the idea of gauge principle. We shall outline Weyl's approach in following subsection.

6.2.2 *Weyl's approach*

While the Dirac spinor is a (two-valued) representation of the Lorentz group $SO(3,1)$, there does not exist a spinor generalization for the group $GL(4)$ of the general coordinate transformations. In a curved space, a Dirac spinor can only be defined with respect to local Cartesian frame, which is arbitrary up to local Lorentz transformations. A connection field is thus needed to relate spinors at different spacetime points so that covariant differentials can be defined. Under local Lorentz transformations of the Cartesian frame, which is represented by the four orthonormal vectors $e^a{}_\mu(x)$, the Dirac spinor transforms according to

$$\psi(x) \to e^{-i\epsilon^{ab}\sigma_{ab}/4}\psi(x), \tag{6.13}$$

where

$$\sigma_{ab} = \frac{i}{2}[\gamma_a, \gamma_b], \quad \{\gamma_a, \gamma_b\} = 2\eta_{ab}, \quad \eta_{ab} = \{1, -1, -1, -1\}. \tag{6.14}$$

The connection field $\omega^{ab}{}_\mu(x)$ is introduced such that

$$D_\mu\psi \equiv \left(\partial_\mu - i\frac{1}{4}\omega^{ab}{}_\mu(x)\sigma_{ab}\right)\psi, \tag{6.15}$$

transforms in a covariant way:

$$D_\mu\psi(x) \to e^{-i\epsilon^{ab}(x)\sigma_{ab}/4}D_\mu\psi(x). \tag{6.16}$$

This requires that the connection field $\omega^{ab}{}_\mu(x)$ transforms according to

$$\omega_\mu \to \omega'{}_\mu = e^{-i\epsilon(x)}\omega_\mu e^{i\epsilon(x)} - \left[i\partial_\mu e^{-i\epsilon(x)}\right]e^{i\epsilon(x)} \tag{6.17}$$

where

$$\omega_\mu \equiv \frac{1}{4}\sigma_{ab}\omega^{ab}{}_\mu, \quad \epsilon(x) \equiv \sigma_{ab}\epsilon^{ab}(x). \tag{6.18}$$

This is essentially what is contained in Weyl's work[8], here rephrased in modern notations. It is seen that the whole approach is that of a gauge theory that has been described in Section 2.8. The connection field $\omega^{ab}{}_\mu$ is the gauge potential.

With the introduction of the *vielbein* field $e^a{}_\mu(x)$ and the Lorentz spin-connection field $\omega^{ab}{}_\mu(x)$, one can make use of the covariant derivative defined by Eq.(6.15) to construct the following scalar combination

$$\bar{\psi}i\gamma^a e_a{}^\mu D_\mu \psi, \tag{6.19}$$

where γ^a is the flat-space Dirac matrices satisfying Eq.(6.13), and $e_a{}^\mu$ is the inverse of $e^a{}_\mu$ given by:

$$e_a{}^\mu e^{a'}{}_\mu = \delta_a^{a'}. \tag{6.20}$$

The combination (6.18) is, however, not Hermitian, and its Hermitian adjoint should be added. The action for a Dirac field in the background of $e^a{}_\mu(x)$ and $\omega^{ab}{}_\mu(x)$ is thus of the *Hermitian* form

$$A = \int d^4x\, h \left[\frac{1}{2}(\bar{\psi}i\gamma^a e_a{}^\mu D_\mu \psi - \bar{\psi}\overleftarrow{D}_\mu i\gamma^a e_a{}^\mu \psi) - \frac{mc}{\hbar}\bar{\psi}\psi\right], \tag{6.21}$$

where

$$D_\mu = \overrightarrow{\partial}_\mu - i\frac{1}{4}\sigma_{ab}\omega^{ab}{}_\mu, \quad \overleftarrow{D}_\mu = \overleftarrow{\partial}_\mu + i\frac{1}{4}\sigma_{ab}\omega^{ab}{}_\mu, \tag{6.22}$$

and

$$h = \det e^a{}_\mu. \tag{6.23}$$

The Lagrangian in Eq.(6.21) is invariant under general coordinate transformations.

Define the metric tensor by

$$g_{\mu\nu} = \eta_{ab}e^a{}_\mu e^b{}_\nu, \tag{6.24}$$

and the $GL(4)$ connection by (see, e.g., [Nieh (1980)])

$$\Gamma^\lambda_{\mu\nu} \equiv e_a{}^\lambda \left(e^a{}_{\mu,\nu} + \omega^a{}_{b\nu}e^b{}_\mu\right). \tag{6.25}$$

We further define covariant derivatives respect to both local Lorentz transformations and general coordinate transformations, such as

$$\chi_a{}^\lambda{}_{;\mu} \equiv \chi_a{}^\lambda{}_{,\mu} - \omega^b{}_{a\mu}\chi_b{}^\lambda + \Gamma^\lambda_{\nu\mu}\chi_a{}^\nu,$$
$$\chi^a{}_{\nu;\mu} \equiv \chi^a{}_{\nu,\mu} + \omega^a{}_{b\mu}\chi^b{}_\nu - \Gamma^\lambda_{\nu\mu}\chi^a{}_\lambda. \tag{6.26}$$

It can be easily verified that

$$e^a{}_{\mu;\nu} = 0, \quad e_a{}^\mu{}_{;\nu} = 0, \tag{6.27}$$

and consequently

$$g^{\mu\nu}{}_{;\lambda} = 0, \quad g_{\mu\nu;\lambda} = 0. \tag{6.28}$$

The connection $\Gamma^\lambda_{\mu\nu}$ as defined by Eq.(6.25) is in general not symmetric:

$$\Gamma^\lambda_{\mu\nu} \neq \Gamma^\lambda_{\nu\mu}, \tag{6.29}$$

giving rise to torsion:

$$C^\lambda{}_{\mu\nu} \equiv \Gamma^\lambda_{\mu\nu} - \Gamma^\lambda_{\nu\mu}. \tag{6.30}$$

In the presence of torsion, the relations (6.28) imply a relation that is a generalization of usual expression for the Christoffel connection:

$$\Gamma^\lambda_{\mu\nu} = \frac{1}{2}g^{\lambda\rho}(g_{\rho\mu,\nu} + g_{\rho\nu,\mu} - g_{\mu\nu,\rho}) + Y^\lambda_{\mu\nu}, \tag{6.31}$$

where the contorsion $Y^\lambda_{\mu\nu}$ is given by

$$Y^\lambda_{\mu\nu} \equiv \frac{1}{2}(C^\lambda{}_{\mu\nu} + C_{\mu\nu}{}^\lambda + C_{\nu\mu}{}^\lambda). \tag{6.32}$$

We are ready to derive Dirac equation in a curved Riemann-Cartan space. The Euler-Lagrange equation for Dirac field that follows from the action (6.21) is

$$\frac{1}{2}hi\gamma^a e_a{}^\mu D_\mu \psi + \frac{1}{2}(\partial_\mu - i\frac{1}{4}\sigma_{cd}\omega^{cd}{}_\mu)(hi\gamma^a e_a{}^\mu \psi) - \frac{mc}{\hbar}\psi = 0, \tag{6.33}$$

which, on account of

$$\partial_\mu(he_a{}^\mu) = h(h^{-1}h_{,a} + e_a{}^\mu{}_{,\mu})$$
$$= h(e_a{}^\mu{}_{,\mu} + C^\lambda{}_{\lambda\mu}e_a{}^\mu + \omega^b{}_{a\mu}e_b{}^\mu),$$

can be written in the form

$$\left[i\gamma^a e_a{}^\mu\left(D_\mu + \frac{1}{2}C^\lambda_{\lambda\mu}\right) - \frac{mc}{\hbar}\right]\psi = 0, \tag{6.34}$$

where

$$D_\mu = \partial_\mu - i\frac{1}{4}\sigma_{ab}\omega^{ab}{}_\mu \ . \tag{6.35}$$

For our purpose, we are interested in the Dirac equation in the torsion-free curved Riemann space. In this case, the torsion vanishes, and we have

$$C^\lambda{}_{\mu\nu} \equiv \Gamma^\lambda_{\mu\nu} - \Gamma^\lambda_{\nu\mu} = 0, \tag{6.36}$$

and then the connection $\Gamma^\lambda_{\mu\nu}$ reduces to the usual Christoffel symbol (2.181). The Dirac equation (6.34) reduces to the following:

$$\left[i\gamma^a e_a{}^\mu D_\mu - \frac{mc}{\hbar}\right]\psi = 0, \tag{6.37}$$

or more explicitly

$$\left[i\gamma^a e_a{}^\mu\left(\partial_\mu - i\frac{1}{4}\omega^{ab}{}_\mu\sigma_{ab}\right) - \frac{mc}{\hbar}\right]\psi = 0, \tag{6.38}$$

where the Ricci coefficients $\omega_{ab\mu}$ can be derived from Eqs.(6.24), (6.25) and (2.181):

$$\omega_{ab\mu} = \frac{1}{2}e^c{}_\mu(\gamma_{cab} - \gamma_{abc} - \gamma_{bca}), \tag{6.39}$$

where

$$\gamma^c{}_{ab} = (e_a{}^\mu e_b{}^\nu - e_b{}^\mu e_a{}^\nu)e^c{}_{\mu,\nu}. \tag{6.40}$$

Equation (6.37) is the desired Dirac equation in Riemann spacetime.

6.2.3 *Dirac equation in Beltrami spacetime*

The relativistic quantum mechanical wave equation for particles with spin-1/2 is Dirac equation. In the dS-SR, that is the Dirac equation in the Beltrami spacetime $\mathcal{B}$ with metric that we have derived in Section 3.1 (see Eq.(3.6)):

$$\begin{aligned}B_{\mu\nu}(x) &= \frac{\eta_{\mu\nu}}{\sigma} + \frac{\eta_{\mu\lambda}x^\lambda\eta_{\nu\rho}x^\rho}{R^2\sigma^2}\\ &\equiv \frac{\eta_{\mu\nu}}{\sigma} + \frac{\bar{x}_\mu\bar{x}_\nu}{R^2\sigma^2},\end{aligned} \tag{6.41}$$

where

$$\bar{x}_\mu \equiv \eta_{\mu\lambda}x^\lambda, \quad x^\lambda \equiv \bar{x}^\lambda = \eta^{\mu\lambda}\bar{x}_\mu, \tag{6.42}$$

$$\sigma = \sigma(x) = 1 - \frac{\eta_{\mu\nu}x^\mu x^\nu}{R^2} = 1 - \frac{\bar{x}_\nu\bar{x}^\nu}{R^2} \equiv 1 - \frac{\bar{x}^2}{R^2}, \tag{6.43}$$

where the notation

$$\bar{x}^2 \equiv \eta_{\mu\nu} x^\mu x^\nu = \bar{x}_\nu \bar{x}^\nu, \tag{6.44}$$

has been used.   We introduce two projection operators in spacetime $\{\bar{x}^\mu, \ \eta_{\mu\nu}\}$:

$$\theta_{\mu\nu} = \eta_{\mu\nu} - \frac{\bar{x}_\mu \bar{x}_\nu}{\bar{x}^2}, \qquad \omega_{\mu\nu} = \frac{\bar{x}_\mu \bar{x}_\nu}{\bar{x}^2}. \tag{6.45}$$

It is easy to check the calculation rules for projection operators:

$$\theta_{\mu\lambda} \theta^\lambda{}_\nu \equiv \theta_{\mu\lambda} \eta^{\lambda\rho} \theta_{\rho\nu} = \theta_{\mu\nu}, \quad \text{or in short } \theta \cdot \theta = \theta, \tag{6.46}$$

$$\omega_{\mu\lambda} \omega^\lambda{}_\nu \equiv \omega_{\mu\lambda} \eta^{\lambda\rho} \omega_{\rho\nu} = \omega_{\mu\nu}, \quad \text{or in short } \omega \cdot \omega = \omega, \tag{6.47}$$

$$\theta_{\mu\lambda} \omega^\lambda{}_\nu \equiv \theta_{\mu\lambda} \eta^{\lambda\rho} \omega_{\rho\nu} = 0, \quad \text{or in short } \theta \cdot \omega = 0, \tag{6.48}$$

$$\theta_{\mu\nu} + \omega_{\mu\nu} = \eta_{\mu\nu}, \quad \text{or in short } \theta + \omega = I. \tag{6.49}$$

Since $B_{\mu\nu}$ is a tensor in the spacetime $\{\bar{x}^\mu, \ \eta_{\mu\nu}\}$, this quantity can be written from Eq.(6.41) as follows:

$$B_{\mu\nu} = \frac{1}{\sigma} \theta_{\mu\nu} + \frac{1}{\sigma^2} \omega_{\mu\nu}. \tag{6.50}$$

Furthermore, by means of $B^{\mu\nu} B_{\nu\lambda} = \delta^\mu_\nu \equiv \eta^\mu_\nu$ and the rules Eqs.(6.45)–(6.49), the above expression of $B_{\mu\nu}$ lead to:

$$B^{\mu\nu} = \sigma \theta^{\mu\nu} + \sigma^2 \omega^{\mu\nu}, \tag{6.51}$$

which is the same as Eq.(3.7).

In the Beltrami spacetime $\mathcal{B}$, the metric is $B_{\mu\nu}$. Equation (6.24) becomes

$$B_{\mu\nu} = \eta_{ab} e^a{}_\mu e^b{}_\nu. \tag{6.52}$$

Generally, we have expansion of $e^a{}_\mu$:

$$e^a{}_\mu = a \delta^a_\nu \theta^\nu{}_\mu + b \delta^a_\nu \omega^\nu{}_\mu, \tag{6.53}$$

where a and b are unknown constants. Substituting Eqs.(6.50) and (6.53) into Eq.(6.52) gives

$$\frac{1}{\sigma} \theta_{\mu\nu} + \frac{1}{\sigma^2} \omega_{\mu\nu} = \eta_{\alpha\beta} (a\theta^\alpha{}_\mu + b\omega^\alpha{}_\mu)(a\theta^\beta{}_\nu + b\omega^\beta{}_\nu)$$
$$= a^2 \theta_{\mu\nu} + b^2 \omega_{\mu\nu}. \tag{6.54}$$

Comparing the left side of Eq.(6.54) with the right side and noting θ and ω being projection operators with properties of Eqs.(6.46)–(6.49), we find that:

$$a = \sqrt{\frac{1}{\sigma}}, \qquad b = \frac{1}{\sigma}. \tag{6.55}$$

Substituting Eq.(6.55) into Eq.(6.53) gives

$$
\begin{aligned}
e^a{}_\mu &= \sqrt{\frac{1}{\sigma}}\delta^a_\nu\theta^\nu{}_\mu + \frac{1}{\sigma}\delta^a_\nu\omega^\nu{}_\mu \\
&= \sqrt{\frac{1}{\sigma}}\delta^a_\mu + \left(\frac{1}{\sigma} - \frac{1}{\sqrt{\sigma}}\right)\frac{\eta_{\mu\nu}\delta^a_\beta x^\beta x^\nu}{(1-\sigma)R^2}.
\end{aligned} \tag{6.56}
$$

With Eqs.(6.20) and (6.52), we have

$$
\begin{aligned}
e_a{}^\mu &= e^b{}_\nu B^{\nu\mu}\eta_{ba} = \delta^b_\beta\left(\sqrt{\frac{1}{\sigma}}\theta^\beta{}_\nu + \frac{1}{\sigma}\omega^\beta{}_\nu\right)(\sigma\theta^{\nu\mu} + \sigma^2\omega^{\nu\mu})\eta_{ab} \\
&= \delta^\nu_a(\sqrt{\sigma}\theta_\nu{}^\mu + \sigma\omega_\nu{}^\mu) \\
&= \sqrt{\sigma}\delta^\mu_a + \frac{\sigma - \sqrt{\sigma}}{1-\sigma}\frac{\eta_{ab}\delta^b_\nu x^\nu x^\mu}{R^2},
\end{aligned} \tag{6.57}
$$

and

$$
\begin{aligned}
e^{a\mu} &= e^a{}_\nu B^{\nu\mu} = \delta^a_\alpha\left(\sqrt{\frac{1}{\sigma}}\theta^\alpha{}_\nu + \frac{1}{\sigma}\omega^\alpha{}_\nu\right)(\sigma\theta^{\nu\mu} + \sigma^2\omega^{\nu\mu}) \\
&= \delta^a_\nu(\sqrt{\sigma}\theta^{\nu\mu} + \sigma\omega^{\nu\mu}) \\
&= \sqrt{\sigma}\delta^a_\nu\eta^{\nu\mu} + \frac{\sigma - \sqrt{\sigma}}{1-\sigma}\frac{\delta^a_\nu x^\nu x^\mu}{R^2},
\end{aligned} \tag{6.58}
$$

$$
\begin{aligned}
e_{a\mu} &= \delta^\nu_a\left(\sqrt{\frac{1}{\sigma}}\theta_{\nu\mu} + \frac{1}{\sigma}\omega_{\nu\mu}\right) \\
&= \sqrt{\frac{1}{\sigma}}\delta^\nu_a\eta_{\nu\mu} + \left(\frac{1}{\sigma} - \frac{1}{\sqrt{\sigma}}\right)\frac{\eta_{\mu\nu}\delta^\lambda_a\eta_{\lambda\beta}x^\beta x^\nu}{(1-\sigma)R^2}.
\end{aligned} \tag{6.59}
$$

To calculate the Ricci coefficients $\omega_{ab\mu}$ of Eq.(6.39), the following formulas are needed:

$$
\theta^\alpha{}_{\mu,\nu} = -\frac{\eta^\alpha_\nu\eta_{\mu\lambda}x^\lambda + x^\alpha\eta_{\mu\nu}}{R^2(1-\sigma)} + \frac{2x^\alpha}{R^2(1-\sigma)}\omega_{\mu\nu},
$$

$$
\omega^\alpha{}_{\mu,\nu} = -\theta^\alpha{}_{\mu,\nu} = \frac{\eta^\alpha_\nu\eta_{\mu\lambda}x^\lambda + x^\alpha\eta_{\mu\nu}}{R^2(1-\sigma)} - \frac{2x^\alpha}{R^2(1-\sigma)}\omega_{\mu\nu},
$$

$$
\sigma_{,\nu} = -\frac{2\eta_{\mu\nu}x^\mu}{R^2}.
$$

with which we have

$$e^c_{\ \mu,\nu} = \delta^c_\alpha \left(\sqrt{\frac{1}{\sigma}} \theta^\alpha_{\ \mu} + \frac{1}{\sigma} \omega^\alpha_{\ \mu} \right)_{,\nu}$$

$$= \delta^c_\alpha \left(\frac{1}{2\sqrt{\sigma}} \theta^\alpha_{\ \mu} + \frac{1}{\sigma} \omega^\alpha_{\ \mu} \right) \frac{2\eta_{\lambda\nu} x^\lambda}{\sigma R^2}$$

$$+ (-\delta^c_\nu \eta_{\mu\lambda} x^\lambda - \delta^c_\alpha x^\alpha \eta_{\mu\nu} + 2\delta^c_\alpha x^\alpha \omega_{\mu\nu}) \frac{1}{R^2(1-\sigma)} \left(\sqrt{\frac{1}{\sigma}} - \frac{1}{\sigma} \right).$$

$$(6.60)$$

Now we are ready to derive the Ricci coefficients (or spin connection) $\omega^{ab}_{\ \ \mu}$. Substituting Eqs.(6.57) and (6.60) into Eq.(6.40) gives

$$\gamma^c_{\ ab} \equiv (e_a^{\ \mu} e_b^{\ \nu} - e_b^{\ \mu} e_a^{\ \nu}) e^c_{\ \mu,\nu}$$

$$= \delta^\alpha_a \delta^\beta_b \{ (\sqrt{\sigma}\theta_\alpha^{\ \mu} + \sigma\omega_\alpha^{\ \mu})(\sqrt{\sigma}\theta_\beta^{\ \nu} + \sigma\omega_\beta^{\ \nu}) - (\sqrt{\sigma}\theta_\beta^{\ \mu} + \sigma\omega_\beta^{\ \mu})(\sqrt{\sigma}\theta_\alpha^{\ \nu} + \sigma\omega_\alpha^{\ \nu}) \}$$

$$\times \left\{ \delta^c_\gamma \left(\frac{1}{2\sqrt{\sigma}} \theta^\gamma_{\ \mu} + \frac{1}{\sigma} \omega^\gamma_{\ \mu} \right) \frac{2\eta_{\lambda\nu} x^\lambda}{\sigma R^2} \right.$$

$$\left. + (-\delta^c_\nu \eta_{\mu\lambda} x^\lambda - \delta^c_\gamma x^\gamma \eta_{\mu\nu} + \delta^c_\gamma 2 x^\gamma \omega_{\mu\nu}) \frac{1}{R^2(1-\sigma)} \left(\sqrt{\frac{1}{\sigma}} - \frac{1}{\sigma} \right) \right\}$$

$$= \frac{1}{R^2(1+\sqrt{\sigma})} (\eta_{bd} \delta^d_\gamma x^\gamma \eta^c_a - \eta_{ad} \delta^d_\gamma x^\gamma \eta^c_b), \tag{6.61}$$

$$\gamma_{cab} = \frac{1}{R^2(1+\sqrt{\sigma})} (\eta_{bd} \delta^d_\mu x^\mu \eta_{ca} - \eta_{ad} \delta^d_\mu x^\mu \eta_{cb}). \tag{6.62}$$

Substituting Eqs.(6.56) and (6.62) into Eq.(6.39) we obtain:

$$\omega_{ab\mu} \equiv \frac{1}{2} e^c_{\ \mu} (\gamma_{cab} - \gamma_{abc} - \gamma_{bca})$$

$$= \delta^c_\gamma \frac{1}{2} \left(\frac{1}{\sqrt{\sigma}} \theta^\gamma_{\ \mu} + \frac{1}{\sigma} \omega^\gamma_{\ \mu} \right) \frac{1}{R^2(1+\sqrt{\sigma})}$$

$$\times \delta^d_\lambda (\eta_{ca}\eta_{db} x^\lambda - \eta_{cb}\eta_{da} x^\lambda - \eta_{ab}\eta_{dc} x^\lambda + \eta_{ca}\eta_{db} x^\lambda - \eta_{bc}\eta_{da} x^\lambda + \eta_{ba}\eta_{dc} x^\lambda)$$

$$= \frac{1}{R^2(1+\sqrt{\sigma})\sqrt{\sigma}} \delta^d_\lambda (\eta_{a\mu}\eta_{db} x^\lambda - \eta_{b\mu}\eta_{da} x^\lambda), \tag{6.63}$$

$$\omega^{ab}_{\ \ \mu} = \frac{1}{R^2(1+\sqrt{\sigma})\sqrt{\sigma}} (\delta^a_\mu \delta^b_\nu - \delta^b_\mu \delta^a_\nu) x^\nu. \tag{6.64}$$

By the above formulae, it is straightforward to verify that (see Exercise of this section):

$$e^a_{\ \mu;\nu} = 0, \quad e_a^{\ \mu}_{\ ;\nu} = 0. \tag{6.65}$$

We have so far given the explicit expressions of $e_a{}^\mu$ and $\omega^{ab}{}_\mu$ which can be substituted into the Dirac equation (6.38) in $\mathcal{B}$ for practical calculations in de Sitter invariant relativistic quantum mechanics.

We address that as with the ordinary Dirac equation (2.123) $(i\hbar\gamma^\mu\partial_\mu - mc)\psi(\mathbf{r}, t) = 0$ in inertial reference frame, Eqs.(6.38) with (6.57) and (6.64) is also the Dirac equation in inertial system. But Eq.(2.123) is in flat spacetime, and Eq.(6.38) is in the spacetime $\mathcal{B}$ with a constant curvature.

Exercise

Problem 1, To verify

$$e^a{}_{\mu;\nu} = 0, \quad e_a{}^\mu{}_{;\nu} = 0. \tag{6.66}$$

Chapter 7

Distant Hydrogen Atom in Cosmology

7.1 Introduction

We have derived the Dirac equation in the Beltrami spacetime in previous chapter. In this chapter we use that equation to discuss the distant Hydrogen atom (i.e., one-electron atom) in cosmology.

Physically, *the locally inertial coordinate system* of the Ricci Friedmann–Robertson–Walker (RFW) Universe described by ΛCDM model is the Beltrami spacetime $\mathcal{B}$ in the real world (about meaning of the *real world*, see: Eqs.(4.11), (5.8), (5.19), (5.29) and (5.91)). And the quantum mechanics in the inertial coordinate system is well defined[1]. Therefore it is a meaningful topic for real world physics to solve the relativistic quantum mechanics problem of the Hydrogen atom in $\mathcal{B}$.

Existence of *local inertial coordinate system* is required by the Equivalence Principle. The principle states that experiments in a sufficiently small falling laboratory, over a sufficiently short time, give results that are indistinguishable from those of the same experiments in an inertial frame in empty space of Special Relativity (SR) (see, e.g., *pp.119* in [Hartle (2003)]). Such a sufficiently small falling laboratory, over a sufficiently short time represents *a local inertial coordinates system*. This principle suggests that the local properties of curved spacetime should be indistinguishable from those of the spacetime with *inertial metric* of special relativity. A concrete

---

[1]To the quantum theory in the no-inertial coordinate system will suffer how to deal with the ambiguities related to the quantum effects of the inertial forces in such kind of coordinate systems. In principle the inertial forces are equivalent to the gravities locally, so the quantum effects of the inertial forces may relate to quantum gravity. However the quantum gravity theory is not yet completely well defined.

expression of this ideal is the requirement that, given a metric $g_{\alpha\beta}$ in one system of coordinates x^α, at each point P of spacetime it is possible to introduce new coordinates x'^α such that

$$g'_{\alpha\beta}(x'_P) = \text{inertial metric of SR at } x'_P, \tag{7.1}$$

and the connection at x'_P is the Christoffel symbols deduced from $g'_{\alpha\beta}(x'_P)$.

In usual Einstein's general relativity (without Λ), the above expression is

$$g'_{\alpha\beta}(x'_P) = \eta_{\alpha\beta}, \quad and \quad \Gamma^\lambda_{\alpha\beta} = 0, \tag{7.2}$$

which satisfies the Einstein equation of the Einstein's general relativity in empty space: $G_{\mu\nu} = 0$.

In de Sitter invariant general relativity (i.e., general relativity with a Λ), *the local inertial coordinate system* at x'^α_P is characterized by

$$g'_{\alpha\beta}(x'_P) = B^{(M)}_{\alpha\beta}(x'_P) \equiv \frac{\eta_{\alpha\beta}}{\sigma(x'_P - M)} + \frac{\eta_{\alpha\lambda}(x'^\lambda_P - M^\lambda)\eta_{\beta\rho}(x'^\rho_P - M^\rho)}{R^2\sigma(x'_P - M)^2}, \tag{7.3}$$

$$\Gamma^\lambda_{\alpha\beta} = \frac{1}{R^2\sigma(x'_P - M)}\left(\delta^\rho_\alpha\eta_{\beta\lambda}(x'^\lambda_P - M^\lambda) + \delta^\rho_\beta\eta_{\alpha\lambda}(x'^\lambda_P - M^\lambda)\right), \tag{7.4}$$

$$\text{with} \quad \sigma(x'_P - M) = 1 - \frac{1}{R^2}\eta_{\mu\nu}(x'^\mu_P - M^\mu)(x'^\nu_P - M^\nu),$$

which satisfies the Einstein equation of de Sitter invariant general relativity in empty spacetime: $G_{\mu\nu} + \Lambda g_{\mu\nu} = 0$ with $\Lambda = 3/R^2$ (see: Eqs.(3.6)–(3.11), and (3.165), (3.166)), and M^μ is the coordinate of Minkowski point position in $\mathcal{B}$ (see Eq.(3.165)). Such a *local inertial coordinate system* will be called *the Beltrami local inertial coordinate system.*

Generally, almost all laboratory experiments on atomic and particle physics are in *the local inertial coordinate system* in our present Universe. As is well known that it plays an essential role in the quantum physics to solve Hydrogen problem. Therefore determining the energy level shifts of a distant Hydrogen atom due to effects arisen with the de Sitter invariant special relativistic quantum mechanics will be meaningful to the cosmology when the curved spacetime is the Ricci Friedmann–Robertson–Walker Universe.

Reference [Yan, Xiao, Huang and Li (2005)] shows that the de Sitter invariant special relativistic dynamical action for free particle associates the dynamics with time- and coordinates-dependent Hamiltonian and Lagrangian (see: Eqs.(3.32), (3.130) and (6.38)). In this chapter, the adiabatic approach [Born, Fock (1928)] (see also: [Messiah (1970)] [Bayfield (1999)]) will be used to deal with the time-dependent Hamiltonian problems in the de Sitter relativistic quantum mechanics. Generally, to a Hamiltonian $H(x,t)$, we may express it as $H(x,t) = H_0(x) + H'(x,t)$. Suppose two eigenstates $|s\rangle$ and $|m\rangle$ of $H_0(x)$ are not generate, i.e., $\Delta E \equiv \hbar(\omega_m - \omega_s) \equiv \hbar\omega_{ms} \neq 0$. The validity of for adiabatic approximation relies on a condition that the variation of the potential $H'(x,t)$ in the Bohr time-period $(\Delta T_{ms}^{(Bohr)})\dot{H}'_{ms} = (2\pi/\omega_{ms})\dot{H}'_{ms}$ is much less than $\hbar\omega_{ms}$, where $H'_{ms} \equiv \langle m|H'(x,t)|s\rangle$. That makes the quantum transition from state $|s\rangle$ to state $|m\rangle$ almost impossible. Thus, the non-adiabatic effect corrections are small enough (or tiny), and the adiabatic approximations are legitimate .

We show in this chapter that the perturbation Hamiltonian $H'(x,t)$ that describes the time evolution of the atom system in de Sitter relativistic quantum mechanics is $\propto (c^2 t^2/R^2)$ (where t is the cosmic time). Since R is cosmologically large and $R >> ct$, the factor $(c^2 t^2/R^2)$ will make the time-evolution of the system is so slow that the adiabatic approximation works. We shall provide a calculation to confirm this point below. By this approach, we solve the stationary de Sitter relativistic Dirac equation for a distant one electron atom, and the spectra of the corresponding Hamiltonian with time-parameter are obtained. Consequently, we find out that the electron mass m_e, the electric charge e, the Planck constant $\hbar$ and the fine structure constant $\alpha = e^2/(\hbar c)$ all vary as cosmic time going by. These are interesting consequences since they indicate that the time-variations of fundamental physics constants are due to well-known time-dependent quantum mechanics, a topic that has been widely discussed for a long history (e.g., see [Bayfield (1999)] and the references within).

The life time of a stable atom, e.g., the Hydrogen atom, is almost infinitely long. We can practically compare the spectra of atoms at nowadays laboratories to those emitted (or absorbed) from the atoms of a distant galaxy, typically a Quasi-Stellar Object (QSO). The time interval could be on the cosmic scales. Such observation of spectra of distant astrophysical objects may encode some cosmological information in the atomic energy levels at the position and time of emission. As is well known that the solutions of ordinary Einstein relativistic Dirac equation of atom are cosmological effects free because the Hamiltonian of it is time-independent, and the solutions at any time are of the same. Thus, after deducting Hubble red shifts, any deviation of cosmology atom spectrum observations from the results of Einstein–Dirac equation could be attributed to some new physics beyond E-SR. The effect arisen from dS-SR is one of the most straightforward answers to such kind of deviations.

7.2 Beltrami local inertial coordinate system in light cone of Ricci Friedmann–Robertson–Walker universe

In Chapter 5, we showed that the isotropic and homogeneous cosmology solution of Einstein equation in General Relativity is Ricci Friedmann–Robertson–Walker (RFW) metric. In this chapter we discuss the Hydrogen atom embedded in RFW Universe and described by the de Sitter invariant special relativistic Dirac equation in the Beltrami local coordinate systems on the light cone of the Universe.

The method to detect the spectrum of distant atom is spectroscopic observations of gas clouds seen in absorption against background Quasi-Stellar Objects (QSO), which can be used to search for level shifts of atom for various purposes (see, e.g., [Webb *et al.* (1999)] [van Weerdenburg, *et al.* (2011)] [Ellingsen, *et al.* (2011)]). In the observations of gas-QSO systems in the expanding Universe, one can observe two species of frequency changes in atomic spectra: the Hubble redshift (z) caused by the usual Doppler effects and a rest frequency change due to the dynamics of atom beyond E-SR. The latter can be found by measuring the relative size of relativistic corrections to the transition frequencies of atoms on the gas-QSO (or on QSO for brevity).

Now, we describe the Earth-QSO reference frame. As illustrated in Fig.7.1, the Earth is located in the origin of frame, the proton (nucleus of Hydrogen atom) is located in $Q = \{Q^0 \equiv ct, Q^1, Q^2 = 0, Q^3 = 0\}$ (for simplicity we take $Q^2 = 0$, $Q^3 = 0$ hereafter). In addition, we assume that the coordinates of Minkowski point $M^\mu = 0$ in this Chapter, which means that the experiments in laboratories on the Earth are in consistent to the predictions of usual special relativity (see Section 3.8). From Eqs.(7.1) and

(7.3), *the local inertial coordinate system* at Q is characterized by

$$B_{\mu\nu}(Q) \equiv B_{\mu\nu}^{(M=0)}(Q) = \frac{\eta_{\mu\nu}}{\sigma(Q)} + \frac{\eta_{\mu\lambda}Q^\lambda \eta_{\nu\rho}Q^\rho}{R^2\sigma(Q)^2}, \qquad (7.5)$$

where

$$\sigma(Q) \equiv \sigma = 1 - \frac{1}{R^2}\left((Q^0)^2 - (Q^1)^2\right). \qquad (7.6)$$

The spacetime coordinate of the electron position in the atom is $L = \{L^0 \equiv ct_L,\ L^1,\ L^2,\ L^3\}$.

To an observable atom in four-dimensional spacetime, the proton has to be located at QSO-light-cone with cosmic metric $g_{\mu\nu}$. Namely, Q must satisfy following light-cone equation (see Fig.7.1 and set $Q^2 = Q^3 = 0$ for simplification)

$$ds^2 = g_{\mu\nu}(Q)dQ^\mu dQ^\nu = 0, \qquad (7.7)$$

which determines $Q^1 = f(Q^0)$. We emphasize that the underlying space-time symmetry for the atom near Q described by de Sitter invariant special relativistic dynamics is de Sitter group instead of a limit it usually known as Poincaré symmetry of E-SR. Poincaré symmetry is only a special limit of de Sitter invariant special relativity's. The corresponding spacetime metric has been shown in Eq.(7.5), of which the expansion of $1/R^2$ reads

$$B_{\mu\nu}(Q) = \eta_{\mu\nu}\left(1 + \frac{(Q^0)^2 - (Q^1)^2}{R^2}\right) + \frac{1}{R^2}\eta_{\mu\lambda}Q^\lambda \eta_{\nu\rho}Q^\rho + \mathcal{O}(1/R^4). \quad (7.8)$$

Note $B_{\mu\nu}(Q)$ is position Q-dependent, and Lorentz metric $\eta_{\mu\nu}$ for E-SR is not. We have explicitly from Eq.(7.8),

$$B_{\mu\nu}(Q) = \begin{pmatrix} 1 + \frac{2(Q^0)^2 - (Q^1)^2}{R^2} & -\frac{Q^0 Q^1}{R^2} & 0 & 0 \\ -\frac{Q^1 Q^0}{R^2} & -1 + \frac{2(Q^1)^2 - (Q^0)^2}{R^2} & 0 & 0 \\ 0 & 0 & -1 - \frac{(Q^0)^2 - (Q^1)^2}{R^2} & 0 \\ 0 & 0 & 0 & -1 - \frac{(Q^0)^2 - (Q^1)^2}{R^2} \end{pmatrix},$$
$$(7.9)$$

$$B^{\mu\nu}(Q) = \begin{pmatrix} 1 - \frac{2(Q^0)^2 - (Q^1)^2}{R^2} & \frac{Q^0 Q^1}{R^2} & 0 & 0 \\ \frac{Q^1 Q^0}{R^2} & -1 - \frac{2(Q^1)^2 - (Q^0)^2}{R^2} & 0 & 0 \\ 0 & 0 & -1 + \frac{(Q^0)^2 - (Q^1)^2}{R^2} & 0 \\ 0 & 0 & 0 & -1 + \frac{(Q^0)^2 - (Q^1)^2}{R^2} \end{pmatrix}.$$
$$(7.10)$$

We will solve Eq.(7.7) to determine Q^1 in RFW Universe model.

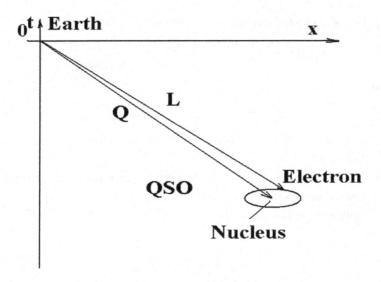

Fig. 7.1 Sketch of the Earth-QSO reference frame. The Earth is located in the origin. The position vector for nucleus of atom on QSO is Q, and for electron is L. The distance between nucleus and electron is r.

The Ricci Friedmann–Robertson–Walker (RFW) metric is (see, Eq.(5.8))

$$ds^2 = c^2dt^2 - a(t)^2 \left\{ \frac{dr^2}{1-kr^2} + r^2d\theta^2 + r^2\sin^2\theta d\phi^2 \right\}$$
$$= (dQ^0)^2 - a(t)^2 \left\{ dQ^i dQ^i + \frac{k(Q^i dQ^i)^2}{1-kQ^iQ^i} \right\}$$
$$\equiv g_{\mu\nu}(Q)dQ^\mu dQ^\nu, \tag{7.11}$$

where $r = \sqrt{Q^iQ^i}$, $Q^1 = r\sin\theta\cos\phi$, $Q^2 = r\sin\theta\sin\phi$, $Q^3 = r\cos\theta$ has been used. As is well-known RFW metric satisfies Homogeneity and Isotropy principle of present day cosmology. When $Q^2 = Q^3 = 0$, We have from Eq.(7.11)

$$g_{\mu\nu}(Q) = \eta_{\mu\nu} - a(t)^2\delta_{\mu 1}\delta_{\nu 1}\left(-\frac{1}{a(t)^2} + 1 + \frac{k(Q^1)^2}{1-k(Q^1)^2} \right)$$
$$- (a(t)^2 - 1)(\delta_{\mu 2}\delta_{\nu 2} + \delta_{\mu 3}\delta_{\nu 3}). \tag{7.12}$$

For simplicity, we take $k = 0$ and $a(t) = 1/(1 + z(t))$ (i.e., $a(t_0) = 1$). The red shift function $z(t)$ is determined by ΛCDM model described in Section 5.4. Using the observation data [Komatsu, *et al.* (2009)], the function $t(z)$ defined by Eq.(5.79) for looking-back cosmologic time is:

$$t = \int_0^z \frac{dz'}{H(z')(1 + z')}, \tag{7.13}$$

where [Riess, *et al.* (1998)] [Jarosik, *et al.* (2011)]

$$H(z') = H_0\sqrt{\Omega_{m0}(1 + z')^3 + \Omega_{R0}(1 + z')^4 + 1 - \Omega_{m0}},$$

$$H_0 = 100\, h \simeq 100 \times 0.705 \text{km} \cdot \text{s}^{-1}/\text{Mpc},$$

$$\Omega_{m0} \simeq 0.274, \quad \Omega_{R0} \sim 10^{-5}. \tag{7.14}$$

Figure of $t(z)$ of Eq.(7.13) is shown in Fig.7.2.

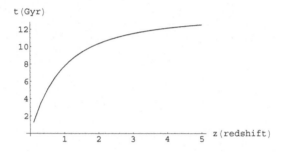

Fig. 7.2   The $t - z$ relation in ΛCDM model (Eq.(7.13)).

From Eq.(7.12), the RFW metric for $k = 0$ reads

$$g_{\mu\nu}(Q) = \eta_{\mu\nu} - (a(t)^2 - 1)(\delta_{\mu1}\delta_{\nu1} + \delta_{\mu2}\delta_{\nu2} + \delta_{\mu3}\delta_{\nu3}). \tag{7.15}$$

Substituting Eq.(7.15) into Eq.(7.7), we have

$$dQ^0 = \sqrt{\frac{-g_{11}}{g_{00}}}dQ^1 = a(t)dQ^1 = \frac{1}{1 + z(t)}dQ^1. \tag{7.16}$$

Consequently, by using Eq.(7.13) and $Q^0 = c\,t$, we get desired result:

$$Q^1 = c\int_0^z \frac{dz'}{H(z')}. \tag{7.17}$$

Figure of $Q^1(z)$ of Eq.(7.17) is shown in Fig.7.3. Ratio of Q^1 over Q^0 is shown in Fig.7.4. Then the location of distant proton is $\{Q^0, Q^1, 0, 0\}$ in the space-time with RFW metric.

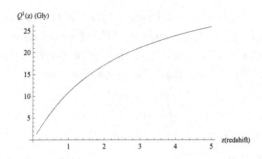

Fig. 7.3   Function $Q^1(z)$ in ΛCDM model (Eq.(7.17)).

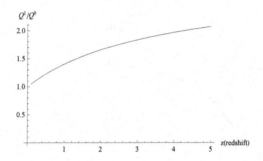

Fig. 7.4   Function of $Q^1(z)/Q^0(z)$. $Q^1(z)$ and $Q^0(z) = ct$ are given in Eqs.(7.17) and (7.13).

7.3   Hydrogen atom at distant locally inertial coordinate system in light cone of RFW universe

We treat Hydrogen atom as a proton-electron bound state described by quantum mechanics under instantaneous approximations (see Fig.7.1). The electron's coordinates are $L = \{L^0 \equiv ct_L \simeq ct,\ L^1,\ L^2,\ L^3\}$, and the relative space coordinates between proton and electron are $x^i = L^i - Q^i$. The magnitude of $r \equiv \sqrt{-\eta_{ij}x^i x^j} \sim a_B$ (where $a_B \simeq 0.5 \times 10^{-10}$m is Bohr radius), and $|x^i| \sim a_B$.

According to gauge principle, the electrodynamic interaction between the nucleus and the electron can be taken into account by replacing the operator D^μ in Eq.(6.38) with the $U(1)$-gauge covariant derivative $\mathcal{D}_L^\mu \equiv D_L^\mu - ie/(c\hbar)A^\mu$, where $A^\mu = \{\phi_B,\ \mathbf{A}\}$. Hence, the de Sitter invariant special relativistic Dirac equation for electron in Hydrogen at QSO reads

$$(ie_{\mu a}\gamma^a \mathcal{D}_L^\mu - \frac{m_e c}{\hbar})\psi = 0, \tag{7.18}$$

where m_e is the mass of electron, $\mathcal{D}_L^\mu = \frac{\partial}{\partial L_\mu} - \frac{i}{4}\omega^{ab\,\mu}\sigma_{ab} - ie/(c\hbar)A^\mu$, e_a^μ and ω_μ^{ab} have been given in Eqs.(6.57) and (6.64) respectively. Through the Taylor-expansion of $1/R^2$, we approximately write e_a^μ and ω_μ^{ab} up to $\mathcal{O}(1/R^2)$ as follows:

$$e_a^\mu = \left(1 - \frac{\eta_{cd}L^c L^d}{2R^2}\right)\eta_a^\mu - \frac{\eta_{ab}L^b L^\mu}{2R^2} + \mathcal{O}(1/R^4), \tag{7.19}$$

$$\omega_\mu^{ab} = \frac{1}{2R^2}(\eta_\mu^a L^b - \eta_\mu^b L^a) + \mathcal{O}(1/R^4). \tag{7.20}$$

In the following sections we are going to solve de Sitter invariant relativistic-Dirac equation for Hydrogen atom on QSO in RFW Universe model. In this quantum system, there are two cosmologic length scales (cosmic radius, say $R \sim 10^{12}$lyr, and the distance between QSO and the Earth is about $\sim ct$, say $R > ct > 10^8$lyr) and two microcosmic length scales(the Compton wave length of electron $a_c = \hbar/(m_e c) \simeq 0.3 \times 10^{-12}$m, and Bohr radius $a_B = \hbar^2/(m_e e^2) \simeq 0.5 \times 10^{-10}$m). The leading order calculations will be accurate up to $\mathcal{O}(c^2 t^2/R^2)$. The terms proportional to $\mathcal{O}(c^4 t^4/R^4)$, $\mathcal{O}(ct a_c/R^2)$, $\mathcal{O}(ct a_B/R^2)$ etc will be omitted. The results beyond the leading order will also be discussed.

7.4 Solutions of ordinary Einstein relativistic Dirac equation for Hydrogen atom at QSO

7.4.1 *Eigenvalues and eigenstates*

At first, we show the solution of ordinary Einstein relativistic Dirac equation in the Earth-QSO reference frame of Fig.7.1, which serves as the leading order of solution for the de Sitter invariant relativistic Dirac equation with $R \to \infty$ in that reference frame. For the Hydrogen, $\partial^\mu \to \mathcal{D}_L^\mu = \partial_L^\mu - ie/(c\hbar)A_M^\mu$ (noting $\omega^{ab\,\mu}|_{R\to\infty} = 0$), where $A_M^\mu \equiv \{\phi_M(x), \mathbf{A}_M(x)\}$ with $x^\mu \equiv L^\mu - Q^\mu$ (see: Fig.7.1), and $\phi_M(x)$ and $\mathbf{A}_M(x)$ are nucleus electric potential and vector potential at x^i in Minkowski space determined by the

Coulomb law and the Ampere law in the electromagnetic dynamics:

$$-\eta^{ij}\partial_i\partial_j\phi_M(x) = \nabla^2\phi_M(x) = -4\pi\rho(x) = -4\pi e\delta^{(3)}(\mathbf{x}), \qquad (7.21)$$

$$\nabla(\partial_\lambda A_M^\lambda) - \partial^2\mathbf{A}_M = -\frac{4\pi}{c}\mathbf{j} = 0. \qquad (7.22)$$

The solutions are $\phi_M(x) = e/r$ and $\mathbf{A}_M = 0$. And hence $\partial_0 \to \mathcal{D}_0^L = \partial_0 - i\eta_{00}e^2/(c\hbar r)$. Then, from Eq.(6.38), the Einstein relativistic Dirac equation reads

$$i\hbar\partial_t\psi = \left(-i\hbar c\vec{\alpha}\cdot\nabla_L + m_e c^2\beta - \frac{e^2}{r}\right)\psi, \qquad (7.23)$$

where $\beta = \gamma^0$, $\alpha^i = \beta\gamma^i$. Noting that the nucleus position $\mathbf{Q}$ =constant, we have

$$\nabla_L = \frac{\partial}{\partial\mathbf{L}} = \frac{\partial}{\partial(\mathbf{Q}+\mathbf{r})} = \frac{\partial}{\partial\mathbf{r}} \equiv \nabla, \qquad (7.24)$$

and Eq.(7.23) becomes the standard E-SR Dirac equation for electron in Hydrogen at its nucleus reference frame. Energy is conserved in Einstein's special relativistic mechanics (hereafter, we use notations of [Strange (2008)]), and the Hydrogen is the stationary states of Dirac equation. The stationary state condition

$$i\hbar\partial_t\psi = W\psi. \qquad (7.25)$$

combined with Eqs.(7.23), (7.24) with (7.25) gives

$$W\psi = \left(-i\hbar c\vec{\alpha}\cdot\nabla + m_e c^2\beta - \frac{e^2}{r}\right)\psi \equiv H_0(r,\hbar,m_e,e)\psi, \qquad (7.26)$$

which is the standard Dirac equation for Hydrogen. The solution is standard and can be found in any textbook on relativistic quantum mechanics. The results are follows (see Eq.(A.98) in Appendix A, or [Rose (1961)][Strange (2008)])

$$W = W_{n,\kappa} = m_e c^2\left(1 + \frac{\alpha^2}{(n - |\kappa| + s)^2}\right)^{-1/2} \qquad (7.27)$$

$$\alpha \equiv \frac{e^2}{\hbar c}, \qquad |\kappa| = (j + 1/2) = 1, 2, 3 \cdots$$

$$s = \sqrt{\kappa^2 - \alpha^2}, \qquad n = 1, 2, 3 \cdots.$$

Its expansion in α is

$$W = m_e c^2\left(1 - \frac{\alpha^2}{2n^2} + \alpha^4\left(\frac{3}{8n^2} - \frac{1}{2n^3|\kappa|}\right) + \cdots\right). \qquad (7.28)$$

The corresponding Hydrogen's wave functions ψ have also already been finely derived (see Appendix A, or [Strange (2008)]). The complete set of commutative observables is $\{H,\ \kappa,\ \mathbf{j}^2,\ j_z\}$, so that $\psi = \psi_{n,\kappa,j,j_z}(\mathbf{r}, \hbar, m_e, \alpha) \equiv \psi_j^{m_j}(\mathbf{r})$, where $\mathbf{j} = \mathbf{L} + \frac{\hbar}{2}\boldsymbol{\Sigma}$, $\hbar\kappa\psi = \beta(\boldsymbol{\Sigma}\cdot\mathbf{L} + \hbar)\psi$, $\boldsymbol{\Sigma} = -\gamma_5\boldsymbol{\alpha}$ and $\alpha = e^2/(\hbar c)$ (see Appendix A). The expression of $\psi_j^{m_j}(\mathbf{r})$ is as follows

$$\psi_j^{m_j}(\mathbf{r}) = \begin{pmatrix} g_\kappa(r)\chi_\kappa^{m_j}(\hat{\mathbf{r}}) \\ if_\kappa(r)\chi_{-\kappa}^{m_j}(\hat{\mathbf{r}}) \end{pmatrix} \tag{7.29}$$

where (see Eqs.(A.46), (A.47) and (A.48) in Appendix A)

$$\chi_\kappa^{m_j}(\hat{\mathbf{r}}) = \sum_{m_s=-1/2}^{1/2} C_{l,m_j-m_s;\,1/2,m_s}^{j\,m_j} Y_l^{m_j-m_s}(\hat{\mathbf{r}})\chi^{m_s} \tag{7.30}$$

$$\text{with} \quad \chi^{m_s=1/2} = \begin{pmatrix} 1 \\ 0 \end{pmatrix}, \quad \chi^{m_s=-1/2} = \begin{pmatrix} 0 \\ 1 \end{pmatrix}$$

$$\chi_{-\kappa}^{m_j}(\hat{\mathbf{r}}) = -\sigma_r \chi_\kappa^{m_j}(\hat{\mathbf{r}}),$$

$$\text{with} \quad \sigma_r = \hat{\mathbf{r}}\cdot\vec{\sigma} = \begin{pmatrix} \cos\theta & \sin\theta e^{-i\phi} \\ \sin\theta e^{i\phi} & -\cos\theta \end{pmatrix}, \tag{7.31}$$

and (see Eqs.(A.93) and (A.94) in Appendix A)

$$g_\kappa(r) = 2\lambda(k_C + W_C)^{1/2}e^{-\lambda r}(2\lambda r)^{s-1}\alpha_0$$
$$\times[n'M(1 - n', 2s + 1, 2\lambda r) \tag{7.32}$$
$$+ \left(\kappa - \frac{\alpha k_C}{\lambda}\right)M(-n', 2s + 1, 2\lambda r)],$$

$$f_\kappa(r) = 2\lambda(k_C - W_C)^{1/2}e^{-\lambda r}(2\lambda r)^{s-1}\alpha_0$$
$$\times[n'M(1 - n', 2s + 1, 2\lambda r) \tag{7.33}$$
$$- \left(\kappa - \frac{\alpha k_C}{\lambda}\right)M(-n', 2s + 1, 2\lambda r)],$$

where

$$k_C = m_e c/\hbar, \quad W_C = W/(c\hbar), \quad \lambda = (k_C^2 - W_C^2)^{1/2},$$
$$s = \sqrt{\kappa^2 - \alpha^2}, \quad n' = n - |\kappa|,$$
$$\kappa = -(j(j+1) - l(l+1) + 1/4), \tag{7.34}$$

with W being given in Eq.(7.27), and $M(a, b, z)$ is the confluent hypergeometric function (see Appendix B):

$$M(a, b, z) = 1 + \frac{az}{b} + \frac{(a)_2 z^2}{2!(b)_2} + \cdots \frac{(a)_n z^n}{n!(b)_n} + \cdots$$
$$(a)_0 = 1, \quad (a)_n = a(a+1)(a+2)\cdots(a+n-1).$$

The normalization constant α_0 is determined by

$$\int_0^\infty r^2(g_\kappa^2(r) + f_\kappa^2(r))dr = 1.$$

For Hydrogen-like one electron atoms with $Z > 1$, the energy level formulae and the eigen-wave functions expressions are all the same as Eqs.(7.27)–(7.34) except $\alpha \to \xi = Z\alpha$. The levels of one electron atom with $Z = 92$ are shown in Fig.8.9. (see Appendix A.5).

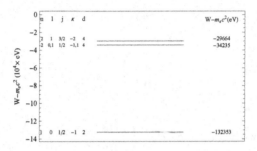

Fig. 7.5 The energy levels of the one-electron atom with $Z = 92$. The figure shows relativistic energy levels calculated using Eq.(7.27) with replacement of $\alpha \to \xi = Z\alpha$. They are labeled by the quantum numbers n, l, j, κ and their degeneracy d, up to $n = 2$.

It is learned from above that since $R \to \infty$, $B_{\mu\nu} \to \eta_{\mu\nu}$, the dS-SR tends to the E-SR. The Hamiltonian of E-SR is cosmic time independent. So, the spectra of Hydrogen at any place in RFW universe are the same, and there is no cosmology information of the universe in the spectrum solutions equation (7.27) of E-SR Dirac equation.

7.4.2 $2s^{1/2}$ *and* $2p^{1/2}$ *states of Hydrogen*

As is well known that the state of $2s^{1/2}$ and state of $2p^{1/2}$ are complete degenerate to all order of α in the E-SR Dirac equation of Hydrogen. Namely, from Eq.(7.27) and $\kappa = -(j(j+1) - l(l+1) + 1/4)$, we have

$$\Delta W(2s^{1/2} - 2p^{1/2}) \equiv W_{(n=2, \, \kappa=-1)} - W_{(n=2, \, \kappa=+1)} = 0. \qquad (7.35)$$

By means of Eqs.(7.29)–(7.33), the wave functions of states $2s^{1/2}$ with $\kappa = -1$ are as follows [Strange (2008)] (see Appendix A.4, (A.104)–(A.107))

$$\psi^{m_j}_{(2s)j=1/2}(\mathbf{r}) = \begin{pmatrix} g_{(2s^{1/2})}(r)\chi^{m_j}_{\kappa}(\hat{\mathbf{r}})_{(2s^{1/2})} \\ if_{(2s^{1/2})}(r)\chi^{m_j}_{-\kappa}(\hat{\mathbf{r}})_{(2s^{1/2})} \end{pmatrix}, \tag{7.36}$$

where

$$g_{(2s^{1/2})}(r) = \sqrt{\frac{(2\lambda)^{2s+1}k_C(2s+1)(k_C+W_C)}{2W_C(2W_C+k_C)\Gamma(2s+1)}}r^{s-1}e^{-\lambda r}$$

$$\times \left(\frac{W_C}{k_C} - \frac{\lambda}{2s+1}\left(1+\frac{2W_C}{k_C}\right)r\right) \tag{7.37}$$

$$f_{(2s^{1/2})}(r) = \sqrt{\frac{(2\lambda)^{2s+1}k_C(2s+1)(k_C-W_C)}{2W_C(2W_C+k_C)\Gamma(2s+1)}}r^{s-1}e^{-\lambda r}$$

$$\times \left(\frac{W_C}{k_C}+1 - \frac{\lambda}{2s+1}\left(1+\frac{2W_C}{k_C}\right)r\right) \tag{7.38}$$

with k_C, W_C, λ and s given in Eq.(7.34) following [Strange (2008)]. Due to Eqs.(7.30) and (7.31)

$$\chi^{1/2}_{\kappa}(\hat{\mathbf{r}})_{(2s^{1/2})} = \begin{pmatrix} Y_0^0 \\ 0 \end{pmatrix}, \ \chi^{-1/2}_{\kappa}(\hat{\mathbf{r}})_{(2s^{1/2})} = \begin{pmatrix} 0 \\ Y_0^0 \end{pmatrix}, \tag{7.39}$$

$$\chi^{1/2}_{-\kappa}(\hat{\mathbf{r}})_{(2s^{1/2})} = \begin{pmatrix} -\cos\theta Y_0^0 \\ -\sin\theta e^{i\phi}Y_0^0 \end{pmatrix}, \tag{7.40}$$

$$\chi^{-1/2}_{-\kappa}(\hat{\mathbf{r}})_{(2s^{1/2})} = \begin{pmatrix} -\sin\theta e^{-i\phi}Y_0^0 \\ \cos\theta Y_0^0 \end{pmatrix}, \ \text{with} \ Y_0^0 = \frac{1}{\sqrt{4\pi}}.$$

For the $2p^{1/2}$ state with $\kappa' = 1$

$$\psi^{m_j}_{(2p)j=1/2}(\mathbf{r}) = \begin{pmatrix} g_{(2p^{1/2})}(r)\chi^{m_j}_{\kappa'}(\hat{\mathbf{r}})_{(2p^{1/2})} \\ if_{(2p^{1/2})}(r)\chi^{m_j}_{-\kappa'}(\hat{\mathbf{r}})_{(2p^{1/2})} \end{pmatrix} \tag{7.41}$$

where

$$g_{(2p^{1/2})}(r) = \sqrt{\frac{(2\lambda)^{2s+1}k_C(2s+1)(k_C+W_C)}{2W_C(2W_C-k_C)\Gamma(2s+1)}}r^{s-1}e^{-\lambda r}$$

$$\times \left(\frac{W_C}{k_C} - 1 + \frac{\lambda}{2s+1}\left(1-\frac{2W_C}{k_C}\right)r\right), \tag{7.42}$$

$$f_{(2p^{1/2})}(r) = \sqrt{\frac{(2\lambda)^{2s+1}k_C(2s+1)(k_C-W_C)}{2W_C(2W_C-k_C)\Gamma(2s+1)}}r^{s-1}e^{-\lambda r}$$

$$\times \left(\frac{W_C}{k_C} + \frac{\lambda}{2s+1}\left(1-\frac{2W_C}{k_C}\right)r\right). \tag{7.43}$$

$$\chi_{\kappa'}^{1/2}(\hat{\mathbf{r}})_{(2p^{1/2})} = \begin{pmatrix} -\sqrt{\frac{1}{3}}Y_1^0(\theta\phi) \\ \sqrt{\frac{2}{3}}Y_1^1(\theta\phi) \end{pmatrix}, \; \chi_{\kappa'}^{-1/2}(\hat{\mathbf{r}})_{(2p^{1/2})} = \begin{pmatrix} -\sqrt{\frac{2}{3}}Y_1^{-1}(\theta\phi) \\ \sqrt{\frac{1}{3}}Y_1^0(\theta\phi) \end{pmatrix},$$

(7.44)

$$\chi_{-\kappa'}^{1/2}(\hat{\mathbf{r}})_{(2p^{1/2})} = \begin{pmatrix} \sqrt{\frac{1}{3}}\cos\theta Y_1^0(\theta\phi) - \sqrt{\frac{2}{3}}\sin\theta e^{-i\phi}Y_1^1(\theta\phi) \\ \sqrt{\frac{1}{3}}\sin\theta e^{i\phi}Y_1^0(\theta\phi) + \sqrt{\frac{2}{3}}\cos\theta Y_1^1(\theta\phi) \end{pmatrix},$$

$$\chi_{-\kappa'}^{-1/2}(\hat{\mathbf{r}})_{(2p^{1/2})} = \begin{pmatrix} \sqrt{\frac{2}{3}}\cos\theta Y_1^{-1}(\theta\phi) - \sqrt{\frac{1}{3}}\sin\theta e^{-i\phi}Y_1^0(\theta\phi) \\ \sqrt{\frac{2}{3}}\sin\theta e^{i\phi}Y_1^{-1}(\theta\phi) + \sqrt{\frac{1}{3}}\cos\theta Y_1^0(\theta\phi) \end{pmatrix}, \quad (7.45)$$

Explicitly $Y_1^0(\theta\phi) = \sqrt{\frac{3}{4\pi}}\cos\theta$, $Y_1^1(\theta\phi) = -\sqrt{\frac{3}{8\pi}}e^{i\phi}\sin\theta$, $Y_1^{-1}(\theta\phi) = \sqrt{\frac{3}{8\pi}}e^{-i\phi}\sin\theta$. (see Appendix B).

Exercise

Problem 1: Verify the normalization of $2s^{1/2}$-radial wave function (7.37):

$$\int_0^\infty \left(g_{(2s^{1/2})}(r)^2 + f_{(2s^{1/2})}(r)^2 \right) r^2 dr = 1$$

numerically.

7.5 De Sitter invariant relativistic Dirac equation for distant Hydrogen atom

Now we turn to discussion for the de Sitter invariant relativistic Dirac equation of distant Hydrogen atom.

From Eqs.(7.18), (7.19) and (7.20), $\partial^\mu \to \mathcal{D}_L^\mu = \frac{\partial}{\partial L_\mu} - \frac{i}{4}\omega^{ab\,\mu}\sigma_{ab} - ie/(c\hbar)A^\mu$, and noting $\mathcal{D}_\mu^L = B_{\mu\nu}\mathcal{D}_L^\nu$, we have the de Sitter invariant relativistic Dirac equation for the electron in Hydrogen at the earth-QSO reference frame as follows

$$\hbar c\beta \left[i\left(1 - \frac{\eta_{ab}L^a L^b}{2R^2}\right)\gamma^\mu \mathcal{D}_\mu^L - \frac{i}{2R^2}\eta_{ab}L^a\gamma^b L^\mu \mathcal{D}_\mu^L - \frac{m_e c}{\hbar} \right]\psi = 0, \quad (7.46)$$

where factor $\hbar c\beta$ in the front of the equation is only for convenience. We expand each terms of Eq.(7.46) in order of $\mathcal{O}(1/R^2)$ as follows:

(1) Since observed QSO must be located at the light cone, $\eta_{ab}L^aL^b \simeq \eta_{ab}Q^aQ^b = (Q^0)^2 - (Q^1)^2$, and the first term of Eq.(7.46) reads

$$\hbar c \beta i \left(1 - \frac{\eta_{ab}L^aL^b}{2R^2}\right) \gamma^\mu \mathcal{D}_\mu^L = \left(1 - \frac{(Q^0)^2 - (Q^1)^2}{2R^2}\right) \hbar c \beta i \gamma^\mu \mathcal{D}_\mu^L \psi \quad (7.47)$$

$$\text{with} \quad \beta\gamma^\mu = \{\beta\gamma^0 = \beta^2 = 1, \ \beta\gamma^i = \alpha^i\} \quad (7.48)$$

$$\hbar c \beta i \gamma^\mu \mathcal{D}_\mu^L \psi = (i\hbar\partial_t + \hbar c \vec{\alpha} \cdot \nabla + \frac{\hbar c \beta}{4}\omega_\mu^{ab}\gamma^\mu\sigma_{ab} + e\beta\gamma^\mu B_{\mu\nu}A^\nu)\psi, \quad (7.49)$$

where $A^\nu = \{A^0 = \phi_B, \ A^i\}$ are determined by Maxwell equations within constant metric $g_{\mu\nu} = B_{\mu\nu}(Q)$ and $j^\nu = \{j^0 \equiv c\rho/\sqrt{B_{00}}, \ j^i = 0\}$ (see Appendix D):

$$-B^{ij}(Q)\partial_i\partial_j\phi_B(x) = \left[\sigma\nabla^2 + \frac{\sigma(Q^1)^2}{R^2}\frac{\partial^2}{\partial(x^1)^2}\right]\phi_B(x) = -\frac{4\pi}{c}j^0$$

$$= -\frac{4\pi}{c}\frac{c\rho_B}{\sqrt{B_{00}}}$$

$$= \frac{-4\pi e}{\sqrt{-\det(B_{\mu\nu}(Q))}}\delta^{(3)}(\mathbf{x}), \quad (7.50)$$

$$\partial^i\partial_\mu A^\mu - B^{\mu\nu}\partial_\mu\partial_\nu A^i = -\frac{4\pi}{c}j^i = 0. \quad (7.51)$$

The solution is (see Appendix D, Eqs.(D.12) and (D.14))

$$\phi_B = \frac{1}{\sqrt{\left[\left(\frac{1}{\sigma} - \frac{(Q^1)^2}{R^2\sigma^2}\right)\left(\frac{1}{\sigma} + \frac{(Q^0)^2}{R^2\sigma^2}\right) + \frac{(Q^0)^2(Q^1)^2}{(R^4\sigma^4)}\right]\left(1 + \frac{(Q^1)^2}{R^2}\right)}}\frac{e}{r_B},$$

$$A^i = 0, \quad (7.52)$$

where $r_B = \sqrt{(\tilde{x}^1)^2 + (x^2)^2 + (x^3)^2}$ with $\tilde{x}^1 \equiv x^1/[1 + \frac{(Q^1)^2}{2R^2}]$. In the follows, we use variables $\{\tilde{x}^1, x^2, x^3\}$ to replace $\{x^1, x^2, x^3\}$. The following notations are employed hereafter:

$$\mathbf{r}_B = \mathbf{i}\tilde{x}^1 + \mathbf{j}x^2 + \mathbf{k}x^3, \quad |\mathbf{r}_B| = r_B, \quad (7.53)$$

$$\nabla_B = \mathbf{i}\frac{\partial}{\partial\tilde{x}^1} + \mathbf{j}\frac{\partial}{\partial x^2} + \mathbf{k}\frac{\partial}{\partial x^3}, \quad \tilde{x}^i \in \{\tilde{x}^1, \ x^2, \ x^3\}. \quad (7.54)$$

Under approximation up to $\mathcal{O}(1/R^2)$, from the full expression of ϕ_B of Eq.(7.52), we get

$$\phi_B \simeq \left(1 - \frac{3(Q^0)^2 - 2(Q^1)^2}{2R^2}\right)\frac{e}{r_B}, \quad A^i = 0. \quad (7.55)$$

Then, noting $\frac{\partial}{\partial x^1} = \frac{\partial \tilde{x}^1}{\partial x^1}\frac{\partial}{\partial \tilde{x}^1} = \frac{\partial}{\partial \tilde{x}^1} - \frac{(Q^1)^2}{2R^2}\frac{\partial}{\partial \tilde{x}^1}$, Eq.(7.49) becomes

$$\hbar c\beta i\gamma^\mu \mathcal{D}_\mu^L \psi = \left(i\hbar\partial_t + i\hbar c\vec{\alpha}\cdot\nabla_B - i\hbar c\frac{(Q^1)^2}{2R^2}\alpha^1\frac{\partial}{\partial\tilde{x}^1} \right.$$

$$\left. + \frac{\hbar c\beta}{4}\omega_\mu^{ab}\gamma^\mu\sigma_{ab} + eB_{00}\phi_B + e\alpha^1 B_{10}\phi_B \right)\psi$$

$$= \left(i\hbar\partial_t + i\hbar c\vec{\alpha}\cdot\nabla_B - i\hbar c\frac{(Q^1)^2}{2R^2}\alpha^1\frac{\partial}{\partial\tilde{x}^1} + \frac{\hbar c\beta}{4}\omega_\mu^{ab}\gamma^\mu\sigma_{ab} \right.$$

$$\left. + \left[1 + \frac{2(Q^0)^2 - (Q^1)^2}{R^2} \right]e\phi_B - \frac{Q^1 Q^0}{R^2}\alpha^1 e\phi_B \right)\psi. \quad (7.56)$$

(2) Estimation of the contributions of the fourth term in RSH of Eq.(7.56) (the spin-connection contributions) is as follows: By Eq.(7.20), the ratio of the fourth term to the first term of Eq.(7.56) is:

$$\left| \frac{\frac{\hbar c\beta}{4}\omega_\mu^{ab}\gamma^\mu\sigma_{ab}\psi}{i\hbar\partial_t\psi} \right| \sim \frac{\hbar c}{4}\frac{ct}{2R^2}\frac{1}{m_e c^2} = \frac{ct}{8R^2}\frac{\hbar}{m_e c} = \frac{1}{8}\frac{cta_c}{R^2} \sim 0, \quad (7.57)$$

where $a_c = \hbar/(m_e c) \simeq 0.3 \times 10^{-12}$m is the Compton wave length of electron. $\mathcal{O}(cta_c/R^2)$-term is negligible. Therefore the 3-rd term in RSH of (7.49) has no contribution to our approximation calculations.

(3) Substituting Eqs.(7.57), (7.56) and (7.48) into Eq.(7.47) and noting $L^a \simeq Q^a$ (see Fig.(7.1)), we get the first term in LHS of Eq.(7.46)

$$\hbar c\beta i\left(1 - \frac{\eta_{ab}L^a L^b}{2R^2} \right)\gamma^\mu\mathcal{D}_\mu^L\psi = \left(1 - \frac{(Q^0)^2 - (Q^1)^2}{2R^2} \right)\left(i\hbar\frac{\partial}{\partial t} + i\hbar c\vec{\alpha}\cdot\nabla_B \right.$$

$$\left. - i\hbar c\frac{(Q^1)^2}{2R^2}\alpha^1\frac{\partial}{\partial\tilde{x}^1} + \left[1 + \frac{2(Q^0)^2 - (Q^1)^2}{R^2} \right]e\phi_B - \frac{Q^1 Q^0}{R^2}\alpha^1 e\phi_B \right)\psi$$

$$= \left\{ \left(1 - \frac{(Q^0)^2 - (Q^1)^2}{2R^2} \right)\left(i\hbar\frac{\partial}{\partial t} + i\hbar c\vec{\alpha}\cdot\nabla_B \right) - i\hbar c\frac{(Q^1)^2}{2R^2}\alpha^1\frac{\partial}{\partial\tilde{x}^1} \right.$$

$$\left. + \left[1 + \frac{3(Q^0)^2 - (Q^1)^2}{2R^2} \right]e\phi_B - \frac{Q^1 Q^0}{R^2}\alpha^1 e\phi_B + \mathcal{O}(1/R^4) \right\}\psi.$$

$$(7.58)$$

(4) The second term of Eq.(7.46) is

$$-\hbar c\beta \frac{i}{2R^2}\eta_{ab}L^a\gamma^b L^\mu \mathcal{D}_\mu^L \psi$$

$$= -\frac{i\hbar c\beta}{2R^2}(\gamma^0 L^0 - \vec{\gamma}\cdot\vec{L})L^\mu \left[\partial_\mu^L - \delta_{\mu 0}\frac{ie}{c\hbar}\phi_B\right]\psi + \mathcal{O}(\frac{1}{R^4})$$

$$= -\frac{i\hbar c}{2R^2}(L^0 - \vec{L}\cdot\vec{\alpha})\left(L^0\partial_0^L - L^0\frac{ie}{c\hbar}\phi_B + L^i\partial_i^L\right)\psi$$

$$\simeq -\frac{ic\hbar}{2R^2}(L^0 - L^1\alpha^1)(L^0\partial_0^L - L^0\frac{ie}{c\hbar}\phi_B + L^1\partial_1^L)\psi$$

$$\simeq -\frac{ic\hbar}{2R^2}\left[((L^0)^2 - L^1L^0\alpha^1)\partial_0 - \frac{ie}{c\hbar}(L^0)^2\phi_B\right.$$

$$\left. +\frac{ie}{c\hbar}L^0L^1\alpha^1\phi_B + L^0L^1\partial_1^L - (L^1)^2\alpha^1\partial_1^L\right]\psi, \qquad (7.59)$$

where the following estimations were used

$$\frac{(L^2)}{R} \sim \frac{(L^3)}{R} \sim \frac{a_B}{R} \sim 0. \qquad (7.60)$$

Noting $L^0 \simeq Q^0$, $L^1 \simeq Q^1$, Eq.(7.59) becomes

$$-\hbar c\beta\frac{i}{2R^2}\eta_{ab}L^a\gamma^b L^\mu\mathcal{D}_\mu^L\psi = \left[\left(\frac{-(Q^0)^2}{2R^2} + \frac{Q^1Q^0}{2R^2}\alpha^1\right)i\hbar\partial_t - \frac{(Q^0)^2}{2R^2}e\phi_B\right.$$

$$\left. +\frac{Q^1Q^0}{2R^2}\alpha^1 e\phi_B - \frac{Q^1Q^0}{2R^2}i\hbar c\frac{\partial}{\partial\tilde{x}^1} + \frac{(Q^1)^2}{2R^2}i\hbar c\alpha^1\frac{\partial}{\partial\tilde{x}^1}\right]\psi. \quad (7.61)$$

(5) Therefore, substituting Eqs.(7.58) and (7.61) into Eq.(7.46), we have

$$i\hbar\left(1 - \frac{2(Q^0)^2 - (Q^1)^2}{2R^2} + \frac{Q^1Q^0}{2R^2}\alpha^1\right)\partial_t\psi$$

$$= \left[-i\hbar c\left(1 - \frac{(Q^0)^2 - (Q^1)^2}{2R^2}\right)\vec{\alpha}\cdot\nabla_B + m_e c^2\beta\right.$$

$$-\left(1 + \frac{2(Q^0)^2 - (Q^1)^2}{2R^2}\right)e\phi_B + \frac{Q^0Q^1}{2R^2}\alpha^1 e\phi_B$$

$$\left. +\frac{Q^1Q^0}{2R^2}i\hbar c\frac{\partial}{\partial\tilde{x}^1}\right]\psi, \qquad (7.62)$$

or

$$i\hbar\partial_t\psi = \left[-\left(1 + \frac{(Q^0)^2}{2R^2}\right) i\hbar c\vec{\alpha}\cdot\nabla_B + \left(1 + \frac{2(Q^0)^2 - (Q^1)^2}{2R^2}\right) m_e c^2\beta \right.$$
$$\left. - \left(1 + \frac{(Q^0)^2}{2R^2}\right) \frac{e^2}{r_B}\right]\psi$$
$$- \frac{Q^1 Q^0}{2R^2}\alpha^1 \left[-i\hbar c\vec{\alpha}\cdot\nabla_B + m_e c^2\beta - \frac{(\sqrt{2}e)^2}{r_B}\right]\psi$$
$$+ \left[\frac{Q^1 Q^0}{2R^2} i\hbar c\frac{\partial}{\partial\tilde{x}^1}\right]\psi, \tag{7.63}$$

where $\phi_B = (1 - \frac{3(Q^0)^2 - 2(Q^1)^2}{2R^2})e/r_B$ is used (see Eq.(7.55)). Equation (7.63) is dS-SR Dirac wave equation to the first order of $\mathcal{O}(\frac{c^2 t^2}{R^2})$. Two remarks on Eq.(7.63) are as follows:

i) When $R \to \infty$, Eq.(7.63) goes back to usual E-SR Dirac equation of Hydrogen, which has been discussed in the last section.

ii) Equation (7.63) is a time-dependent wave equation. It is somehow difficult to deal with the time-dependent problems in quantum mechanics. Generally, there are two approximate approaches to discuss two extreme cases respectively: (i) The modification in states obtained by the wave equation depends critically on the time T during which the modification of the system's "Hamiltonian" take place. For this case, one would use the sudden approach; And, (ii), for case of a very slow modification of Hamiltonian, the adiabatic approach works [Messiah (1970)]. As discussed previously, since $|R|$ is cosmologically large and $|R| >> ct$, factor $\{(Q^0)^2/R^2, (Q^1)^2/R^2\} \propto (c^2 t^2/R^2)$ makes the perturbed part of the time-evolution of the system described by wave equation of (7.63) is so slow that the adiabatic approximation [Born, Fock (1928)] may legitimately works. We will provide a calculations to confirm this point below (Section 7.7).

7.6 De Sitter relativistic quantum mechanics spectrum equation for Hydrogen

In order to discuss the spectra of Hydrogen by de Sitter invariant special relativistic Dirac equation, we need to find out its solutions with certain physics energy E. From Eq.(6.10), the energy operator in the dS-SR can be derived by means of the operator expression of momentum:

$$p^0 = \frac{E}{c} = i\hbar \left[\frac{1}{c}\partial_t - \frac{ct}{R^2}x^\nu \partial_\nu^L + \frac{5ct}{2R^2} \right]$$

$$E = i\hbar \left[\partial_t - \frac{c^2t^2}{R^2}\partial_t - \frac{c^2tL^i}{R^2}\frac{\partial}{\partial L^i} + \frac{5c^2t}{2R^2} \right]$$

$$E\psi \simeq i\hbar \left(1 - \frac{c^2t^2}{R^2} \right) \partial_t\psi - i\hbar c\frac{ctL^1}{R^2}\frac{\partial}{\partial L^1}\psi$$

$$= i\hbar \left(1 - \frac{(Q^0)^2}{R^2} \right) \partial_t\psi - i\hbar c\frac{Q^0Q^1}{R^2}\frac{\partial}{\partial x^1}\psi, \qquad (7.64)$$

where a estimation for the ratio of the 3-rd term to the 2-nd of $E\psi$ was used:

$$\frac{|i\hbar\frac{5c^2t}{2R^2}\psi|}{|\frac{-c^2t^2}{R^2}i\hbar\partial_t\psi|} \sim \frac{|i\hbar\frac{5c^2t}{2R^2}|}{|\frac{-2c^2t^2}{R^2}E|} \sim \frac{5\hbar}{2tm_ec^2} \equiv \frac{5}{2}\frac{a_c}{ct},$$

where $a_c \simeq 0.3 \times 10^{-12}$m is the Compton wave length of electron and ct is about the distance between earth and QSO. In our approximate calculations $a_c/(ct)$ is negligible. For instance, to a QSO with $ct \sim 10^9$ly, $a_c/(ct) \sim 10^{-38} << (ct)^2/R^2 \sim 10^{-5}$. Hence the 3-rd term of $E\psi$ were ignored.

Inserting Eq.(7.64) into Eq.(7.63), we obtain the de Sitter relativistic spectra equation of Hydrogen

$$\left(1 + \frac{(Q^0)^2}{R^2} \right) E\psi = \left[-\left(1 + \frac{(Q^0)^2}{2R^2} \right) i\hbar c\vec{\alpha} \cdot \nabla_B \right.$$

$$+ \left(1 + \frac{2(Q^0)^2 - (Q^1)^2}{2R^2} \right) m_ec^2\beta - \left(1 + \frac{(Q^0)^2}{2R^2} \right) \frac{e^2}{r_B} \Bigg] \psi$$

$$- \frac{Q^1Q^0}{2R^2}\alpha^1 \left[-i\hbar c\vec{\alpha} \cdot \nabla_B + m_ec^2\beta - \frac{(\sqrt{2}e)^2}{r_B} \right] \psi - \left[\frac{Q^1Q^0}{2R^2}i\hbar c\frac{\partial}{\partial \tilde{x}^1} \right] \psi,$$

or

$$E\psi = \left[-\left(1 - \frac{(Q^0)^2}{2R^2}\right) i\hbar c\vec{\alpha} \cdot \nabla_B + \left(1 - \frac{(Q^1)^2}{2R^2}\right) m_e c^2 \beta \right.$$
$$\left. - \left(1 - \frac{(Q^0)^2}{2R^2}\right) \frac{e^2}{r_B} \right] \psi$$
$$- \frac{Q^1 Q^0}{2R^2} \alpha^1 \left[-i\hbar c\vec{\alpha} \cdot \nabla_B + m_e c^2 \beta - \frac{(\sqrt{2}e)^2}{r_B} \right] \psi - \left[\frac{Q^1 Q^0}{2R^2} i\hbar c \frac{\partial}{\partial \tilde{x}^1} \right] \psi$$
$$= \left[-i\hbar_t c\vec{\alpha} \cdot \nabla_B + m_e^{(t)} c^2 \beta - \frac{e_t^2}{r_B} \right] \psi$$
$$- \frac{Q^1 Q^0}{2R^2} \alpha^1 \left[-i\hbar c\vec{\alpha} \cdot \nabla_B + m_e c^2 \beta - \frac{(\sqrt{2}e)^2}{r_B} \right] \psi - \left[\frac{Q^1 Q^0}{2R^2} i\hbar c \frac{\partial}{\partial \tilde{x}^1} \right] \psi$$
$$\equiv (\ H_0(r_B, \hbar_t, m_e^{(t)}, e_t) + H'\ )\psi \equiv H_{(dS-SR)}\psi, \qquad (7.65)$$

which is up to $\mathcal{O}(c^2 t^2/R^2)$ (once again, $\mathcal{O}(1/R^4)$, $\mathcal{O}(cta_B/R^2)$, $\mathcal{O}(cta_c/R^2)$ terms have been ignored), and where

$$H_0(r_B, \hbar_t, m_e^{(t)}, e_t) = -i\hbar_t c\vec{\alpha} \cdot \nabla_B + m_e^{(t)} c^2 \beta - \frac{e_t^2}{r_B}, \qquad (7.66)$$

$$H' = \frac{Q^1 Q^0}{2R^2} \left(-\alpha^1 H_0(r_B, \hbar_t, m_e^{(t)}, \sqrt{2}e_t) - i\hbar c \frac{\partial}{\partial \tilde{x}^1} \right), \qquad (7.67)$$

with

$$\hbar_t = \left(1 - \frac{(Q^0)^2}{2R^2}\right) \hbar = \left(1 - \frac{c^2 t^2}{2R^2}\right) \hbar, \qquad (7.68)$$

$$m_e^{(t)} = \left(1 - \frac{(Q^1)^2}{2R^2}\right) m_e, \qquad (7.69)$$

$$e_t = \left(1 - \frac{(Q^0)^2}{4R^2}\right) e = \left(1 - \frac{c^2 t^2}{4R^2}\right) e, \qquad (7.70)$$

and

$$\alpha_t \equiv \frac{e_t^2}{\hbar_t c} = \frac{e^2}{\hbar c} = \alpha. \qquad (7.71)$$

Using notation in [Strange (2008)], $E_0 = W$ and the unperturbed eigenstate equation is

$$W\psi = H_0(\hbar_t, m_e^{(t)}, e_t)\psi = \left(-i\hbar_t c\vec{\alpha} \cdot \nabla_B + m_e^{(t)} c^2 \beta - \frac{e_t^2}{r_B} \right) \psi, \qquad (7.72)$$

which is the same as Eq.(7.26) except $\hbar$, m_e, e being replaced by $\hbar_t$, $m_e^{(t)}$, e_t. The time t is dynamic variable in the time-dependent Hamiltonian system, yet we do not know whether t can be approximately treated as a parameter in the system at present. Hence, we cannot yet conclude $\hbar$, m_e, e are time variations described by Eqs.(7.68), (7.69) and (7.70) at this stage. In the following section, we pursue this subject.

7.7 Adiabatic approximation solution to de Sitter–Dirac spectra equation and time variation of physical constants

Comparing Eq.(7.72) with Eq.(7.26), we can see that there are three correction terms in Eq.(7.72), which are proportional to (c^2t^2/R^2), accounting for the effects of dS-SR QM. Since $R \gg ct$, we argue that the corrections due to these effects should be small, and the adiabatic approach works quite well for solving this QM problem. In order to be sure on this point, we examine the corrections beyond adiabatic approximations below by calculating them explicitly for a certain z. Suppose $z \simeq 3 \sim 4$, Fig.7.4 indicates $Q^1 \approx 1.7Q^0 = 1.7\,ct$, and hence $(Q^1)^2 \approx 3(Q^0)^2 = 3\,c^2t^2$ in (7.72). Rewriting the spectral equation (7.72) in the form of wave equation like Eq.(7.23) via $E \Rightarrow i\hbar\partial_t$, we have

$$i\hbar\partial_t\psi = H(t)\psi = [H_0(r_B, \hbar, m_e, e) + H_0'(t)]\psi, \tag{7.73}$$

where

$$H_0(r_B, \hbar, m_e, e) = -i\hbar c\vec{\alpha}\cdot\nabla_B + m_ec^2\beta - \frac{e^2}{r_B} \quad \text{(see Eq.(7.26))} \tag{7.74}$$

$$H_0'(t) = -\left(\frac{c^2t^2}{2R^2}\right)H_0(r_B, \hbar, 3m_e, e). \tag{7.75}$$

Suppose the initial state of the atom is $\psi(t=0) = \psi_s(r_B, \hbar, m_e, \alpha)$ where $s = \{n_s,\ \kappa_s,\ \mathbf{j}_s^2,\ j_{sz}\}$. By Eqs.(7.73)–(7.75) and taking into account the time-evolution effects, we have (see Appendix E, or/and Chapter XVII of Vol II of [Messiah (1970)])

$$\psi(t) \simeq \psi_s(\mathbf{r}_B, \hbar_t, m_e^{(t)}, e_t)e^{-i\frac{W_s}{\hbar}t} + \sum_{m\neq s}\frac{\dot{H}_0'(t)_{ms}}{i\hbar\omega_{ms}^2}\left(e^{i\omega_{ms}t} - 1\right)\psi_m(\mathbf{r}_B, \hbar_t, m_e^{(t)}, e_t)$$

$$\times\ e^{\left(-i\int_0^t \frac{W_m(\theta)}{\hbar}d\theta\right)}, \tag{7.76}$$

where $\psi_s(\mathbf{r}_B, \hbar_t, m_e^{(t)}, e_t)$ is the adiabatic wave function, $\hbar_t$, $m_e^{(t)}$ e_t are given in Eqs.(7.68)–(7.70), and

$$
\begin{aligned}
\dot{H}_0'(t)_{ms}|_{(m\neq s)} &= \langle m|\dot{H}_0'(t)|s\rangle|_{(m\neq s)} = \frac{-c^2 t}{R^2}\langle m|H_0(r_B, \hbar, 3m_e, e)|s\rangle|_{(m\neq s)} \\
&= \frac{-c^2 t}{R^2}\langle m|\left(H_0(r_B, \hbar, 3\mu, e) - H_0(r_B, \hbar, m_e, e)\right)|s\rangle|_{(m\neq s)} \\
&\simeq \frac{-c^2 t}{R^2}\langle n_m, \kappa_m, \mathbf{j}_m^2, j_{mz}|2m_e c^2 \beta|n_s, \kappa_s, \mathbf{j}_s^2, j_{sz}\rangle e^{-i(\omega_s - \omega_m)t} \\
&= \frac{-2m_e c^4 t}{R^2}\langle m|\beta|s\rangle e^{-i(\omega_s - \omega_m)t},
\end{aligned}
\tag{7.77}
$$

$$
\omega_{ms} = \omega_m - \omega_s, \quad \omega_m = \frac{W_m}{\hbar}.
\tag{7.78}
$$

Note, formula $\langle m|H_0(r,e)|s\rangle|_{m\neq s} = 0$ has been used in the calculations of (7.77). The second term of Right-Hand-Side (RHS) of Eq.(7.76) represents the quantum transition amplitudes from ψ_s-state to ψ_m, which belong to the correction effects beyond adiabatic approximations. Now for showing the order of magnitude of such corrections, we estimate $|\dot{H}_0'(t)_{ms}/\hbar\omega_{ms}^2|$ for $s = 1 \equiv (1s^{1/2}, \kappa = -1, m_j = 1/2)$, $m = 2 \equiv (2s^{1/2}, \kappa = -1, m_j = 1/2)$. Noting $W_n \approx m_e c^2 - m_e c^2 \alpha^2/(2n^2)$, and the Compton wave length of electron $a_c = \hbar/(m_e c) \simeq \hbar/(m_e c) \simeq 0.3 \times 10^{-12} m$, from Eqs.(7.77) and (7.78) we have

$$
\left|\frac{\dot{H}_0'(t)_{21}}{\hbar\omega_{21}^2}\right| = \left|\frac{128}{9\alpha^4}\langle 2|\beta|1\rangle\right|\frac{a_c}{R}\frac{ct}{R}, \quad with \quad \beta = \begin{pmatrix} I & 0 \\ 0 & -I \end{pmatrix},
\tag{7.79}
$$

where the state $\langle 2|$ has been given in Eqs.(7.36)–(7.43) and the state $|1\rangle$ is as follows: [Strange (2008)]

$$
|1\rangle = \psi_{(1s)j=1/2}^{m_j=1/2}(\mathbf{r}) = \begin{pmatrix} g_{(1s^{1/2})}(r)\chi_\kappa^{1/2}(\hat{\mathbf{r}})_{(1s^{1/2})} \\ if_{(1s^{1/2})}(r)\chi_{-\kappa}^{1/2}(\hat{\mathbf{r}})_{(1s^{1/2})} \end{pmatrix},
\tag{7.80}
$$

where

$$
g_{(1s^{1/2})}(r) = \sqrt{\frac{(2\lambda_1)^{2s+1}(k_C + W_{C1})}{2k_C\Gamma(2s+1)}}\, r^{s-1}e^{-\lambda_1 r}
\tag{7.81}
$$

$$
f_{(1s^{1/2})}(r) = -\sqrt{\frac{(2\lambda_1)^{2s+1}(k_C - W_{C1})}{2k_C\Gamma(2s+1)}}\, r^{s-1}e^{-\lambda_1 r}
\tag{7.82}
$$

with $W_{C1} \simeq m_e c^2 (1 - \alpha^2/2)/(c\hbar)$, $\lambda_1 = \sqrt{k_C^2 - W_{C1}^2}$ and

$$\chi_\kappa^{1/2}(\hat{\mathbf{r}})_{(1s^{1/2})} = \begin{pmatrix} Y_0^0 \\ 0 \end{pmatrix}, \quad \chi_{-\kappa}^{1/2}(\hat{\mathbf{r}})_{(1s^{1/2})} = \begin{pmatrix} -\cos\theta Y_0^0 \\ -\sin\theta e^{i\phi} Y_0^0 \end{pmatrix}. \quad (7.83)$$

By means of expressions of $\psi_{(1s)j=1/2}^{m_j=1/2}(\mathbf{r})$ (Eqs.(7.80)–(7.83)) and $\psi_{(2s)j=1/2}^{m_j=1/2}(\mathbf{r})$ (Eqs.(7.36)–(7.43)), we have

$$\langle 2|\beta|1 \rangle = \int_0^\infty dr r^2 \left(g_{(1s^{1/2})}(r) g_{(2s^{1/2})}(r) - f_{(1s^{1/2})}(r) f_{(2s^{1/2})}(r) \right)$$

$$\simeq -1.12 \times 10^{-5}. \quad (7.84)$$

Substituting Eq.(7.84) into Eq.(7.79), we obtain

$$\left| \frac{\dot{H}_0'(t)_{21}}{\hbar \omega_{21}^2} \right| = 5.6 \times 10^4 \times \frac{a_c}{R} \frac{ct}{R} \quad (7.85)$$

Considering Compton wave length of electron $a_c = \frac{\hbar}{m_e c} \simeq 0.3 \times 10^{-12} \text{m} \simeq 0.3 \times 10^{-28} ly$ and both $Q^0 = ct$ and R are of cosmological large length scales as well as $R > Q^0$, we have

$$\left| \frac{\dot{H}_0'(t)_{21}}{\hbar \omega_{21}^2} \right| \simeq \frac{(1.7 \times 10^{-24} ly)}{R} \times \frac{Q^0}{R} << 1. \quad (7.86)$$

For generic $\langle m|$ and $|s\rangle$, similar to the calculations of Eq.(7.85), we always have

$$\left| \frac{\dot{H}_0'(t)_{ms}}{\hbar \omega_{ms}^2} \right| \simeq (constant) \cdot \frac{a_c}{R} \times \frac{Q^0}{R}. \quad (7.87)$$

Since the $(constant)$ at last is about $\sim 10^{10}$, then $\left| \frac{\dot{H}_0'(t)_{ms}}{\hbar \omega_{ms}^2} \right| \simeq \frac{(1.7 \times 10^{-18} ly)}{R} \times \frac{Q^0}{R}$, we finally obtain

$$\left| \frac{\dot{H}_0'(t)_{ms}}{\hbar \omega_{ms}^2} \right| << 1, \quad (7.88)$$

which indicates the second term of adiabatic expansion expression Eq.(7.76) can be ignored, or the corrections from beyond adiabatic approximations are quite small (or tiny), and hence the adiabatic approximation is legitimate for solving this dS-SR Dirac equation of the atom.

Thus, we arrive at an interesting conclusion that the fundamental physical constants vary adiabatically along with cosmologic time in dS-SR quantum mechanics framework. As is well known, the quantum evolution in the time-dependent quantum mechanics has been widely accepted and studied during past several decades (see, e.g., [Bayfield (1999)]). It is remarkable that the time-variations of m_e, $\hbar$ and e (see Eqs.(7.68)–(7.70)) belong to such quantum evolution effects.

7.8 $2s^{1/2}$–$2p^{1/2}$ splitting in the dS-SR Dirac equation of Hydrogen

In the subsection 7.4.2, we have pointed that the state of $2s^{1/2}$ and state of $2p^{1/2}$ are complete degenerate to all orders of α in the E-SR Dirac equation of Hydrogen described in Hamiltonian $H_0(r, \hbar, \mu, e)$ of Eq.(7.26) (see also Eq.(7.27)). In this section, we calculate the $(2S^{1/2} - 2p^{1/2})$-splitting duo to dS-SR effects.

7.8.1   *Energy levels shifts of Hydrogen in dS-SR QM as perturbation effects of E-SR Dirac equation of atom*

The dS-SR Dirac spectrum equation for Hydrogen atom has been derived in Section 7.6 (7.65), which is as follows

$$H_{(dS-SR)}\psi = (\ H_0(r, \hbar_t, m_e^{(t)}, e_t) + H'\ )\psi = E\psi, \qquad (7.89)$$

where

$$H_0(r, \hbar_t, m_e^{(t)}, e_t) = -i\hbar_t c\vec{\alpha} \cdot \nabla + m_e^{(t)} c^2 \beta - \frac{e_t^2}{r}, \qquad (7.90)$$

$$H' = \frac{1}{2}(H'^{\dagger} + H') \equiv H_1' + H_2', \qquad (7.91)$$

and[2]

$$H_1' = -\frac{Q^1 Q^0}{4R^2}\left(\alpha^1 H_0(r, \hbar_t, m_e^{(t)}, \sqrt{2}e_t) + H_0(r, \hbar_t, m_e^{(t)}, \sqrt{2}e_t)\alpha^1\right), (7.92)$$

$$H_2' = -\frac{Q^1 Q^0}{4R^2}\left(i\hbar c\overrightarrow{\frac{\partial}{\partial x^1}} - i\hbar c\overleftarrow{\frac{\partial}{\partial x^1}}\right). \qquad (7.93)$$

Comparing above equations with Eqs.(7.65)–(7.67), hereafter we have removed the subscript B in r_B and ∇_B, and the tilde notation $\sim$ in $\tilde{x}^1$ for

---

[2]Equations (7.91) with (7.92) and (7.93) is similar to equation (74) (or (76)) in [Yan (2012)] except 2 mistakes in writing in it: (i) e in (74) of [Yan (2012)] should be $\sqrt{2}e$; (ii) the sign of the second term in it should be minus.

simplicity. The perturbation Hamiltonian H' is also rewritten to be explicitly Hermitian. In the spherical coordinates system the operator $\frac{\partial}{\partial x^1} \equiv \partial_1$ in Eq.(7.92) is as follows

$$\partial_1 = \frac{\partial}{\partial x^1} = \vec{i} \cdot \nabla = \sin\theta\cos\phi\frac{\partial}{\partial r} + \cos\theta\cos\phi\frac{1}{r}\frac{\partial}{\partial\theta} - \frac{\sin\phi}{r\sin\theta}\frac{\partial}{\partial\phi}. \quad (7.94)$$

Comparing dS-SR QM with ordinary E-SR Dirac equation of Hydrogen, we see two distinguishing effects in the dS-SR QM descriptions of distant Hydrogen atom (or one-electron atom) in the Earth-QSO reference frame: (i) The physical constants variations with cosmic time adiabatically, which have been discussed in the previous section (see Eqs.(7.68)−(7.70)); (ii) Perturbation effects arising from H' of Eq.(7.91) in $H_{(dS-SR)}$ of Eq.(7.89). In this section, we focus on the latter.

For adiabatic quantum system, the states are quasi-stationary in all instants. Hence at all instants the quasi-stationary perturbation theory works. When $H_{(dS-SR)} = H_0(r, \hbar_t, m_e^{(t)}, e_t) + H'$, the unperturbed quasi-stationary solutions of $H_0(r, \hbar_t, m_e^{(t)}, e_t)\psi = W\psi$ are the same as Eqs.(7.26)−(7.34) except $\hbar \to \hbar_t$, $m_e \to m_e^{(t)}$, $e \to e_t$. Then the energy levels shifts due to H' of Eq.(7.91) are computable in practice by the perturbation approach in QM.

Those shifts $\Delta E^i \equiv W_i - E$ due to H' are determined by

$$\det\left(\langle H_{(dS-SR)}\rangle_{ii'} - E\delta_{ii'}\right) = 0, \quad (7.95)$$

where $i = \{n,\ l,\ j,\ \kappa,\ m_j\}$, $H_{(dS-SR)}$ has been given in Eq.(7.89) and the elements are

$$\langle H_{(dS-SR)}\rangle_{ii'} = \langle i|H_0|i'\rangle + \langle i|H'|i'\rangle = W_i\delta_{ii'} + \langle H'\rangle_{ii'}, \quad (7.96)$$

where $W_i = W_{n,\kappa}$ are shown in Eq.(7.27).

Firstly, we compute the elements $\langle H'\rangle_{ii'} \equiv \langle(H_1' + H_2')\rangle_{ii'}$ with $i = i'$. From Eq.(7.92), we have

$$\langle H_1'\rangle_{ii} = -\frac{Q^1 Q^0}{2R^2}W_i\langle i|\alpha^1|i\rangle + \frac{Q^1 Q^0 e^2}{2R^2}\langle i|\frac{1}{r}\alpha^1|i\rangle$$

$$= -\frac{Q^1 Q^0}{2R^2}W_i\int dr\, r^2 \int d\Omega$$

$$\times \left(g_\kappa(r)\chi_\kappa^{m_j\dagger}(\hat{\mathbf{r}}), -if_\kappa(r)\chi_{-\kappa}^{m_j\dagger}(\hat{\mathbf{r}}) \right) \begin{pmatrix} 0 & \sigma^1 \\ \sigma^1 & 0 \end{pmatrix} \begin{pmatrix} g_\kappa(r)\chi_\kappa^{m_j}(\hat{\mathbf{r}}) \\ if_\kappa(r)\chi_{-\kappa}^{m_j}(\hat{\mathbf{r}}) \end{pmatrix}$$

$$+\frac{Q^1 Q^0 e^2}{2R^2} \int drr \int d\Omega$$

$$\times \left(g_\kappa(r)\chi_\kappa^{m_j\dagger}(\hat{\mathbf{r}}), -if_\kappa(r)\chi_{-\kappa}^{m_j\dagger}(\hat{\mathbf{r}}) \right) \begin{pmatrix} 0 & \sigma^1 \\ \sigma^1 & 0 \end{pmatrix} \begin{pmatrix} g_\kappa(r)\chi_\kappa^{m_j}(\hat{\mathbf{r}}) \\ if_\kappa(r)\chi_{-\kappa}^{m_j}(\hat{\mathbf{r}}) \end{pmatrix}$$

$$\propto \int d\Omega \left(\chi_\kappa^{m_j\dagger}(\hat{\mathbf{r}})\sigma^1\chi_{-\kappa}^{m_j}(\hat{\mathbf{r}}) - \chi_{-\kappa}^{m_j\dagger}(\hat{\mathbf{r}})\sigma^1\chi_\kappa^{m_j}(\hat{\mathbf{r}}) \right). \tag{7.97}$$

Substituting Eqs.(7.30) and (7.31) into Eq.(7.97), we get

$$\int d\Omega \chi_\kappa^{m_j\dagger}(\hat{\mathbf{r}})\sigma^1\chi_{-\kappa}^{m_j}(\hat{\mathbf{r}}) = 0 \tag{7.98}$$

$$\int d\Omega \chi_{-\kappa}^{m_j\dagger}(\hat{\mathbf{r}})\sigma^1\chi_\kappa^{m_j}(\hat{\mathbf{r}}) = 0, \tag{7.99}$$

and hence

$$\langle H_1' \rangle_{ii} = 0. \tag{7.100}$$

It is easy to check the validness of Eqs.(7.98) and (7.99). Considering, for instance, the case of state $|i\rangle = |(2p^{1/2})^{m_j=1/2}\rangle$ (see Eqs.(7.44) and (7.45)), we can calculate the left sides of Eqs.(7.98) and (7.99) explicitly:

$$\int d\Omega \chi_{\kappa=1}^{1/2\dagger}(\hat{\mathbf{r}})_{(2p^{1/2})}\sigma^1\chi_{-\kappa=-1}^{1/2}(\hat{\mathbf{r}})_{(2p^{1/2})}$$

$$= \int d\Omega \left(-\sqrt{\frac{1}{3}}Y_1^0(\theta\phi)^\dagger, \sqrt{\frac{2}{3}}Y_1^1(\theta\phi)^\dagger \right) \begin{pmatrix} 0 & 1 \\ 1 & 0 \end{pmatrix}$$

$$\times \begin{pmatrix} \sqrt{\frac{2}{3}}\cos\theta Y_1^{-1}(\theta\phi) - \sqrt{\frac{1}{3}}\sin\theta e^{-i\phi}Y_1^0(\theta\phi) \\ \sqrt{\frac{2}{3}}\sin\theta e^{i\phi}Y_1^{-1}(\theta\phi) + \sqrt{\frac{1}{3}}\cos\theta Y_1^0(\theta\phi) \end{pmatrix}$$

$$= \frac{-i}{2\pi}\int_0^{2\pi} d\phi \sin\phi \int_0^\pi d\theta \sin\theta \cos^2\theta(\sin\theta - \cos\theta) = 0, \tag{7.101}$$

$$\int d\Omega \chi_{-\kappa=-1}^{1/2\dagger}(\hat{\mathbf{r}})_{(2p^{1/2})}\sigma^1\chi_{\kappa=1}^{1/2}(\hat{\mathbf{r}})_{(2p^{1/2})}$$

$$= \left(\int d\Omega \chi_{\kappa=1}^{1/2\dagger}(\hat{\mathbf{r}})_{(2p^{1/2})}\sigma^1\chi_{-\kappa=-1}^{1/2}(\hat{\mathbf{r}})_{(2p^{1/2})} \right)^\dagger = 0. \tag{7.102}$$

Therefore Eqs.(7.98) and (7.99) hold for state of $|i\rangle = |(2p^{1/2})^{m_j=1/2}\rangle$. Similarly, for cases of $|i\rangle = |(2s^{1/2})^{m_j=1/2}\rangle$, $|i\rangle = |(2s^{1/2})^{m_j=-1/2}\rangle$ and $|i\rangle = |(2p^{1/2})^{m_j=-1/2}\rangle$, Eqs.(7.98) and (7.99) can also be verified.

From Eq.(7.93), we have

$$\langle H_2' \rangle_{ii} = \frac{Q^1 Q^0}{4R^2} \langle i | \left(i\hbar c \frac{\overrightarrow{\partial}}{\partial x^1} - i\hbar c \frac{\overleftarrow{\partial}}{\partial x^1} \right) | i \rangle = -\frac{Q^1 Q^0}{2R^2} c \langle i | \hat{p}^1 | i \rangle = 0,$$

(7.103)

where $\hat{p}^1 = -i \frac{\partial}{\partial x^1}$, and the fact that the average value for p^1 in the stationary bound state $|i\rangle$ must vanish has been used. Equation (7.103) can also be checked by explicit calculations based on the known wave functions Eqs.(7.29)–(7.31).

Combining Eq.(7.103) with Eq.(7.101), we find that

$$\langle H' \rangle_{ii} = \langle H_1' \rangle_{ii} + \langle H_2' \rangle_{ii} = 0,$$

(7.104)

which means $\langle H' \rangle_{ii'}$ is an off-diagonal matrix in the Hilbert space. As is well known that the energy shifts for non-degenerate levels due to H' are expressed as

$$\Delta E^i \equiv W_i - E = \langle H' \rangle_{ii} + \sum_{i' \neq i}{}' \frac{|\langle H' \rangle_{i'i}|^2}{W_i - W_{i'}} + \cdots,$$

(7.105)

which could be considered the perturbation solution of Eq.(7.95) for non-degenerate case. Thus, noting $H' \propto \mathcal{O}(1/R^2)$ and Eq.(7.104), the non-degenerate level shifts ΔE^i are of order $\mathcal{O}(1/R^4)$, which is beyond the considerations of this paper. We will show in next subsection the meaningful $\mathcal{O}(1/R^2)$-energy level shifts due to off-diagonal perturbation interaction H' for degeneration levels. Typical example is $2S^{1/2} - 2p^{1/2}$ splitting due to H'.

7.8.2 $2s^{1/2} - 2p^{1/2}$ *splitting caused by* $\mathbf{H'}$

In Section 7.4, we have shown that the state of $2s^{1/2}$ and state of $2p^{1/2}$ are complete degenerate to all order of α in the E-SR Dirac equation of Hydrogen (see Eq.(7.35)). The degeneracy will be broken by the effects of H'. In this subsection we calculate the $2s^{1/2} - 2p^{1/2}$ splitting caused by H'.

By using the explicit expressions of $2s^{1/2}$- and $2p^{1/2}$ wave functions Eqs.(7.36)–(7.45), all matrix elements of $H' = H_1' + H_2'$ between them can be calculated, i.e.,

$$\langle H' \rangle_{2L^{1/2}, 2L'^{1/2}}^{m_j, m_j'} = \langle H_1' \rangle_{2L^{1/2}, 2L'^{1/2}}^{m_j, m_j'} + \langle H_2' \rangle_{2L^{1/2}, 2L'^{1/2}}^{m_j, m_j'},$$

(7.106)

where

$$\langle H_i' \rangle_{2L^{1/2},2L'^{1/2}}^{m_j,\, m_j'} = \langle (2L^{1/2})^{m_j} | H_i' | (2L'^{1/2})^{m_j'} \rangle$$

$$= \int dr\, r^2 \int d\Omega \; \psi_{(2L)j=1/2}^{m_j \dagger}(\mathbf{r})\, H_i' \, \psi_{(2L')j=1/2}^{m_j'}(\mathbf{r}),$$

$$(7.107)$$

here $\{L, L'\} = \{s, p\}$, $i = 1, 2$, and $\psi_{(2L)j=1/2}^{m_j}(\mathbf{r})$ are given in Eqs.(7.36)–(7.45). The matrix element calculations are presented in step by step in Appendix F, and the results are listed as follows:

(1) H_1'-matrix elements: H_1' is given in Eq.(7.92), i.e.,

$$H_1' = -\frac{Q^1 Q^0}{4R^2} \left(\alpha^1 H_0(r, \hbar, m_e, \sqrt{2}e) + H_0(r, \hbar, m_e, \sqrt{2}e)\alpha^1 \right),$$

$$(7.108)$$

where the subscript t of $\hbar, \mu, e$ has been removed because there is already a factor of $(1/R^2)$ in the H_1'-expression and $1/R^4$-terms are ignorable. From Eqs.(7.98), (7.99) and (7.108), and noting $\int_0^{2\pi} d\phi \, \exp(\pm in\phi) = 0$, $\int_0^{\pi} d\theta \sin\theta \cos^{2n+1}\theta = 0$ etc., the explicit calculations of $\langle H_1' \rangle_{ii'}$ show

$$-i\Theta_1 \equiv \langle H_1' \rangle_{2s^{1/2},2p^{1/2}}^{1/2,\,-1/2} = \langle H_1' \rangle_{2s^{1/2},2p^{1/2}}^{-1/2,\,1/2} = -\langle H_1' \rangle_{2p^{1/2},2s^{1/2}}^{1/2,\,-1/2}$$

$$= -\langle H_1' \rangle_{2p^{1/2},2s^{1/2}}^{-1/2,\,1/2} = \text{unknown}, \qquad (7.109)$$

and others $= 0$.

Equivalently and explicitly, the matrix form of $\{\langle H_1' \rangle_{ii'}\}$ is:

$$\{\langle H_1' \rangle_{ii'}\} = \begin{array}{c} {\scriptstyle (2L^j, L'^{j'})^{m_j, m_j'} : \; (2s^{\frac{1}{2}})^{\frac{1}{2}} \;\; (2s^{\frac{1}{2}})^{\frac{-1}{2}} \;\; (2p^{\frac{1}{2}})^{\frac{1}{2}} \;\; (2p^{\frac{1}{2}})^{\frac{-1}{2}}} \\ \begin{array}{c} (2s^{1/2})^{1/2} \\ (2s^{1/2})^{-1/2} \\ (2p^{1/2})^{1/2} \\ (2p^{1/2})^{-1/2} \end{array} \left(\begin{array}{cccc} 0 & 0 & 0 & -i\Theta_1 \\ 0 & 0 & -i\Theta_1 & 0 \\ 0 & i\Theta_1 & 0 & 0 \\ i\Theta_1 & 0 & 0 & 0 \end{array} \right) \end{array},$$

$$(7.110)$$

where the matrix row and column indices have been labeled explicitly. Compactly, Eq.(7.110) can be written as follows

$$\{\langle H_1' \rangle_{ii'}\} = \begin{pmatrix} 0 & -i\Theta_1 \sigma^1 \\ i\Theta_1 \sigma^1 & 0 \end{pmatrix}, \quad \text{with } \sigma^1 = \begin{pmatrix} 0 & 1 \\ 1 & 0 \end{pmatrix}. \quad (7.111)$$

(2) By means of the wave functions $\psi_j^{m_j}(\mathbf{r})$ of Eq.(7.29) the Θ_1 can be calculated. Details are shown in Appendix by using the known wave functions with $n = 2$ and $\kappa = \pm 1$. Explicit calculations for this particular input result in

$$\Theta_1 = 0. \qquad (7.112)$$

(see Eq.(F.11) in Appendix F).

(3) H_2' matrix elements: H_2' is shown in Eq.(7.93):

$$H_2' = -\frac{Q^1 Q^0}{4R^2}\left(i\hbar c \overrightarrow{\frac{\partial}{\partial x^1}} - i\hbar c \overleftarrow{\frac{\partial}{\partial x^1}}\right), \qquad (7.113)$$

where $\dfrac{\partial}{\partial x^1} = \sin\theta\cos\phi\dfrac{\partial}{\partial r} + \cos\theta\cos\phi\dfrac{1}{r}\dfrac{\partial}{\partial\theta} - \dfrac{\sin\phi}{r\sin\theta}\dfrac{\partial}{\partial\phi}.$ (7.114)

By means of straightforward calculations (see Appendix F), we obtain all elements of H_2':

$$\langle H_2'\rangle_{2s^{1/2},2p^{1/2}}^{1/2,\,-1/2} = \langle H_2'\rangle_{2s^{1/2},2p^{1/2}}^{-1/2,\,1/2} = -\langle H_2'\rangle_{2p^{1/2},2s^{1/2}}^{1/2,\,-1/2} = -\langle H_2'\rangle_{2p^{1/2},2s^{1/2}}^{-1/2,\,1/2}$$

$$= -i\frac{Q^1 Q^0}{2R^2}\frac{\hbar c\lambda}{6\sqrt{4W_C^2 - k_C^2}}\left(\frac{k_C^2}{W_C} - 2(\frac{1}{s}+1)k_C - W_C + \frac{2}{k_C s}W_C^2\right)$$

$$\equiv -i\Theta_2,$$

and others $= 0$, $\qquad (7.115)$

where notations of $k_C, W_C, \lambda, s, \kappa$ have been given in Eq.(7.34). The matrix form of Eq.(7.115) is as follows:

$$\{\langle H_2'\rangle_{ii'}\} = \begin{pmatrix} 0 & -i\Theta_2\sigma^1 \\ i\Theta_2\sigma^1 & 0 \end{pmatrix}, \qquad (7.116)$$

where

$$\Theta_2 = \frac{Q^1 Q^0}{2R^2}\frac{\hbar c\lambda}{6\sqrt{4W_C^2 - k_C^2}}\left(\frac{k_C^2}{W_C} - 2(\frac{1}{s}+1)k_C - W_C + \frac{2}{k_C s}W_C^2\right). \qquad (7.117)$$

(4) Since $H' = H_1' + H_2'$ (see Eq.(7.91)), the matrix form of H' is:

$$\{\langle H'\rangle_{ii'}\} = \{\langle H_1'\rangle_{ii'}\} + \{\langle H_2'\rangle_{ii'}\} = \begin{pmatrix} 0 & -i\Theta\sigma^1 \\ i\Theta\sigma^1 & 0 \end{pmatrix}, \quad (7.118)$$

where

$$\Theta = \Theta_1 + \Theta_2 = 0 + \Theta_2$$
$$= \frac{Q^1 Q^0}{2R^2}\frac{\hbar c\lambda}{6\sqrt{4W_C^2 - k_C^2}}\left(\frac{k_C^2}{W_C} - 2(\frac{1}{s}+1)k_C - W_C + \frac{2}{k_C s}W_C^2\right). \qquad (7.119)$$

Obviously, $\Theta \propto \mathcal{O}(1/R^2)$.

(5) Substituting Eq.(7.118) into Eqs.(7.96) and (7.95), we get the secular equation for eigenvalue E in the degenerate perturbation calculations:

$$
\begin{vmatrix}
W - E & 0 & 0 & -i\Theta \\
0 & W - E & -i\Theta & 0 \\
0 & i\Theta & W - E & 0 \\
i\Theta & 0 & 0 & W - E
\end{vmatrix} = 0. \tag{7.120}
$$

The real energy solutions are

$$
W - E = \pm\Theta, \quad \text{or} \quad E^{(+)} = W + \Theta, \quad E^{(-)} = W - \Theta, \tag{7.121}
$$

and hence we obtain the desired expression of energy level splitting of $(2s^{1/2} - 2p^{1/2})$ due to H'

$$
(\Delta E)_{(2p^{1/2} - 2s^{1/2})} \equiv E^{(-)} - E^{(+)} = -2\Theta
$$

$$
= -\frac{Q^1 Q^0}{R^2} \frac{\hbar c \lambda}{6\sqrt{4W_C^2 - k_C^2}} \left(\frac{k_C^2}{W_C} - 2(\frac{1}{s} + 1)k_C - W_C + \frac{2}{k_C s}W_C^2 \right), \tag{7.122}
$$

which is of order $\mathcal{O}(1/R^2)$. Equation (7.122) represents an important effect of dS-SR for one-electron atom and it is the main result of this chapter.

(6) The eigenstates with eigenvalues $E^{(\pm)}$: Generally, the eigenstates of H' are

$$
|E^{(\pm)}\rangle = C_1^{(\pm)}|2s^{1/2}\rangle^{1/2} + C_2^{(\pm)}|2s^{1/2}\rangle^{-1/2} + C_3^{(\pm)}|2p^{1/2}\rangle^{1/2} + C_4^{(\pm)}|2p^{1/2}\rangle^{-1/2}, \tag{7.123}
$$

where $\{|2s^{1/2}\rangle^{m_j}, |2s^{1/2}\rangle^{m_j}\} \in |nL^j\rangle^{m_j}$, and $C_i^{(\pm)}$ $(i = 1, 2, 3, 4)$ satisfy the following eigenequation corresponding to Eq.(7.120):

$$
\begin{pmatrix}
W & 0 & 0 & -i\Theta \\
0 & W & -i\Theta & 0 \\
0 & i\Theta & W & 0 \\
i\Theta & 0 & 0 & W
\end{pmatrix}
\begin{pmatrix}
C_1^{(\pm)} \\
C_2^{(\pm)} \\
C_3^{(\pm)} \\
C_4^{(\pm)}
\end{pmatrix} = E^{(\pm)}
\begin{pmatrix}
C_1^{(\pm)} \\
C_2^{(\pm)} \\
C_3^{(\pm)} \\
C_4^{(\pm)}
\end{pmatrix}. \tag{7.124}
$$

Substituting Eq.(7.121) into Eq.(7.124), we get the eigenstates with eigenvalues $E^{(\pm)}$:

$$
|E^{(+)}\rangle = \frac{1}{2}(|2s^{1/2}\rangle^{1/2} + |2s^{1/2}\rangle^{-1/2} + i|2p^{1/2}\rangle^{1/2} + i|2p^{1/2}\rangle^{-1/2}), \tag{7.125}
$$

$$
|E^{(-)}\rangle = \frac{1}{2}(|2s^{1/2}\rangle^{1/2} + |2s^{1/2}\rangle^{-1/2} - i|2p^{1/2}\rangle^{1/2} - i|2p^{1/2}\rangle^{-1/2}), \tag{7.126}
$$

which satisfy $\langle E^{(+)}|E^{(+)}\rangle = \langle E^{(-)}|E^{(-)}\rangle = 1$, and $\langle E^{(-)}|E^{(+)}\rangle = 0$.

(7) Numerical discussions: In order to have a comprehensive understanding of Eq.(7.122), we now discuss $(\Delta E)_{(2p^{1/2}-2s^{1/2})} \equiv E^{(-)} - E^{(+)} = -2\Theta$ numerically. For Hydrogen's $2s^{1/2}$- and $2p^{1/2}$-states, $m_e = 510998.910$ eV/c^2, $Z = 1$, $n = 2$, $\kappa = \pm 1$. Substituting them into Eq.(7.27), we get $W = 510995.51$ eV/c^2. And by Eq.(7.34), we further have k_C, W_C, λ and s. Inserting all of them into Eq.(7.122), we obtain

$$\Delta E(z) \equiv (\Delta E)_{(2p^{1/2}-2s^{1/2})} = \frac{Q^1(z)Q^0(z)}{R^2} \times 358.826 \text{ eV}$$

$$= \frac{Q^1(z)Q^0(z)}{R^2} \times 8.36 \times 10^7 \text{ (Lamb shift)}, \qquad (7.127)$$

where $Q^0(z) \equiv ct(z)$ and $Q^1(z)$ have been given in Eqs.(7.13) and (7.17) respectively (see also Fig.7.2 and Fig.7.3). In the expression of function $\Delta E(z)$ of Eq.(7.127), when z were fixed, the only unknown number is R which is the universal parameter of dS-SR. Therefore a way to determine R may be through such kind of observations of the level spectrum shifts of atoms on distant galaxy. The curves of $\Delta E(z)$ of Eq.(7.127) with $|R| = \{10^5\text{Gly}, \ 10^6\text{Gly}, \ \times 10^7\text{Gly}\}$ are shown in Fig.7.6. In the Table 7.1, the $\Delta E(z)$ for $|R| = \{10^3\text{Gly}, \ 10^4\text{Gly}, \ 10^5\text{Gly}\}$ and $z = \{1, \ 2\}$ is listed.

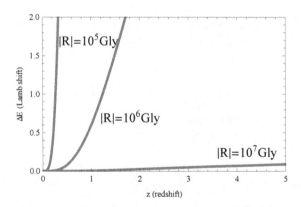

Fig. 7.6 Functions $\Delta E(z)$ of Eq.(7.122) with $|R| = 0.5 \times 10^5\,\text{Gly}$, $10^5\,\text{Gly}$, $2 \times 10^5\,\text{Gly}$ are shown. The unit of the energy splitting $\Delta E(z)$ is (Lamb shift)$\simeq 4.3 \times 10^{-6}$eV.

Table 7.1 The energy level splitting of Hydrogen's $(2p^{1/2} - 2s^{1/2})$, $\Delta E(z) \equiv (\Delta E)_{(2p^{1/2}-2s^{1/2})}$ (see Eqs.(7.122) and (7.127)): R is the radius of de Sitter pseudo-sphere in dS-SR, which is a universal parameter in the theory and $|R| > R_H \equiv 13.7$Gly (Horizon of the Universe). Gly $\equiv 10^9$ light years. z is the red shift. (Lamb shift)$\simeq 4.3 \times 10^{-6}$eV.

$|R|$	10^3 Gly		10^4 Gly		10^5 Gly	
z	1	2	1	2	1	2
$\Delta E(z)$ (eV)	2.6	11	2.6×10^{-2}	0.11	2.6×10^{-4}	1.1×10^{-3}
(Lamb shift)	6×10^5	2.7×10^6	5977	26541	59.77	265.41

7.9 High-order calculations of $1/R^2$-expansions

Substituting Eqs.(6.57), (6.64), (D.12) and (D.16) into Eq.(7.18) gives full expression of the de Sitter invariant special relativistic Dirac equation for the electron in Hydrogen in the earth-QSO reference frame:

$$\hbar c\beta \left[i\sqrt{\sigma}\gamma^\mu \mathcal{D}_\mu^L + i\frac{\sigma - \sqrt{\sigma}}{(1-\sigma)R^2}\eta_{ab}Q^a\gamma^b Q^\mu \mathcal{D}_\mu^L - \frac{m_e c}{\hbar} \right]\psi = 0, \qquad (7.128)$$

where factor $\hbar c\beta$ in the front of the equation is only for convenience, $L^\mu \simeq Q^\mu$ has been used (see Fig.7.1 in Section 7.2), and

$$\mathcal{D}_\mu^L \equiv \frac{\partial}{\partial L^\mu} - \frac{i}{4}\omega^{ab}_{\ \ \mu}\sigma_{ab} - i\frac{e}{ch}B_{\mu\nu}A^\nu$$

$$= \frac{\partial}{\partial L^\mu} - \frac{i}{4}\frac{1}{R^2(1+\sqrt{\sigma})\sqrt{\sigma}}(\eta_\mu^a Q^b - \eta_\mu^b Q^a)\sigma_{ab}$$

$$-i\frac{e}{ch}\frac{\frac{1}{\sigma} + \frac{(Q^0)^2}{R^2\sigma^2} - \frac{Q^0 Q^1}{R^2\sigma^2}}{\sqrt{\left[\left(\frac{1}{\sigma} - \frac{(Q^1)^2}{R^2\sigma^2}\right)\left(\frac{1}{\sigma} + \frac{(Q^0)^2}{R^2\sigma^2}\right) + \frac{(Q^0)^2(Q^1)^2}{(R^4\sigma^4)}\right]\left(1 + \frac{(Q^1)^2}{R^2}\right)}}\frac{e}{r_B},$$

$$(7.129)$$

We expand each term of Eq.(7.128) in order of $1/R^2$ as follows:

(1) Since observed QSO must be located on the light cone, we have $\eta_{ab}L^a L^b \simeq \eta_{ab}Q^a Q^b = (Q^0)^2 - (Q^1)^2$, and the first term of Eq.(7.128) reads

$$\hbar c\beta i\sqrt{\sigma}\gamma^\mu \mathcal{D}_\mu^L = \sqrt{1 - \frac{(Q^0)^2 - (Q^1)^2}{R^2}}\,\hbar c\beta i\gamma^\mu \mathcal{D}_\mu^L\psi \qquad (7.130)$$

with $\beta\gamma^\mu = \{\beta\gamma^0 = \beta^2 = 1,\ \beta\gamma^i = \alpha^i\}$ $\qquad (7.131)$

$$\hbar c\beta i\gamma^\mu \mathcal{D}_\mu^L\psi = (i\hbar\partial_t + i\hbar c\vec{\alpha}\cdot\nabla + \frac{\hbar c\beta}{4}\omega^{ab}_{\ \ \mu}\gamma^\mu\sigma_{ab} + e\beta\gamma^\mu B_{\mu\nu}A^\nu)\psi.$$

$$(7.132)$$

In the following, we use variables $\{\tilde{x}^1, x^2, x^3\}$ (where $\tilde{x}^1 \equiv \frac{1}{\sqrt{1+(Q^1)^2/R^2}} x^1$, see Eq.(A.37)) to replace $\{x^1, x^2, x^3\}$. Using notations of (8.97) and (8.98), and noting $\frac{\partial}{\partial x^1} = \frac{\partial \tilde{x}^1}{\partial x^1} \frac{\partial}{\partial \tilde{x}^1} = \frac{1}{\sqrt{1+(Q^1)^2/R^2}} \frac{\partial}{\partial \tilde{x}^1}$, Eq.(7.130) becomes

$$
\hbar c \beta i \gamma^\mu \mathcal{D}_\mu^L \psi = \left(i\hbar\partial_t + i\hbar c\vec{\alpha} \cdot \nabla_B + i\hbar c \left[\frac{1}{\sqrt{1+(Q^1)^2/R^2}} - 1 \right] \alpha^1 \frac{\partial}{\partial \tilde{x}^1} \right.
$$
$$
\left. + \frac{\hbar c \beta}{4} \omega_\mu^{ab} \gamma^\mu \sigma_{ab} + eB_{00}\phi_B + e\alpha^1 B_{10}\phi_B \right)\psi
$$
$$
= \left(i\hbar\partial_t + i\hbar c\vec{\alpha} \cdot \nabla_B + i\hbar c \left[\frac{1}{\sqrt{1+(Q^1)^2/R^2}} - 1 \right] \alpha^1 \frac{\partial}{\partial \tilde{x}^1} \right.
$$
$$
\left. + \frac{\hbar c \beta}{4} \omega_\mu^{ab} \gamma^\mu \sigma_{ab} + \left[\frac{1}{\sigma} + \frac{(Q^0)^2}{R^2\sigma^2} \right] e\phi_B - \frac{Q^1 Q^0}{R^2\sigma^2} \alpha^1 e\phi_B \right)\psi. \quad (7.133)
$$

(2) Estimation of the contributions of the fourth term in RSH of (7.133) (the spin-connection contributions) is as follows. From Eq.(6.64), the ratio of the fourth term to the first term of Eq.(7.133) is:

$$
\left| \frac{\frac{\hbar c \beta}{4} \omega_\mu^{ab} \gamma^\mu \sigma_{ab}\psi}{i\hbar\partial_t\psi} \right| \sim \frac{\hbar c}{4} \frac{ct}{2R^2} \frac{1}{m_e c^2} = \frac{ct}{8R^2} \frac{\hbar}{m_e c} = \frac{1}{8} \frac{cta_c}{R^2} \sim 0, \quad (7.134)
$$

where $a_c = \hbar/(m_e c) \simeq 0.3 \times 10^{-12}$m is the Compton wave length of electron. $\mathcal{O}(cta_c/R^2)$-term is negligible. Therefore the 3-rd term in RSH of (7.132) has no contribution to our approximation calculations.

(3) Substituting Eqs.(7.134), (7.133) and (7.131) into Eq.(7.130) and noting $L^a \simeq Q^a$ (see Fig.7.1), we get the first term in LHS of Eq.(7.128)

$$
\hbar c \beta i \sqrt{\sigma} \gamma^\mu \mathcal{D}_\mu^L \psi = \sqrt{\sigma} \left(i\hbar\partial_t + i\hbar c\vec{\alpha} \cdot \nabla_B + i\hbar c \left[\frac{1}{\sqrt{1+(Q^1)^2/R^2}} - 1 \right] \right.
$$
$$
\left. \times \alpha^1 \frac{\partial}{\partial \tilde{x}^1} + \left[\frac{1}{\sigma} + \frac{(Q^0)^2}{R^2\sigma^2} \right] e\phi_B - \frac{Q^1 Q^0}{R^2\sigma^2} \alpha^1 e\phi_B \right)\psi.
$$
$$
(7.135)
$$

(4) The second term of Eq.(7.128) is

$$\hbar c \beta i \frac{\sigma - \sqrt{\sigma}}{(1-\sigma)R^2} \eta_{ab} L^a \gamma^b L^\mu \mathcal{D}_\mu^L \psi$$

$$= \hbar c \beta i \frac{\sigma - \sqrt{\sigma}}{(1-\sigma)R^2} (\gamma^0 L^0 - \vec{\gamma} \cdot \vec{L}) \left[L^0 \mathcal{D}_0 + L^i \mathcal{D}_i \right] \psi$$

$$\simeq \hbar c i \frac{\sigma - \sqrt{\sigma}}{(1-\sigma)R^2} (L^0 - L^1 \alpha^1)$$

$$\times \left[L^0 \left(\partial_0^L - \frac{\gamma^0 \gamma^1 L^1}{2R^2(1+\sqrt{\sigma})\sqrt{\sigma}} - \frac{ie}{c\hbar} (\frac{1}{\sigma} + \frac{(Q^0)^2}{R^2 \sigma^2}) \phi_B \right) \right.$$

$$\left. + L^1 \left(\partial_1^L + \frac{\gamma^0 \gamma^1 L^0}{2R^2(1+\sqrt{\sigma})\sqrt{\sigma}} + \frac{ie}{c\hbar} \frac{Q^0 Q^1}{R^2 \sigma^2} \phi_B \right) \right] \psi$$

$$\simeq \hbar c i \frac{\sigma - \sqrt{\sigma}}{(1-\sigma)R^2} (L^0 - L^1 \alpha^1)$$

$$\times \left[L^0 \partial_0^L + L^1 \partial_1^L - \frac{ie}{c\hbar} \left(\frac{L^0}{\sigma} + \frac{L^0 (Q^0)^2 - L^1 Q^0 Q^1}{R^2 \sigma^2} \right) \phi_B \right] \psi,$$

$$(7.136)$$

where the following approximation estimations is used

$$\frac{(L^2)}{R} \sim \frac{(L^3)}{R} \sim \frac{a_B}{R} \sim 0. \tag{7.137}$$

Noting $L^0 \simeq Q^0$, $L^1 \simeq Q^1$, $L^2 \simeq 0$, $L^3 \simeq 0$, Eq.(7.136) becomes

$$\hbar c \beta i \frac{\sigma - \sqrt{\sigma}}{(1-\sigma)R^2} \eta_{ab} L^a \gamma^b L^\mu \mathcal{D}_\mu^L \psi$$

$$= \frac{\sigma - \sqrt{\sigma}}{(1-\sigma)R^2} \left[[(Q^0)^2 - Q^1 Q^0 \alpha^1] i\hbar \partial_t + Q^1 Q^0 i\hbar c \frac{\partial}{\partial \tilde{x}^1} - (Q^1)^2 i\hbar c \alpha^1 \frac{\partial}{\partial \tilde{x}^1} \right.$$

$$\left. + e[Q^0 - Q^1 \alpha^1] \left(\frac{Q^0}{\sigma} + \frac{(Q^0)^3 - Q^0 (Q^1)^2}{R^2 \sigma^2} \right) \phi_B \right] \psi$$

$$= \frac{\sigma - \sqrt{\sigma}}{(1-\sigma)R^2} \left[[(Q^0)^2 - Q^1 Q^0 \alpha^1] i\hbar \partial_t + Q^1 Q^0 i\hbar c \frac{\partial}{\partial \tilde{x}^1} - (Q^1)^2 i\hbar c \alpha^1 \frac{\partial}{\partial \tilde{x}^1} \right.$$

$$\left. + e[Q^0 - Q^1 \alpha^1] \frac{Q^0}{\sigma^2} \phi_B \right] \psi, \tag{7.138}$$

where

$$\frac{Q^0}{\sigma} + \frac{(Q^0)^3 - Q^0 (Q^1)^2}{R^2 \sigma^2} = Q^0 \left(\frac{1}{\sigma} + \frac{1-\sigma}{\sigma^2} \right) = \frac{Q^0}{\sigma^2}$$

was used.

(5) Therefore, substituting Eqs.(7.135) and (7.138) into Eq.(7.128), we have

$$i\hbar \left(\sqrt{\sigma} + \frac{\sigma - \sqrt{\sigma}}{(1-\sigma)R^2} [(Q^0)^2 - Q^0 Q^1 \alpha^1] \right) \partial_t \psi$$

$$= \left\{ -i\hbar\sqrt{\sigma}c\vec{\alpha} \cdot \nabla_B - \left[\frac{1}{\sqrt{\sigma}} \right. \right.$$

$$\left. + \frac{(Q^0)^2}{R^2 \sigma \sqrt{\sigma}} + \frac{\sigma - \sqrt{\sigma}}{(1-\sigma)R^2} \frac{(Q^0)^2}{\sigma^2} \right] e\phi_B + m_e c^2 \beta \right\} \psi$$

$$+ \left\{ -i\hbar c\sqrt{\sigma} \left(\frac{1}{\sqrt{1+(Q^1)^2/R^2}} - 1 \right) \alpha^1 \frac{\partial}{\partial \tilde{x}^1} + \frac{Q^1 Q^0}{R^2 \sigma \sqrt{\sigma}} \alpha^1 e\phi_B \right.$$

$$\left. - \frac{\sigma - \sqrt{\sigma}}{(1-\sqrt{\sigma})R^2} \left[i\hbar c(Q^1 Q^0 - (Q^1)^2 \alpha^1) \frac{\partial}{\partial \tilde{x}^1} - \frac{Q^1 (Q^0)^2}{\sigma^2} \alpha^1 e\phi_B \right] \right\} \psi.$$

$$(7.139)$$

In order to discuss the spectra of Hydrogen atom in the dS-SR Dirac equation, we need to find its solutions with certain energy E for the electron in the atom. From Eq.(7.64) in Section 7.6, we have

$$i\hbar\partial_t\psi = \frac{E}{1-(Q^0)^2/R^2}\psi + \frac{i\hbar c Q^1 Q^0}{R^2(1-(Q^0)^2/R^2)} \frac{1}{\sqrt{1+(Q^1)2/R^2}} \frac{\partial}{\partial \tilde{x}^1}\psi.$$

Then substituting Eq.(8.106) into Eq.(8.104) gives

$$E\psi = H_0\psi + H'\psi \qquad (7.140)$$

$$\text{with } H_0 = \left(1 - \frac{(Q^0)^2}{R^2} \right) \left[\sqrt{\sigma} + \frac{(\sigma - \sqrt{\sigma})(Q^0)^2}{(1-\sigma)R^2} \right]^{-1}$$

$$\times \left\{ -i\hbar\sqrt{\sigma}c\vec{\alpha} \cdot \nabla_B - \left[\frac{1}{\sqrt{\sigma}} + \frac{(Q^0)^2}{R^2 \sigma \sqrt{\sigma}} \right. \right.$$

$$\left. \left. + \frac{\sigma - \sqrt{\sigma}}{(1-\sigma)R^2} \frac{(Q^0)^2}{\sigma^2} \right] e\phi_B + m_e c^2 \beta \right\}, \qquad (7.141)$$

$$H' = \left(1 - \frac{(Q^0)^2}{R^2} \right) \left[\sqrt{\sigma} + \frac{(\sigma - \sqrt{\sigma})(Q^0)^2}{(1-\sigma)R^2} \right]^{-1}$$

$$\left\{ -i\hbar c\sqrt{\sigma} \left(\frac{1}{\sqrt{1+(Q^1)^2/R^2}} - 1 \right) \alpha^1 \frac{\partial}{\partial \tilde{x}^1} + \frac{Q^1 Q^0}{R^2 \sigma \sqrt{\sigma}} \alpha^1 e\phi_B \right.$$

$$-\frac{\sigma - \sqrt{\sigma}}{(1-\sigma)R^2}\left(i\hbar c(Q^1 Q^0 - (Q^1)^2\alpha^1)\frac{\partial}{\partial \tilde{x}^1} - \frac{Q^1(Q^0)^2}{\sigma^2}\alpha^1 e\phi_B\right)$$

$$+\frac{(\sigma - \sqrt{\sigma})Q^0 Q^1 \alpha^1}{(1-(Q^0)^2/R^2)(1-\sigma)R^2}\left(E + i\hbar c\frac{Q^0 Q^1}{R^2\sqrt{1+(Q^1)^2/R^2}}\frac{\partial}{\partial \tilde{x}^1}\right)\Bigg\}$$

$$-\frac{i\hbar c Q^1 Q^0}{R^2\sqrt{1+(Q^1)^2/R^2}}\frac{\partial}{\partial \tilde{x}^1}, \tag{7.142}$$

where ϕ_B was given in Eq.(D.12). Equation (7.140) is dS-SR Dirac wave equation to order of $\mathcal{O}(\frac{c^2 t^2}{R^2})$.

(6) From Eqs.(7.142) and (7.6), the H' up to order of $\mathcal{O}(\frac{1}{R^4})$ reads

$$H' \simeq \frac{i}{8}c\hbar\left(\frac{(Q^0)^2(Q^1)^2 + 2(Q^1)^4}{R^4}\right)\alpha^1\frac{\partial}{\partial \tilde{x}^1}$$

$$+\frac{Q^0 Q^1}{2R^2}\left(1 - \frac{(Q^0)^2 + 3(Q1)^2}{4R^2}\right)\frac{e^2}{r_B}\alpha^1$$

$$-\frac{Q^0 Q^1}{2R^2}\left(1 + \frac{3(Q^0)^2 - (Q^1)^2}{4R^2}\right)\alpha^1 H_0(e_t)$$

$$-\frac{ic\hbar Q^0 Q^1}{2R^2}\left(1 + \frac{(Q^0)^2 - 3(Q^1)^2}{4R^2}\right)\frac{\partial}{\partial \tilde{x}^1}. \tag{7.143}$$

A further approximation of neglecting $\mathcal{O}(\frac{1}{R^4})$-terms in Eq.(7.143) gives

$$H' \simeq \frac{Q^1 Q^0}{2R^2}\left(-\alpha^1 H_0(r_B, \hbar, m_e, \sqrt{2}e) - i\hbar c\frac{\partial}{\partial \tilde{x}^1}\right). \tag{7.144}$$

This is the same as the H'-expression of Eq.(7.91) with Eqs.(7.92) and (7.93).

7.10 Universal constant variations over cosmological time

Now we base on the Dirac equation describing the distance Hydrogen atom in cosmology Eq.(7.140) with Eq.(7.141) to discuss the time-variations of $\hbar$, e, m_e and $\alpha \equiv e^2/(\hbar c)$. Under the adiabatic approximation, the unperturbed Dirac equation for a distant Hydrogen atom should be

$$i\hbar\frac{\partial}{\partial t}\psi \equiv E_0\psi = \left\{-i\hbar_t c\vec{\alpha}\cdot\nabla_B + m_e^{(t)}c^2\beta - \frac{e_t^2}{r_B}\right\}\psi, \quad \text{with } \alpha_t = \frac{e_t^2}{\hbar_t c}, \tag{7.145}$$

where the subscript (or superscript) t indicates the back-looking time in cosmology. Substituting Eqs.(7.141) and (D.12) into Eq.(7.140) gives

$$
E_0\psi = H_0\psi = \left(1 - \frac{(Q^0)^2}{R^2}\right)\left[\sqrt{\sigma} + \frac{(\sigma - \sqrt{\sigma})(Q^0)^2}{(1-\sigma)R^2}\right]^{-1}
$$

$$
\times \Bigg\{ -i\hbar\sqrt{\sigma}c\vec{\alpha}\cdot\nabla_B + m_e c^2\beta
$$

$$
- \frac{\left[\frac{1}{\sqrt{\sigma}} + \frac{(Q^0)^2}{R^2\sigma\sqrt{\sigma}} + \frac{\sigma-\sqrt{\sigma}}{(1-\sigma)R^2}\frac{(Q^0)^2}{\sigma^2}\right]}{\sqrt{\left[\left(\frac{1}{\sigma} - \frac{(Q^1)^2}{R^2\sigma^2}\right)\left(\frac{1}{\sigma} + \frac{(Q^0)^2}{R^2\sigma^2}\right) + \frac{(Q^0)^2(Q^1)^2}{(R^4\sigma^4)}\right]\left(1 + \frac{(Q^1)^2}{R^2}\right)}} \frac{e^2}{r_B}\Bigg\}\psi
$$

$$
(7.146)
$$

Comparing Eq.(7.145) with Eq.(7.146), we obtain

$$
\hbar_t = \left(1 - \frac{(Q^0)^2}{R^2}\right)\left[\sqrt{\sigma} + \frac{(\sigma - \sqrt{\sigma})(Q^0)^2}{(1-\sigma)R^2}\right]^{-1}\sqrt{\sigma}\,\hbar\,,
\tag{7.147}
$$

$$
m_e^{(t)} = \left(1 - \frac{(Q^0)^2}{R^2}\right)\left[\sqrt{\sigma} + \frac{(\sigma - \sqrt{\sigma})(Q^0)^2}{(1-\sigma)R^2}\right]^{-1}m_e\,,
\tag{7.148}
$$

$$
e_t = \left(\frac{\left(1-\frac{(Q^0)^2}{R^2}\right)\left[\sqrt{\sigma}+\frac{(\sigma-\sqrt{\sigma})(Q^0)^2}{(1-\sigma)R^2}\right]^{-1}\left[\frac{1}{\sqrt{\sigma}} + \frac{(Q^0)^2}{R^2\sigma\sqrt{\sigma}} + \frac{\sigma-\sqrt{\sigma}}{(1-\sigma)R^2}\frac{(Q^0)^2}{\sigma^2}\right]}{\sqrt{\left[\left(\frac{1}{\sigma} - \frac{(Q^1)^2}{R^2\sigma^2}\right)\left(\frac{1}{\sigma} + \frac{(Q^0)^2}{R^2\sigma^2}\right) + \frac{(Q^0)^2(Q^1)^2}{(R^4\sigma^4)}\right]\left(1 + \frac{(Q^1)^2}{R^2}\right)}}\right)^{1/2} e\,,
\tag{7.149}
$$

and

$$
\alpha_t = \frac{\frac{1}{\sqrt{\sigma}} + \frac{(Q^0)^2}{R^2\sigma\sqrt{\sigma}} + \frac{\sigma-\sqrt{\sigma}}{(1-\sigma)R^2}\frac{(Q^0)^2}{\sigma^2}}{\sqrt{\sigma}\sqrt{\left[\left(\frac{1}{\sigma} - \frac{(Q^1)^2}{R^2\sigma^2}\right)\left(\frac{1}{\sigma} + \frac{(Q^0)^2}{R^2\sigma^2}\right) + \frac{(Q^0)^2(Q^1)^2}{(R^4\sigma^4)}\right]\left(1 + \frac{(Q^1)^2}{R^2}\right)}}\,\alpha\,.
\tag{7.150}
$$

We address that the un-perturbed part $H_0\psi$ (7.146) of full Dirac equation (7.140) has included all $1/R$-expansion terms. So Eqs.(7.147)–(7.150) are complete expressions for $\hbar_t$, $m_e^{(t)}$, e_t and α_t respectively, which have included all terms in the $1/R$-expansions. Noting Eq.(7.6), it is straightforward to complete the $1/R^2$-taylor expansions for $\hbar_t$, $m_e^{(t)}$, e_t and α_t

respectively. Up to order-$1/R^4$ terms, the results are follows

$$\hbar_t = \left[1 - \frac{(Q^0)^2}{2R^2} + \frac{1}{8R^4}\left(-(Q^0)^4 - (Q^0)^2(Q^1)^2\right) + \mathcal{O}(1/R^6)\right]\hbar, \quad (7.151)$$

$$m_e^{(t)} = \left[1 - \frac{(Q^1)^2}{2R^2} + \frac{1}{8R^4}\left(-5(Q^0)^2(Q^1)^2 + 3(Q^1)^4\right) + \mathcal{O}(1/R^6)\right]m_e, \quad (7.152)$$

$$e_t = \left[1 - \frac{(Q^0)^2}{4R^2} - \frac{3(Q^0)^4}{32R^4} + \mathcal{O}(1/R^6)\right]e, \quad (7.153)$$

$$\alpha_t = \left[1 + \frac{1}{8R^4}(Q^0)^2(Q^1)^2 + \mathcal{O}(1/R^6)\right]\alpha. \quad (7.154)$$

(The calculations of these equations can be easily checked by "*Mathematica*". See Appendix A.) If we dropped all $\mathcal{O}(1/R^4)$-terms in Eqs.(7.151)–(7.154), the above results are exactly the same as Eqs.(7.68)–(7.71). These are the main results of this chapter. Especially, Eq.(7.154) indicates that $\alpha_t \equiv \alpha_z \neq \alpha$ up to $\mathcal{O}(1/R^4)$. This conclusion is different from the $\mathcal{O}(1/R^2)$-result $\alpha_t = \alpha$ of Eq.(7.71). Namely, when one observes the absorbing (or radiating) spectra of Hydrogen or Hydrogen-like atoms at distant galaxies with redshift z, it will be found that the observed fine-structure constant α_z will not be equal to the value of α in the laboratory on the earth.

Chapter 8

Temporal and Spatial Variation of the Fine Structure Constant

This chapter focuses the space-time variation of the fine-structure constant in the cosmology, a phenomenon exhibiting one of the physical effects of de Sitter invariant special relativity (dS-SR).

8.1 Methods to search for changes of fine-structure constant

Quasar absorption system: A Quasar (or Quasi-Stellar Object (QSO)) is a celestial body with very high luminosity and large red-shifts in sky. Some quasar sight lines observed by us may pass through some gas clouds in the Universe. In this case, spectroscopic observations of gas clouds seen in absorption against background quasars can be used to search for the *absorption spectrums* of atoms, ions and molecules in the gas clouds. Comparing these observational spectra with the corresponding spectra in the Earth laboratories, we can learn that (i) the Hubble's red shift of the gas-cloud $z \equiv \Delta\nu/\nu$ (see Eq.(5.16)) which is the same for any frequencies; (ii) the relative frequency shifts between the energy levels of atoms (or ions, molecules) in the gas-cloud. The latter is related to measurements of the space-time variation of the fine-structure constant in the cosmology. Such systems are called *quasar absorption systems*. Quasar absorption systems present ideal laboratories for seeking any temporal or spatial variation of fundamental constants by comparing atomic spectra from the distant objects with laboratory spectra on earth.

Methods to search α-variation via astro-observations:

We note that one needs to measure the *relative size* of relativistic corrections to atomic energy levels in order to determine the α-variation. To see this point, let us explain in the following. The energy scale of atomic spectra, in the nonrelativistic limit, is given by the atomic unit $me^4/\hbar^2$. In this limit, all atomic spectra are proportional to this quantity and no change of the fundamental constants can be detected. Indeed, any change in the atomic unit will be absorbed in the determination of the redshift parameter $1 + z = \omega/\omega_z$ (ω_z is the redshifted frequency of the atomic transition, and ω is this frequency in the laboratory). However, any change in the fundamental constants can be found by measuring the *relative size* of relativistic corrections, which are functions of α^2, where $\alpha \equiv e^2/(\hbar c)$ is the fine structure constant.

The most straightforward way to look for the variation of α is to measure the ratio of some fine structure interval to an optical transition frequency, such as $\omega(np_{1/2} \to np_{3/2})$ and $\omega(n's_{1/2} \to np_{3/2})$. This ratio can be roughly estimated as $0.2\alpha^2 Z^2$, where Z is the nuclear charge [Sobelman (1979)]. Therefore, any difference in this ratio between a laboratory experiment and a measurement for some distant astrophysical object can be easily converted into the spacetime variation of α. Among them, fine structure interval frequency $\omega(np_{1/2} \to np_{3/2})$ could be is found as a difference: $\omega(np_{1/2} \to np_{3/2}) = \omega(n's_{1/2} \to np_{3/2}) - \omega(n's_{1/2} \to np_{1/2})$. Yet, the accuracy of this method is relatively low. Namely the sensitivity of the ratio to α^2 (i.e., $0.2Z^2$) seems small. Consequently, the measurement of the ratio is not very sensitive to α-variation. So, it is less-useful in the detections of α-variations in practice.

A great improvement on the above method is achieved by the Many-Multiplet (MM) method proposed by [Dzuba, Flambaum, Webb (1999a)] [Dzuba, Flambaum, Webb (1999b)]. One can gain about an order of magnitude in the sensitivity to the α-variation by comparing optical transitions for different atoms. Now we explain it. In this case the frequency of each transition can be expanded in a series in α^2 :

$$\omega_i = \omega_i^{(0)} + \omega_i^{(2)}\alpha^2 + \cdots , \tag{8.1}$$

$$= \omega_{i\,\text{lab}} + q_i x + \cdots , \qquad x \equiv (\alpha/\alpha_0)^2 - 1, \tag{8.2}$$

where α_0 stands for the laboratory value of the fine structure constant.

Note that Eq.(8.1) corresponds to the expansion at $\alpha = 0$, while Eq.(8.2) at $\alpha = \alpha_0$. In both cases, parameters $\omega_i^{(2)}$ and q_i appear due to relativistic corrections. We note that both $\omega_{i\,\text{lab}}$ and q_i are calculable by the Dirac-Hartree-Fock method and the quantum many body perturbation theories (and $\omega_{i\,\text{lab}}$ are of course measurable in laboratories).

For a fine structure transition the first coefficient on the right hand side of Eq.(8.1) turns to zero, while for the optical transitions it does not. Thus, for the case of a fine structure and an optical transition one can write:

$$\frac{\omega_{\text{fs}}}{\omega_{\text{op}}} = \frac{\omega_{\text{fs}}^{(2)}}{\omega_{\text{op}}^{(0)}}\alpha^2 + \mathcal{O}(\alpha^4), \tag{8.3}$$

while for two optical transitions i and k the ratio is:

$$\frac{\omega_i}{\omega_k} = \frac{\omega_i^{(0)}}{\omega_k^{(0)}} + \left(\frac{\omega_i^{(2)} - \omega_k^{(2)}}{\omega_k^{(0)}}\right)\alpha^2 + \mathcal{O}(\alpha^4). \tag{8.4}$$

More often than not, the coefficients $\omega_i^{(2)}$ for optical transitions are about an order of magnitude larger than corresponding coefficients for the fine structure transitions $\omega_{\text{fs}}^{(2)}$ (this is because the relativistic correction to a ground state electron energy is substantially larger than the spin-orbit splitting in an excited state [Dzuba, Flambaum, Webb (1999a)] [Dzuba, Flambaum, Webb (1999b)]). Therefore, the ratio (8.4) is, in general, more sensitive to the variation of α than the ratio (8.3).

Let's illustrate how to determine the distant $\alpha_z \equiv \alpha$ by using MM method. Suppose i stand for transition of Mg I ($3s^2\ {}^1S_0 \to 3s3p\ {}^1P_1$), from expansion expression (8.2) and the calculations in [Dzuba, Flambaum, Webb (1999b)] we have

$$\omega_i \equiv \omega_{i\,\text{obs}} = \omega_{i\,\text{lab}} + q_i x + q_i' x^2 + \dots$$

$$= 35051.277(1) + 106x - 10y, \quad y = \left(\frac{\alpha}{\alpha_l}\right)^4 - 1. \tag{8.5}$$

According to this relation, once $\omega_{i\,\text{obs}} = \omega_{\text{obs}}(\text{Mg I }(3s^2\ {}^1S_0 \to 3s3p\ {}^1P_1))$ is known by means of astro-observations, desired $\alpha_z \equiv \alpha$ will be determined via $x = (\alpha/\alpha_l)^2 - 1$. Since i can be any transitions from different species, we should combine them to get reliable α_z. Part of calculation results for different i are follows [Webb *et al.* (1999)]

$$\text{Mg II } {}^2P \; J = 1/2 : \; \omega = 35669.286(2) + 119.6x,$$
$$J = 3/2 : \; \omega = 35760.835(2) + 211.2x,$$
$$\text{Fe II } {}^6D \; J = 9/2 : \; \omega = 38458.9871(20) + 1394x + 38y,$$
$$J = 7/2 : \; \omega = 38660.0494(20)11632x10y,$$
$${}^6F \, J = 11/2 : \; \omega = 41968.0642(20) + 1622x + 3y,$$
$$J = 9/2 : \; \omega = 42114.8329(20) + 1772x + 0y,$$
$${}^6P \; J = 7/2 : \; \omega = 42658.2404(20) + 1398x - 13y. \qquad (8.6)$$

The key of MM method relies on the combination of transitions from different species. In particular, as can be seen from Eq.(8.6), some transitions are fairly insensitive to a change of the fine-structure constant (e.g., Mg II, hence providing good anchors) while others such as Fe II are more sensitive. The first implementation [Webb *et al.* (1999)] of the method was based on a measurement of the shift of the Fe II (the rest wavelengths of which are very sensitive to EM) spectrum with respect to the one of Mg II. This comparison increases the sensitivity compared with methods using only to measure the ratio of some fine structure interval to an optical transition frequency (e.g., see Eq.(8.3)).

8.2 Spacetime variations of fine-structure constant estimated via combining Keck /HIRES- and VLT /UVES-observation data

The Keck telescope (Mauna Kea, Hawaii) and VLT telescope (Paranal, Chile) locations on Earth are separated by 45^0 in latitude and hence, on average, observe different directions on the sky.

During more than one decade, analysis relied on 143 absorption systems observed in the Keck shows [Webb *et al.* (1999)] [Murphy *et al.* (2003)] that the robust estimates in MM-method is the weighted mean

$$\frac{\Delta\alpha}{\alpha_0} = (-0.57 \pm 0.11) \times 10^{-5}, \quad 0.2 < z < 4.2, \qquad (8.7)$$

where $\Delta\alpha \equiv \alpha_z - \alpha_0$ and $\alpha_0 \equiv \alpha_l$. This result indicates that α varies temporally, and could be smaller in the past time.

Further MM-method studies afterward using 140 absorption systems from the Keck telescope and 153 from the Very Large Telescope (VLT) suggest that α vary also spatially [Webb *et al.* (2011)] [King *et al.* (2012)]. That is, α could be smaller in one direction in the sky yet larger in the opposite direction at the time of absorption.

Specifically, authors of [Webb *et al.* (2011)] and [King *et al.* (2012)] found out that the angular distribution of $\Delta\alpha/\alpha_0$ in sky is:

$$\frac{\Delta\alpha}{\alpha_0} = A\cos\Theta, \qquad (8.8)$$

where amplitude $A \simeq (1.02\pm0.21)\times10^{-5}$, and Θ is the angle in sky between quasar sight line and the direction in the sky equatorial coordinates: {*right-ascension* : 17.4 ± 0.9h; *declination* : -58 ± 9 deg} (see Fig.8.1 and its captions). Distribution of Eq.(8.8) is called *dipole*, which is statistically preferred over a monopole-only model at the 4.1σ level in the above Keck plus VLT observations.

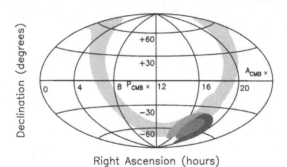

Fig. 8.1 (color online). All-sky plot in equatorial coordinates showing the independent Keck (green, leftmost) and VLT (blue, rightmost) best-fit dipoles, and the combined sample (red, center), for the dipole model, $\Delta\alpha/\alpha_0 = A\cos\Theta$, with $A = (1.02 \pm 0.21) \times 10^{-5}$. Approximate 1σ confidence contours are from the covariance matrix. The best fit dipole is at right ascension 17.4 ± 0.9h, declination -58 ± 9 deg, statistically preferred over a monopole-only model at the 4.1σ level. In this model, a bootstrap analysis shows the probability of the dipole alignments being as good or closer than observed is 6%. For a dipole + monopole this increases to 14%. The cosmic microwave background dipole are illustrated for comparison.

Figure 8.2 illustrates the $\Delta\alpha/\alpha_0$ binned data and the best-fit dipole + monopole model. Figure 8.3 illustrates $\Delta\alpha/\alpha_0$ vs look-back time distance projected onto the dipole axis, $r \cos \Theta$, using the best-fit dipole parameters for this model. This model seems to represent the data reasonably well and $\Delta\alpha/\alpha_0$ appears distance dependent, the correlation being significant at the 4.2σ level.

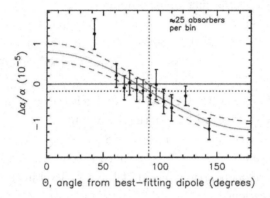

Fig. 8.2   (color online). $\Delta\alpha/\alpha_0$ for the combined Keck and VLT data vs angle Θ from the best-fit dipole position (best-fit parameters given in Fig.8.1 caption). Dashed lines illustrate $\pm1\sigma$ errors.

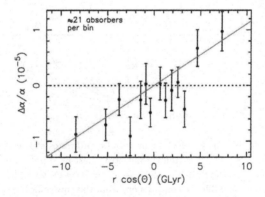

Fig. 8.3   (color online). $\Delta\alpha/\alpha_0$ vs $Ar \cos \Theta$ showing an apparent gradient in α along the best-fit dipole. The best-fit direction is at right ascension 17.4 ± 0.9h, declination -58 ± 9 deg, for which $A = (1.1\pm0.25) \times 10^{-6}\,\mathrm{GLyr}^{-1}$. A spatial gradient is statistically preferred over a monopole-only model at the 4.2σ level. A cosmology with parameters $(H_0, \Omega_{m0}, \Omega_{\Lambda0}) = (70.5, 0.274, 0.726)$ was used in Eq.(7.14).

Conclusions: Quasar spectra obtained using two separate observatories show a spatial variation in the relative spacings of absorption lines which could be attributed to an a dipole variation of α. A fit to the dipole gives a significance of $\geq 4.2\sigma$.

8.3 An overview of implication of spacetime variations of fine-structure constant

In Chapter 7, the dS-SR QM spectrum equation of distant Hydrogen atom in Cosmology has been discussed and solved, and especially the time variation of α was revealed via such discussions in Section 7.10. This is supposed to be a great hint of how to understand the implications of spatial and temporal variations of the fine-structure constant reported by observations [Webb *et al.* (2011)] [King *et al.* (2012)] (see the descriptions in Sections 8.1 and 8.2).

There are very few observations which can be directly and unambiguously related to new physics. The study of relative wavelength shifts in quasar absorption spectra at high redshift is indeed one of them as systematic achromatic shifts in these spectra can be attributed to changes in fundamental constants, and, in particular, in the fine-structure constant, α. This would most certainly call for new physics beyond the Standard Model.

Let's recall the reasons why the fine-structure constant α is unvarying over spacetime in the Standard Model (SM) of physics including cosmology. We examine the relativistic wave equation of an electron in Hydrogen in SM of physics. First, we consider the laboratory atom. From the viewpoint of cosmology, the energy level E of a free Hydrogen atom in laboratory is determined by the Dirac equation in a local inertial coordinates system located at the Earth in the Universe described by Ricci Friedmann–Robertson–Walker (RFW) metric. The spacetime metric of the local inertial system is Minkowski:

$$\{\eta_{\mu\nu}\} = \begin{pmatrix} 1 & 0 & 0 & 0 \\ 0 & -1 & 0 & 0 \\ 0 & 0 & -1 & 0 \\ 0 & 0 & 0 & -1 \end{pmatrix}, \tag{8.9}$$

which is spacetime independent. E satisfies Dirac spectrum equation:

$$E\psi = \left(-i\hbar c\boldsymbol{\alpha} \cdot \nabla - \frac{e^2}{r} + m_e c^2 \beta\right)\psi, \qquad (8.10)$$

where $\{\boldsymbol{\alpha} \equiv \alpha^1 \mathbf{i} + \alpha^2 \mathbf{j} + \alpha^3 \mathbf{k}, \ \beta\}$ are Dirac matrices, $\nabla \equiv (\partial/\partial x^1)\mathbf{i} + (\partial/\partial x^2)\mathbf{j} + (\partial/\partial x^3)\mathbf{k}$ and $r = \sqrt{(x^1)^2 + (x^2)^2 + (x^3)^2}$. This matrix-differential equation is integrable and the solution of the eigenvalue E is (see, e.g., [Strange (2008)])

$$E \equiv W_{n,\kappa} = m_e c^2 \left(1 + \frac{\alpha^2}{(n - |\kappa| + s)^2}\right)^{-1/2} \qquad (8.11)$$

$$\alpha \equiv \frac{e^2}{\hbar c}, \qquad |\kappa| = (j + 1/2) = 1, \ 2, \ 3 \cdots$$

$$s = \sqrt{\kappa^2 - \alpha^2}, \qquad n = 1, \ 2, \ 3 \cdots.$$

We keep in mind that the coefficients of operators $-i\boldsymbol{\alpha} \cdot \nabla$ and $-1/r$ in Eq.(8.10) are $\hbar c$ and e^2 respectively, and their ratio is the definition of α (see Eq.(8.11)).

Next, we consider a distant atom of Hydrogen located on the light-cone of RFW-Universe (see Fig.8.4), i.e., the nucleus coordinate is $Q^\mu(z) \equiv \{Q^0, \ \mathbf{Q}\}$, and electron's is $L^\mu(z) \equiv \{L^0, \ \mathbf{L}\}$. Noting that the metric of the local inertial coordinate system at Q^μ in RFW-Universe is still $\eta_{\mu\nu}$ (8.9) because of the spacetime-independency of $\eta_{\mu\nu}$ and denoting $L^\mu(z) - Q^\mu(z) \equiv x'^\mu$, the electron wave equation in the distant atom reads

$$E\psi = \left(-i\hbar c\boldsymbol{\alpha} \cdot \nabla' - \frac{e^2}{r'} + m_e c^2 \beta\right)\psi, \qquad (8.12)$$

where $\nabla' \equiv (\partial/\partial x'^1)\mathbf{i} + (\partial/\partial x'^2)\mathbf{j} + (\partial/\partial x'^3)\mathbf{k}$ and $r' = \sqrt{(x'^1)^2 + (x'^2)^2 + (x'^3)^2}$. Then we find out that

$$\alpha' \equiv \alpha_z = \frac{e^2}{\hbar c}. \qquad (8.13)$$

Comparing Eq.(8.13) with Eq.(8.11), we conclude that

$$\alpha_z = \alpha, \qquad (8.14)$$

which indicates that the fine-structure constant is unvarying indeed in SM of physics.

The argument above on α-unvarying in SM made by comparing Dirac equation for laboratory Hydrogen atom with that for a distant Hydrogen atom is deeply related to aspects of Special Relativity (SR), General Relativity (GR) and Cosmology. In other words, the α-varying phenomena reported in [Webb *et al.* (2011); King *et al.* (2012); Webb *et al.* (1999); Dzuba, Flambaum, Webb (1999a); Webb *et al.* (2001); Murphy *et al.* (2003)] implies some new physics beyond SM. Further remarks on this issue are follows:

(1) The spectrum equations (8.10) and (8.12) come from the following inhomogeneous-Lorentz (or Poincaré) invariant (i.e., $ISO(3,1)$) Dirac equation and Maxwell equation in local inertial systems of RFW Universe:

$$(i\gamma^\mu D_\mu^L - \frac{m_e c}{\hbar})\psi = 0, \tag{8.15}$$

$$F^{\mu\nu}{}_{,\nu} = j^\mu = -\delta^{\mu 0} 4\pi e \delta^{(3)}(\mathbf{x}), \tag{8.16}$$

where $D_\mu^L = \frac{\partial}{\partial L^\mu} - ie/(c\hbar)\eta_{\mu\nu}A^\nu$, and the electromagnetic potential $A^\nu \equiv \{\phi = e/r, \mathbf{A}\}$. So, the operator structure of Eqs.(8.10), (8.12) and a dimensionless combination of universal constants $\alpha = e^2/(\hbar c)$ are rooted in the symmetry assumption of the theory.

(2) The point that α is unvarying is deduced from the constancy of the adopted metric of local inertial coordinate system in the RFW Universe (i.e., $\{\eta_{\mu\nu}\}$ =const.). So, the fact of the α-varying in real world reported in [Webb *et al.* (2011); King *et al.* (2012); Webb *et al.* (1999); Dzuba, Flambaum, Webb (1999a); Webb *et al.* (2001); Murphy *et al.* (2003)] indicates that the metric of local inertial coordinate system in the real Universe may be spacetime-dependent.

(3) Minkowski metric $\eta_{\mu\nu} = \text{diag}\{+,-,-,-\}$ is the basic spacetime metric of E-SR in SM. The most general transformation to preserve metric $\eta_{\mu\nu}$ is Poincaré group (or inhomogeneous Lorentz group $ISO(1,3)$). It is well known that the Poincaré group is the limit of the de Sitter group with pseudo-sphere radius $|R| \to \infty$. Therefore, E-SR may possibly be extended to a SR theory with de Sitter spacetime symmetry. Since P.A.M. Dirac's pioneering work in 1935 [Dirac (1935)] many discussions (e.g., E. Inönü and E. P. Wigner in 1968 [Inönü, Wigner (1953)]; F. Gürsey and T.D. Lee in 1963 [Gürsey, Lee (1963)], etc.) advocated such a possible extension of E-SR. In 1970's, K.H. Look (Qi-Keng Lu) and his collaborators Z.L. Zou, H.Y. Guo suggested the

de Sitter Invariant Special Relativity (dS-SR) [Lu (1970)][Lu, Zou and Guo (1974)] (see [Guo, Huang, Xu and Zhou (2004a); Yan, Xiao, Huang and Li (2005)] and Appendix in [Sun, Yan, Deng, Huang, Hu (2013)] for the English version). It has been proved that Lu-Zou-Guo's dS-SR is a satisfying and self-consistent special relativity theory. In 2005, one of us (MLY) and Xiao, Huang, Li suggested dS-SR Quantum Mechanics (QM) [Yan, Xiao, Huang and Li (2005)].

(4) Beltrami metric (see Appendix A)

$$B_{\mu\nu}(x) = \eta_{\mu\nu}/\sigma(x) + \eta_{\mu\lambda}x^\lambda\eta_{\nu\rho}x^\rho/(R^2\sigma(x)^2), \qquad (8.17)$$

$$\text{with } \sigma(x) = 1 - \eta_{\mu\nu}x^\mu x^\nu/R^2 > 0$$

is the basic metric of dS-SR with Minkowski point coordinates $M^\mu = 0$ (i.e., $B_{\mu\nu}(x)|_{x=M=0} = \eta_{\mu\nu}$) [Yan, Xiao, Huang and Li (2005)]. Both $\eta_{\mu\nu}$ and $B_{\mu\nu}$ lead to the inertial motion law for free particles $\ddot{\mathbf{x}} = 0$, which is the precondition to define inertial reference systems required by special relativity theories. However, the $B_{\mu\nu}$-preserving coordinate transformation group is de Sitter group $SO(4,1)$ (or $SO(3,2)$) rather than E-SR's inhomogeneous Lorentz group $ISO(1,3)$ [Lu (1970); Lu, Zou and Guo (1974); Guo, Huang, Xu and Zhou (2004a); Yan, Xiao, Huang and Li (2005)], different from the case of $\eta_{\mu\nu}$. In addition, $\eta_{\mu\nu}$ does not satisfy the Einstein equation with Λ (Einstein cosmology constant) in vacuum, but $B_{\mu\nu}(x)$ does satisfy it (see below). Generally, when $M^\mu \neq 0$, the basic metric of dS-SR is modified to be

$$B_{\mu\nu}^{(M)}(x) \equiv B_{\mu\nu}(x - M) = \frac{\eta_{\mu\nu}}{\sigma^{(M)}(x)} + \frac{\eta_{\mu\lambda}(x^\lambda - M^\lambda)\eta_{\nu\rho}(x^\rho - M^\rho)}{R^2\sigma^{(M)}(x)^2},$$

$$(8.18)$$

where

$$\sigma^{(M)}(x) \equiv \sigma(x - M) = 1 - \frac{\eta_{\mu\nu}(x^\mu - M^\mu)(x^\nu - M^\nu)}{R^2}, \qquad (8.19)$$

which will be called Modified Beltrami metric, or M-Beltrami metric. Based on $B_{\mu\nu}^{(M)}(x)$, the dS-invariant special relativity with $M^\mu \neq 0$ can be built. The procedures and formulation are similar to ordinary dS-SR in [Lu (1970); Lu, Zou and Guo (1974); Guo, Huang, Xu and Zhou (2004a); Yan, Xiao, Huang and Li (2005)], which is actually a slight extension of usual dS-SR (see Appendix B). It is essential, however, that $B_{\mu\nu}^{(M)}(x)$ is spacetime dependent and has more parameters $\{R, M^\mu\}$, which may provide a possible clue to solve the puzzle of α-varying.

(5) Our strategy in this chapter for solving this puzzle is to pursue the following dS-SR Dirac equation for both electron in laboratory Hydrogen and electron in distant Hydrogen in the RFW Universe (see Eq.(25) in [Yan (2012)]):

$$(ie_a{}^\mu \gamma^a \mathcal{D}^L_\mu - \frac{m_e c}{\hbar})\psi = 0, \qquad (8.20)$$

where $\mathcal{D}^L_\mu = \frac{\partial}{\partial L^\mu} - \frac{i}{4}\omega^{ab}{}_\mu \sigma_{ab} - ie/(c\hbar)B^{(M)}_{\mu\nu}A^\nu$, $e_a{}^\mu$ is the tetrad, $\omega^{ab}{}_\mu$ is spin-connection, and the electromagnetic potential $A^\nu \equiv \{\phi_B, \mathbf{A}\}$. Unlike E-SR Dirac equation (8.15), the spacetime symmetry of (8.20) is de Sitter invariant group $SO(4,1)$ (or $SO(3,2)$) instead of former $ISO(3,1)$. Specifically, the dS-SR Dirac spectrum equation can be deduced from (8.20). It is essential that the result will be different form E-SR equation (8.10). Following the method used in Eqs.(8.10) and (8.11), the coefficients of resulting $(-i\boldsymbol{\alpha} \cdot \nabla)$-type and $(-1/r)$-type operator terms in the dS-SR Dirac spectrum equation are of $\hbar_z(\boldsymbol{\Omega})c$ and $e_z(\boldsymbol{\Omega})^2$ respectively. Then their ratio yields prediction of $\alpha_z(\boldsymbol{\Omega}) \equiv e_z(\boldsymbol{\Omega})^2/(\hbar_z(\boldsymbol{\Omega})c)$. The adjustable parameters in this model are R and the position of Minkowski point M^μ. For simplicity, we take $M^\mu = \{M^0, M^1, 0, 0\}$. It turns out to be a good a choice for solution to the puzzle of α-varying.

(6) Different from Quantum Mechanics (QM) wave equation (8.15) deduced from $\eta_{\mu\nu}$, Eq.(8.20) is actually a time-dependent Hamiltonian problems in QM. This is because $B^{(M)}_{\mu\nu}(x)$ is time-dependent. Therefore the corresponding Lagrangian L_{dS} (see Eq.(3.32)) and hence Hamiltonian is time-dependent [Yan, Xiao, Huang and Li (2005)]. In this chapter, the adiabatic approach [Born, Fock (1928)][Messiah (1970)][Bayfield (1999)] will be used to deal with the time-dependent Hamiltonian problems in dS-SR QM. Generally, to a $H(x,t)$, we may express it as $H(x,t) = H_0(x) + H'(x,t)$. Suppose two eigenstates $|s\rangle$ and $|m\rangle$ of $H_0(x)$ are not degenerate, i.e., $\Delta E \equiv \hbar(\omega_m - \omega_s) \equiv \hbar\omega_{ms} \neq 0$. The validity of adiabatic approximation relies on the fact that the variation of the potential $H'(x,t)$ in the the Bohr time-period $(\Delta T^{(Bohr)}_{ms})\dot{H}'_{ms} = (2\pi/\omega_{ms})\dot{H}'_{ms}$ is much less than $\hbar\omega_{ms}$, where $H'_{ms} \equiv \langle m|H'(x,t)|s\rangle$. That makes the quantum transition from state $|s\rangle$ to state $|m\rangle$ almost impossible. Thus, the non-adiabatic effect corrections are small enough (or tiny), and the adiabatic approximations are legitimate. For

the wave equation of dS-SR QM of atoms discussed in this chapter, we show that the perturbation Hamiltonian describes the time evolutions of the system $H'(x, t) \propto (c^2 t^2 / R^2)$ (where t is the cosmic time). Since R is cosmologically large and $R >> ct$, the factor $(c^2 t^2 / R^2)$ will make the time-evolution of the system so slow that the adiabatic approximation works. In Chapter 7 we have provided a explicit calculations to confirm this point (see Section 7.7). By this approach, we solve the stationary dS-SR Dirac equation for one electron atom, and the spectra of the corresponding Hamiltonian with time-parameter are obtained. Consequently, we find out that the electron mass m_e, the electric charge e, the Planck constant $\hbar$ and the fine structure constant $\alpha = e^2/(\hbar c)$ vary as cosmic time goes by. These are interesting consequences since they indicate that the time-variations of fundamental physics constants might be attributed to quantum evolutions of time-dependent quantum mechanics that has been widely discussed for a long history (e.g., see [Bayfield (1999)] and the references within).

(7) Finally, we argue that it is reasonable to assume that the Beltrami metric is the appropriate metric for the spacetime of the local inertial system in real world. If we express the total energy momentum tensor $T_{\mu\nu}$ as the sum of a possible vacuum term $-\rho_{(v)} g_{\mu\nu}$ and a term $T_{\mu\nu}^M$ arising from matter (including radiation), then the complete Einstein equation is [Peebles (2009)][Padmanabhan (2009)][Yan, Hu, Huang (2012)]:

$$\mathcal{R}_{\mu\nu} - \frac{1}{2} g_{\mu\nu} \mathcal{R} + \Lambda g_{\mu\nu} = -8\pi G T_{\mu\nu}^M - 8\pi G \rho_{(v)} g_{\mu\nu}, \qquad (8.21)$$

where $\rho_{(v)}$ is the dark energy density, so $\Lambda_{\text{dark energy}} = 8\pi G \rho_{(v)}$, and Λ is originally introduced by Einstein in 1917, and serves as a universal constant in physics. We call it the Einstein (or geometry) cosmological constant . The effective cosmologic constant $\Lambda_{eff} = \Lambda + \Lambda_{\text{dark energy}} \simeq 1.26 \times 10^{-56}$ cm^{-2} is the observed value determined via effects of accelerated expansion of the universe [Riess, *et al.* (1998)] and the recent WMAP data [Jarosik, *et al.* (2011)]. We have no any *a priori* reason to assume the geometry cosmologic constant to be zero, so the vacuum Einstein equation is:

$$\mathcal{R}_{\mu\nu} - \frac{1}{2} g_{\mu\nu} \mathcal{R} + \Lambda g_{\mu\nu} = 0, \qquad (8.22)$$

instead of $G_{\mu\nu} = 0$, and hence the vacuum solution to Eq.(8.22) is $g_{\mu\nu} = B_{\mu\nu}^{(M)}(x)$ with $|R| = \sqrt{3/\Lambda}$ instead of $g_{\mu\nu} = \eta_{\mu\nu}$. Therefore, we conclude that the metric of the local inertial coordinate system in real world should be Beltrami metric rather than Minkowski metric. Thus, the dS-SR Dirac equation (8.20) (instead of E-SR Dirac equation (8.15)) is legitimate to characterize the spectra in the real world, and then the α-varying over the real world space-time would occur naturally.

Exercise

Problem 1: Calculate the turn time Δt of electron to revolve around the proton in Hydrogen. (call Δt the basic Bohr-time interval).

Problem 2: Estimate Hubble red-shift change Δz during the basic Bohr-time interval Δt for $z = 3$.

8.4 Light-cone of Ricci Friedmann–Robertson–Walker universe

In Section 7.2, we described for simplicity the Light Cone of Ricci Friedmann–Robertson–Walker (RFW) Universe with $Q^0 \neq 0$, $Q^1 \neq 0$, $Q^2 = Q^3 = 0$. Now we describe it in 4-dimension spacetime.

We have already shown that the isotropic and homogeneous cosmology solution of Einstein equation in General Relativity (GR) is RFW metric (see Section 5.1). All visible quasars in sky must be located on the light-cone of RFW Universe (see Fig.8.4).

The Ricci Friedmann–Robertson–Walker (RFW) metric is (see, Eq.(5.8))

$$ds^2 = c^2 dt^2 - a(t)^2 \left\{ \frac{dr^2}{1 - kr^2} + r^2 d\theta^2 + r^2 \sin^2\theta d\phi^2 \right\}$$
$$= (dQ^0)^2 - a(t)^2 \left\{ dQ^i dQ^i + \frac{k(Q^i dQ^i)^2}{1 - kQ^i Q^i} \right\}$$
$$\equiv g_{\mu\nu}(Q) dQ^\mu dQ^\nu, \tag{8.23}$$

where $a(t)$ is scale (or expansion) factor and $r = \sqrt{Q^i Q^i} \equiv Q$, $Q^1 = Q \sin\theta\cos\phi$, $Q^2 = Q\sin\theta\sin\phi$, $Q^3 = Q\cos\theta$ and $(Q^0)^2 = c^2 t^2$. For

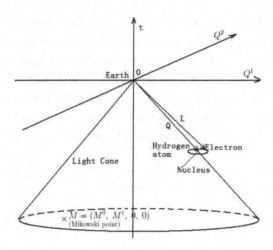

Fig. 8.4 Sketch of the light cone of the Ricci Friedmann–Robertson–Walker universe. Only 3 coordinate axes $\{Q^0 = ct,\ Q^1,\ Q^2\}$ are shown in this figure. The Q^3 axis could be imagined. The Earth is located at the origin. The position vector for nucleus of atom between the QSO and the Earth is $\mathbf{Q}$, and for electron is $\mathbf{L}$. The distance between nucleus and electron is $\bar{r} \sim |\mathbf{L} - \mathbf{Q}|$. The location of the Minkowski point of Beltrami metric is denoted by notation " $\times$ " with $M = (M^0, M^1, 0, 0)$.

the sake of convenience, we take t to be looking-back cosmological time, so that $t < 0$. As is well known RFW metric satisfies Homogeneity and Isotropy principle of present day cosmology. For simplicity, we take $k = 0$ and $a(t) = 1/(1 + z(t))$ (i.e., $a(t_0) = 1$). And the red shift function z is determined by ΛCDM model described in Section 5.4. The function $t(z)$ defined by Eq.(5.79) for looking-back cosmological time is:

$$t(z) = \int_z^0 \frac{dz'}{H(z')(1 + z')}, \tag{8.24}$$

where [Riess, *et al.* (1998)] [Jarosik, *et al.* (2011)]

$$H(z') = H_0 \sqrt{\Omega_{m0}(1 + z')^3 + \Omega_{R0}(1 + z')^4 + 1 - \Omega_{m0}},$$

$$H_0 = 100\,h \simeq 100 \times 0.705\,\text{km} \cdot \text{s}^{-1}/\text{Mpc},$$

$$\Omega_{m0} \simeq 0.274, \quad \Omega_{R0} \sim 10^{-5}. \tag{8.25}$$

Figure of $t(z)$ of Eq.(8.24) is shown in Fig.8.5.

The light-cone of RFW Universe is defined by $ds^2 = 0$. From Eq.(8.23), we have the light-cone equation:

$$(dQ^0)^2 - a(t)^2(dQ)^2 = 0, \quad \text{or} \quad -cdt = a(t)dQ = \frac{1}{1 + z(t)}dQ. \tag{8.26}$$

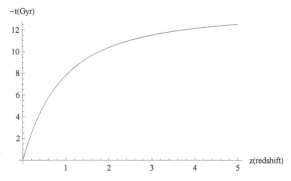

Fig. 8.5 The $t - z$ relation in ΛCDM model (Eq.(7.13)).

Substituting Eq.(8.24) into Eq.(8.26) gives

$$Q(z) = c \int_0^z \frac{dz'}{H(z')}. \tag{8.27}$$

Figure of $Q(z)$ of Eq.(8.27) is shown in Fig.8.6. Ratio of Q over Q^0 is shown in Fig.8.7.

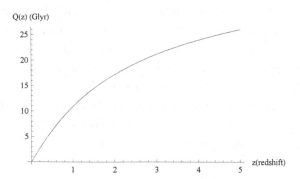

Fig. 8.6 Function $Q(z)$ in ΛCDM model (Eq.(8.27)).

8.5 Local inertial coordinate system in light-cone of RFW universe with Einstein cosmology constant

In principle, almost all calculations on quantum spectrums in atomic physics are achieved in the *inertial coordinate systems*. From the cosmological point of view, the phenomena of atomic spectrums should be described in the local inertial coordinate system s of RFW Universe. Therefore, we

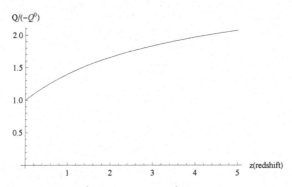

Fig. 8.7   Function of $Q(z)/Q^0(z)$. $Q(z)$ and $Q^0(z) = ct$ are given in Eqs.(8.27) and (7.13).

are interested in how to determine the local inertial coordinate system in light-cone of RFW Universe when the Einstein cosmology constant Λ is present.

Existence of *local inertial coordinate system* is required by the Equivalence Principle. The principle states that experiments in a sufficiently small falling laboratory, over a sufficiently short time, give results that are indistinguishable from those of the same experiments in an inertial frame in empty space of special relativity (see, e.g., *pp.119* in [Hartle (2003)]). Such a sufficiently small falling laboratory, over a sufficiently short time represents *a local inertial coordinates system*. This principle suggests that the local properties of curved spacetime should be indistinguishable from those of the spacetime with *inertial metric* of special relativity. A concrete expression of this ideal is the requirement that, given a metric $g_{\alpha\beta}$ in one system of coordinates x^α, at each point P of spacetime it is possible to introduce new coordinates x'^α such that

$$g'_{\alpha\beta}(x'_P) = \text{inertial metric of SR at } x'_P, \qquad (8.28)$$

and the connection at x'_P is the Christoffel symbols deduced from $g'_{\alpha\beta}(x'_P)$.

In usual Einstein's general relativity (without Λ), the above expression is

$$g'_{\alpha\beta}(x'_P) = \eta_{\alpha\beta}, \quad and \quad \Gamma^\lambda_{\alpha\beta} = 0, \qquad (8.29)$$

which satisfies the Einstein equation of E-GR in empty space: $G_{\mu\nu} = 0$.

In dS-GR (GR with a Λ), the local inertial coordinate system at $x_P'^{\alpha}$ is characterized by

$$g_{\alpha\beta}'(x_P') = B_{\alpha\beta}^{(M)}(x_P') \equiv \frac{\eta_{\mu\nu}}{\sigma^{(M)}(x_P')} + \frac{\eta_{\mu\lambda}(x'^{\lambda} - M^{\lambda})\eta_{\nu\rho}(x_P'^{\rho} - M^{\rho})}{R^2\sigma^{(M)}(x_P')^2},$$

$$\text{with } \sigma^{(M)}(x_P') = 1 - \frac{\eta_{\mu\nu}(x_P'^{\mu} - M^{\mu})(x_P'^{\nu} - M^{\nu})}{R^2}, \qquad (8.30)$$

$$\Gamma_{\alpha\beta}^{\lambda} = \frac{1}{2}(B^{(M)})^{\lambda\rho}(\partial_{\alpha}B_{\rho\beta}^{(M)} + \partial_{\beta}B_{\rho\alpha}^{(M)} - \partial_{\rho}B_{\alpha\beta}^{(M)})$$

$$= \frac{1}{R^2\sigma^{(M)}(x_P')}(\delta_{\mu}^{\lambda}\eta_{\nu\rho} + \delta_{\nu}^{\lambda}\eta_{\mu\rho})(x_P'^{\rho} - M^{\rho}), \qquad (8.31)$$

or

$$\frac{g_{\rho\beta}'(x_P')}{\partial x_P'^{\alpha}} = \frac{x_P'^{\nu} - M^{\nu}}{R^2\sigma^{(M)}(x_P')^2}\Big[2\eta_{\rho\beta}\eta_{\alpha\nu} + \eta_{\rho\alpha}\eta_{\beta\nu} + \eta_{\beta\alpha}\eta_{\rho\nu}$$

$$+\frac{4\eta_{\rho\mu}\eta_{\beta\nu}\eta_{\alpha\lambda}(x_P'^{\mu} - M^{\mu})(x_P'^{\lambda} - M^{\lambda})}{R^2\sigma^{(M)}(x_P')}\Big],$$

where $B_{\alpha\beta}^{(M)}(x_P')$ was given in Eq.(8.18), which satisfies the Einstein equation of dS-GR in empty spacetime: $G_{\mu\nu} + \Lambda g_{\mu\nu} = 0$ with $\Lambda = 3/R^2$. (Note $\eta_{\mu\nu}$ does not satisfy that equation, i.e., $G_{\mu\nu}(\eta) + \Lambda\eta_{\mu\nu} \neq 0$. So it cannot be the metric of the local inertial system in dS-GR with Λ).

In order to show the differences between the Local Inertial Systems (LIS) in E-GR and the LIS in dS-GR visually, we draw a sketch map Fig.8.8. In this figure, for simplicity, we use bold lines to represent a curved spacetime with $g_{\mu\nu}$ which satisfy $G_{\mu\nu} = \kappa T_{\mu\nu}$ for E-GR or $G_{\mu\nu} + \Lambda g_{\mu\nu} = \kappa T_{\mu\nu}$ for dS-GR , and use fine lines (or fine curves) to represent the tangent line ($\eta_{\mu\nu}$) or externally tangent circle ($B_{\mu\nu}$). The tangent point is in x_P. As usual, we could always think the tangent spacetime metric at the tangent point as the metric of the local inertial system at the tangent point. We can see that when $\Lambda \neq 0$, the solution of $G_{\mu\nu} + \Lambda g_{\mu\nu} = 0$ is $g_{\mu\nu} = B_{\mu\nu}$ instead of $g_{\mu\nu} = \eta_{\mu\nu}$, so the metric of tangent spacetime in Fig.8.8(b) is $B_{\mu\nu}$ rather than $\eta_{\mu\nu}$.

For the light cone of RFW Universe with Λ, the coordinate-components $Q^0(z) = ct(z)$ and $Q(z)$ have been shown in Eqs.(8.24) (or Fig.8.5) and (8.27) (or Fig.8.6) respectively. Therefore from Eq.(8.18), the spacetime metric of the local inertial coordinate system at position $Q(z)$ of the light

$$\mathbf{E-GR}: \Lambda = 0 \qquad \mathbf{dS-GR}: \Lambda \neq 0$$

$$g_{\mu\nu}(x): \ G_{\mu\nu} = \kappa T_{\mu\nu} \qquad g_{\mu\nu}(x): \ G_{\mu\nu} + \Lambda g_{\mu\nu} = \kappa T_{\mu\nu}$$

$$\eta_{\mu\nu} \ : \ G_{\mu\nu} = 0 \qquad B_{\mu\nu}(x): \ G_{\mu\nu} + \Lambda g_{\mu\nu} = 0$$

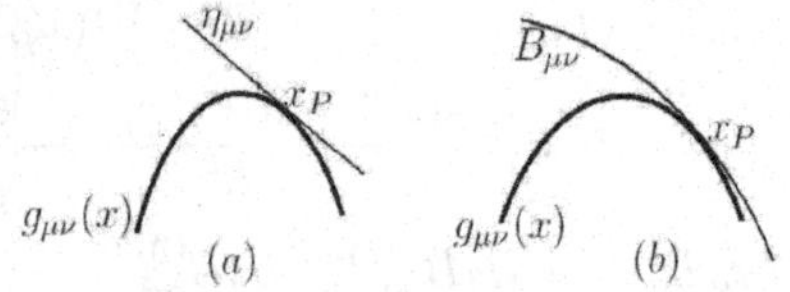

Fig. 8.8   Sketch map for tangent spacetime both in E-GR and in dS-GR. We could think the tangent spacetime metric at the tangent point as the metric of the local inertial system at the tangent point. Fig.(a): Bold line represents a curved spacetime with $g_{\mu\nu}$ satisfying $G_{\mu\nu} = \kappa T_{\mu\nu}$ in E-GR. The fine line is the tangent line with $(\eta_{\mu\nu})$. Tangent point is in x_P. The metric of the local inertial system for E-GR is $\eta_{\mu\nu}$. Fig.(b): Bold line represents a curved spacetime with $g_{\mu\nu}$ satisfying $G_{\mu\nu} + \Lambda g_{\mu\nu} = \kappa T_{\mu\nu}$ in dS-GR. We can see that when $\Lambda \neq 0$, the solution of $G_{\mu\nu} + \Lambda g_{\mu\nu} = 0$ is $g_{\mu\nu} = B_{\mu\nu}$ instead of $g_{\mu\nu} = \eta_{\mu\nu}$, so the metric of tangent spacetime in Fig.8.8(b) is $B_{\mu\nu}$ rather than $\eta_{\mu\nu}$.

cone is determined to be

$$B_{\mu\nu}^{(M)}(\mathcal{Q}) \equiv B_{\mu\nu}(\mathcal{Q}-M) = \frac{\eta_{\mu\nu}}{\sigma^{(M)}(\mathcal{Q})} + \frac{\eta_{\mu\lambda}(Q^\lambda - M^\lambda)\eta_{\nu\rho}(Q^\rho - M^\rho)}{R^2 \sigma^{(M)}(\mathcal{Q})^2}, \tag{8.32}$$

where

$$\sigma^{(M)}(\mathcal{Q}) \equiv \sigma(\mathcal{Q}-M) = 1 - \frac{\eta_{\mu\nu}(Q^\mu - M^\mu)(Q^\nu - M^\nu)}{R^2}. \tag{8.33}$$

We see from Fig.8.4 that the visible atom is embedded into the light cone at Q-point. Since $Q \simeq L$ (i.e., comparing with the Universe, atoms are very very small), we can reasonably treat the metric of the spacetime in the atomic region as a constant. This is just the adiabatic approximation adopted in [Yan (2012)]. When $Q^\mu \to M^\mu$, we have $B_{\mu\nu}^{(M)}(\mathcal{Q}) \Rightarrow \eta_{\mu\nu}$. So, $Q^\mu = M^\mu$ is the Minkowski point of the Beltrami metric $B_{\mu\nu}^{(M)}(\mathcal{Q})$.

Now let's derive $e_a{}^\mu$ and $\omega^{ab}{}_\mu$ from Eq.(8.32). Setting

$$q^\mu \equiv Q^\mu - M^\mu, \tag{8.34}$$

then Eqs.(8.32) and (8.33) become

$$B_{\mu\nu}^{(M)} = \frac{\eta_{\mu\nu}}{\sigma^{(M)}} + \frac{\eta_{\mu\lambda}q^\lambda \eta_{\nu\rho}q^\rho}{R^2 (\sigma^{(M)})^2}, \tag{8.35}$$

where

$$\sigma^{(M)} \equiv \sigma^{(M)}(q) = 1 - \frac{\eta_{\mu\nu} q^\mu q^\nu}{R^2}. \tag{8.36}$$

(denote the spacetime with metric $B_{\mu\nu}^{(M)}$ as $\mathcal{B}^M$). We introduce notations:

$$\bar{q}_\mu \equiv \eta_{\mu\lambda} q^\lambda, \quad q^\lambda \equiv \bar{q}^\lambda = \eta^{\mu\lambda} \bar{q}_\mu, \tag{8.37}$$

$$\bar{q}^2 \equiv \eta_{\mu\nu} q^\mu q^\nu = \bar{q}_\nu \bar{q}^\nu, \tag{8.38}$$

and construct two projection operators in spacetime $\{\bar{q}^\mu\}$ with metric $\eta_{\mu\nu}$:

$$\bar{\theta}_{\mu\nu} \equiv \eta_{\mu\nu} - \frac{\bar{q}_\mu \bar{q}_\nu}{\bar{q}^2}, \quad \bar{\omega}_{\mu\nu} \equiv \frac{\bar{q}_\mu \bar{q}_\nu}{\bar{q}^2}. \tag{8.39}$$

It is easy to check the calculation rules for projection operators:

$$\bar{\theta}_{\mu\lambda} \bar{\theta}^\lambda{}_\nu \equiv \bar{\theta}_{\mu\lambda} \eta^{\lambda\rho} \bar{\theta}_{\rho\nu} = \bar{\theta}_{\mu\nu}, \quad \text{or in short } \bar{\theta} \cdot \bar{\theta} = \bar{\theta}, \tag{8.40}$$

$$\bar{\omega}_{\mu\lambda} \bar{\omega}^\lambda{}_\nu \equiv \bar{\omega}_{\mu\lambda} \eta^{\lambda\rho} \bar{\omega}_{\rho\nu} = \bar{\omega}_{\mu\nu}, \quad \text{or in short } \bar{\omega} \cdot \bar{\omega} = \bar{\omega}, \tag{8.41}$$

$$\bar{\theta}_{\mu\lambda} \bar{\omega}^\lambda{}_\nu \equiv \bar{\theta}_{\mu\lambda} \eta^{\lambda\rho} \bar{\omega}_{\rho\nu} = 0, \quad \text{or in short } \bar{\theta} \cdot \bar{\omega} = 0, \tag{8.42}$$

$$\bar{\theta}_{\mu\nu} + \bar{\omega}_{\mu\nu} = \eta_{\mu\nu}, \quad \text{or in short } \bar{\theta} + \bar{\omega} = I. \tag{8.43}$$

$B_{\mu\nu}^{(M)}(Q)$ can be written as

$$B_{\mu\nu}^{(M)} = \frac{\eta_{\mu\nu}}{\sigma^{(M)}} + \frac{\bar{q}_\mu \bar{q}_\nu}{R^2 (\sigma^{(M)})^2}, \tag{8.44}$$

where

$$\sigma^{(M)} = 1 - \frac{\eta_{\mu\nu} q^\mu q^\nu}{R^2} = 1 - \frac{\bar{q}_\nu \bar{q}^\nu}{R^2} \equiv 1 - \frac{\bar{q}^2}{R^2}. \tag{8.45}$$

Since $B_{\mu\nu}^{(M)}$ is a tensor in the spacetime $\{\bar{q}^\mu, \eta_{\mu\nu}\}$, it can be written as follows from Eq.(8.44):

$$B_{\mu\nu}^{(M)} = \frac{1}{\sigma^{(M)}} \bar{\theta}_{\mu\nu} + \frac{1}{(\sigma^{(M)})^2} \bar{\omega}_{\mu\nu}. \tag{8.46}$$

Furthermore, by means of $(B^{(M)})^{\mu\nu} B_{\nu\lambda}^{(M)} = \delta_\nu^\mu \equiv \eta_\nu^\mu$ and the rules (8.39)–(8.43), the above expression of $B_{\mu\nu}^{(M)}$ leads to:

$$(B^{(M)})^{\mu\nu} = \sigma^{(M)} \bar{\theta}^{\mu\nu} + (\sigma^{(M)})^2 \bar{\omega}^{\mu\nu}. \tag{8.47}$$

Or explicitly in matrix form:

$$\{(B^{(M)})^{\mu\nu}\} = \begin{pmatrix} \sigma^{(M)}(1 - \frac{(q^0)^2}{R^2}) & \frac{-q^0 q^1 \sigma^{(M)}}{R^2} & \frac{-q^0 q^2 \sigma^{(M)}}{R^2} & \frac{-q^0 q^3 \sigma^{(M)}}{R^2} \\ \frac{-q^1 q^0 \sigma^{(M)}}{R^2} & -\sigma^{(M)}(1 + \frac{(q^1)^2}{R^2}) & \frac{-q^1 q^2 \sigma^{(M)}}{R^2} & \frac{-q^1 q^3 \sigma^{(M)}}{R^2} \\ \frac{-q^2 q^0 \sigma^{(M)}}{R^2} & \frac{-q^2 q^1 \sigma^{(M)}}{R^2} & -\sigma^{(M)}(1 + \frac{(q^2)^2}{R^2}) & \frac{-q^2 q^3 \sigma^{(M)}}{R^2} \\ \frac{-q^3 q^0 \sigma^{(M)}}{R^2} & \frac{-q^3 q^1 \sigma^{(M)}}{R^2} & \frac{-q^3 q^2 \sigma^{(M)}}{R^2} & -\sigma^{(M)}(1 + \frac{(q^3)^2}{R^2}) \end{pmatrix}$$

$$\tag{8.48}$$

In the Beltrami spacetime $\mathcal{B}$ with the metric $B^{(M)}_{\mu\nu}$, the tetrad $e^a{}_\mu$ is defined via the following equation

$$B^{(M)}_{\mu\nu} = \eta_{ab} e^a{}_\mu e^b{}_\nu. \tag{8.49}$$

Generally, we have expansion of $e^a{}_\mu$:

$$e^a{}_\mu = a\bar{\theta}^a{}_\mu + b\bar{\omega}^a{}_\mu, \tag{8.50}$$

where a and b are unknown constants. Substituting Eqs.(8.46) and (8.50) into Eq.(8.49) gives

$$\frac{1}{\sigma^{(M)}}\bar{\theta}_{\mu\nu} + \frac{1}{(\sigma^{(M)})^2}\bar{\omega}_{\mu\nu} = \eta_{ab}(a\bar{\theta}^a{}_\mu + b\bar{\omega}^a{}_\mu)(a\bar{\theta}^b{}_\nu + b\bar{\omega}^b{}_\nu)$$
$$= a^2\bar{\theta}_{\mu\nu} + b^2\bar{\omega}_{\mu\nu}. \tag{8.51}$$

Comparing the left side of Eq.(8.51) with the right side, and noting $\bar{\theta}$ and $\bar{\omega}$ being projection operators with properties of Eqs.(8.40)–(8.43), we find that:

$$a = \sqrt{\frac{1}{\sigma^{(M)}}}, \qquad b = \frac{1}{\sigma^{(M)}}. \tag{8.52}$$

Substituting Eq.(8.52) into Eq.(8.50) gives

$$e^a{}_\mu = \sqrt{\frac{1}{\sigma^{(M)}}}\bar{\theta}^a{}_\mu + \frac{1}{\sigma^{(M)}}\bar{\omega}^a{}_\mu$$
$$= \sqrt{\frac{1}{\sigma^{(M)}}}\delta^a_\mu + \left(\frac{1}{\sigma^{(M)}} - \frac{1}{\sqrt{\sigma^{(M)}}}\right)\frac{\eta_{\mu\nu}\delta^a_\lambda(Q^\lambda - M^\lambda)(Q^\nu - M^\nu)}{(1 - \sigma^{(M)})R^2}. \tag{8.53}$$

$e_a{}^\mu$ is the inverse of $e^a{}_\mu$ given by:

$$e_a{}^\mu e^{a'}{}_\mu = \delta^{a'}_a. \tag{8.54}$$

With the Eqs.(8.54) and (8.49), we have

$$e_a{}^\mu = e^b{}_\nu(B^{(M)})^{\nu\mu}\eta_{ba} = \left(\sqrt{\frac{1}{\sigma^{(M)}}}\bar{\theta}^b{}_\nu + \frac{1}{\sigma^{(M)}}\bar{\omega}^b{}_\nu\right)(\sigma^{(M)}\bar{\theta}^{\nu\mu} + (\sigma^{(M)})^2\bar{\omega}^{\nu\mu})\eta_{ab}$$
$$= \sqrt{\sigma^{(M)}}\bar{\theta}_a{}^\mu + \sigma^{(M)}\bar{\omega}_a{}^\mu$$
$$= \sqrt{\sigma^{(M)}}\delta^\mu_a + \frac{\sigma^{(M)} - \sqrt{\sigma^{(M)}}}{1 - \sigma^{(M)}}\frac{\eta_{ab}\delta^b_\lambda(Q^\lambda - M^\lambda)(Q^\mu - M^\mu)}{R^2}, \tag{8.55}$$

and

$$e^{a\mu} = e^a_{\ \nu}(B^{(M)})^{\nu\mu} = \left(\sqrt{\frac{1}{\sigma^{(M)}}}\,\bar\theta^a_{\ \nu} + \frac{1}{\sigma^{(M)}}\bar\omega^a_{\ \nu}\right)(\sigma^{(M)}\bar\theta^{\nu\mu} + (\sigma^{(M)})^2\bar\omega^{\nu\mu})$$

$$= \sqrt{\sigma^{(M)}}\,\bar\theta^{a\mu} + \sigma^{(M)}\bar\omega^{a\mu}$$

$$= \sqrt{\sigma^{(M)}}\,\delta^\mu_b\eta^{ab} + \frac{\sigma^{(M)} - \sqrt{\sigma^{(M)}}}{1-\sigma}\frac{\delta^a_\mu(Q^\mu - M^\mu)(Q^\mu - M^\mu)}{R^2}, \tag{8.56}$$

$$e_{a\mu} = \sqrt{\frac{1}{\sigma^{(M)}}}\,\bar\theta_{a\mu} + \frac{1}{\sigma^{(M)}}\bar\omega_{a\mu}$$

$$= \sqrt{\frac{1}{\sigma^{(M)}}}\,\delta^\nu_a\eta_{\nu\mu} + \left(\frac{1}{\sigma^{(M)}} - \frac{1}{\sqrt{\sigma^{(M)}}}\right)\frac{\eta_{\mu\nu}\eta_{ab}\delta^b_\lambda(Q^\lambda - M^\lambda)(Q^\nu - M^\nu)}{(1-\sigma^{(M)})R^2}. \tag{8.57}$$

Next we derive spin-connection $\omega^{ab}_{\ \ \mu}$. From identity

$$e^\mu_{a\ ;\nu} = \partial_\nu e^\mu_a + \omega_a^{\ b}_{\ \nu}e^\mu_b + \Gamma^\mu_{\lambda\nu}e^\lambda_a = 0, \tag{8.58}$$

$$\Gamma^\rho_{\lambda\mu} = \frac{1}{2}(B^{(M)})^{\rho\nu}(\partial_\lambda B^{(M)}_{\nu\mu} + \partial_\mu B^{(M)}_{\nu\lambda} - \partial_\nu B^{(M)}_{\lambda\mu}), \tag{8.59}$$

we have

$$\omega^{ab}_{\ \ \mu} = \frac{1}{2}(e^{a\rho}\partial_\mu e^b_\rho - e^{b\rho}\partial_\mu e^a_\rho) - \frac{1}{2}\Gamma^\rho_{\lambda\mu}(e^{a\lambda}e^b_\rho - e^{b\lambda}e^a_\rho). \tag{8.60}$$

Substituting Eqs.(8.35) and (8.53) into Eqs.(8.59) and (8.60) gives

$$\omega^{ab}_{\ \ \mu} = \frac{1}{R^2\left(1 + \sqrt{\sigma^{(M)}}\right)\sqrt{\sigma^{(M)}}}(\delta^a_\mu\delta^b_\lambda - \delta^b_\mu\delta^a_\lambda)(Q^\lambda - M^\lambda). \tag{8.61}$$

8.6 Electric Coulomb law at light-cone of RFW universe

The Hydrogen atom is a bound state of a proton and an electron. The electric Coulomb potential binds them together. The action for deriving that potential of proton located at $\mathcal{Q} \equiv Q^\mu = \{Q^0 = ct,\ Q^1,\ Q^2,\ Q^3\}$ with background space-time metric $g_{\mu\nu} \equiv B^{(M)}_{\mu\nu}(\mathcal{Q})$ of Eq.(8.32) (see Fig.8.4) in the Gaussian system of units reads

$$S = -\frac{1}{16\pi c}\int d^4L\sqrt{-g}F_{\mu\nu}F^{\mu\nu} - \frac{e}{c}\int d^4L\sqrt{-g}j^\mu A_\mu, \tag{8.62}$$

where $g = \det(B^{(M)}_{\mu\nu})$, $F_{\mu\nu} = \frac{\partial A_\nu}{\partial L^\mu} - \frac{\partial A_\mu}{\partial L^\nu}$ and $j^\mu \equiv \{j^0 = c\rho_{\text{proton}}/\sqrt{B^{(M)}_{00}},\ \mathbf{j}\}$ is the 4-current density vector of proton (see, e.g., Ref.[Landau, Lifshitz (1987)]: *Chapter 4; Chapter 10, Eq.(90.3)*). Making space-time variable change of $L^\mu \to (L^\mu - Q^\mu) \equiv x^\mu = \{x^0 = ct_L - ct, \ x^i = L^i - Q^i\}$ and noting $L^\mu \simeq Q^\mu$, we have action S as

$$
S = -\frac{1}{16\pi c} \int d^4x \sqrt{-\det(B_{\mu\nu}^{(M)}(Q))} F_{\mu\nu} F^{\mu\nu}
$$
$$
-\frac{e}{c} \int d^4x \sqrt{-\det(B_{\mu\nu}^{(M)}(Q))} j^\mu A_\mu
$$
$$
= \left(-\frac{1}{16\pi c} \ B_{\mu\lambda}^{(M)}(Q) B_{\nu\rho}^{(M)}(Q) \int d^4x F^{\mu\nu}(x) F^{\lambda\rho}(x) \right.
$$
$$
\left. -\frac{e}{c} \int d^4x j^\mu(x) A_\mu(x) \right) \sqrt{-\det(B_{\mu\nu}^{(M)}(Q))}, \quad (8.63)
$$

and the equation of motion $\delta S / \delta A_\mu(x) = 0$ as follows (see, e.g., [Landau, Lifshitz (1987)], Eq.(90.6), pp.257)

$$
\partial_\nu F^{\mu\nu} = (B^{(M)})^{\nu\lambda} \partial_\nu F^\mu{}_\lambda = -\frac{4\pi}{c} j^\mu. \quad (8.64)
$$

In Beltrami space, $A^\mu = \{\phi_B, \ \mathbf{A}\}$ (see, e.g., [Landau, Lifshitz (1987)], Eq.(16.2) in pp.45) and 4-charge current $j^\mu = \{c\rho_{\text{proton}}/\sqrt{B_{00}^{(M)}}, \ \mathbf{j}\}$. According to the expression of charge density in curved space in Ref. [Landau, Lifshitz (1987)], (pp.256, Eq.(90.4)), $\rho_{\text{proton}} \equiv \rho_B = \frac{e}{\sqrt{\gamma}} \delta^{(3)}(\mathbf{x})$ and $\mathbf{j} = 0$, where

$$
\gamma = \det(\gamma_{ij}), \quad (8.65)
$$
$$
dl^2 = \gamma_{ij} dx^i dx^j = \left(-B_{ij}^{(M)} + \frac{B_{0i}^{(M)} B_{j0}^{(M)}}{B_{00}^{(M)}} \right) dx^i dx^j \quad (8.66)
$$

(see Eq.(84.7) in [Landau, Lifshitz (1987)]).

Thus, we have

(1) When $\mu = i$ ($i = 1, 2, 3$) in Eq.(8.64), we have

$$
\partial^i \partial_\mu A^\mu - (B^{(M)})^{\mu\nu} \partial_\mu \partial_\nu A^i = -\frac{4\pi}{c} j^i = 0. \quad (8.67)
$$

By means of the gauge condition

$$
\partial_\mu A^\mu = 0, \quad (8.68)
$$

we have

$$
(B^{(M)})^{\mu\nu} \partial_\mu \partial_\nu A^i = 0. \quad (8.69)
$$

Then

$$
A^i = 0 \quad (8.70)
$$

is a solution that satisfies the gauge condition (8.68) (noting $\partial_0 A^0 = \frac{\partial}{\partial x^0}\phi_B(r_B) = 0$ due to $\frac{\partial Q^0}{\partial x^0} = \frac{\partial Q^0}{\partial L^0} = 0$). Equation (8.70) is the vector potential.

(2) When $\mu = 0$ in Eq.(8.64), we have the Coulomb's law:

$$-(B^{(M)})^{ij}(Q)\partial_i\partial_j\phi_B(x) = -\frac{4\pi}{c}j^0 = -\frac{4\pi}{c}\frac{c\rho_B}{\sqrt{B_{00}^{(M)}}} = \frac{-4\pi e}{\sqrt{B_{00}^{(M)}\gamma}}\delta^{(3)}(\mathbf{x})$$

$$= \frac{-4\pi e}{\sqrt{-\det(B_{\mu\nu}^{(M)}(Q))}}\delta^{(3)}(\mathbf{x}), \qquad (8.71)$$

where $B_{00}^{(M)}\gamma = -\det(B_{\mu\nu}^{(M)})$ has been used, and $B_{\mu\nu}^{(M)}$ (and $(B^{(M)})^{ij}$) was given in Eq.(8.35) (and (8.48)), i.e.,

$$\left\{(B^{(M)})^{ij}\right\} = \begin{pmatrix} -\sigma^{(M)}(1+\frac{(q^1)^2}{R^2}) & \frac{-q^1q^2\sigma^{(M)}}{R^2} & \frac{-q^1q^3\sigma^{(M)}}{R^2} \\ \frac{-q^2q^1\sigma^{(M)}}{R^2} & -\sigma^{(M)}(1+\frac{(q^2)^2}{R^2}) & \frac{-q^2q^3\sigma^{(M)}}{R^2} \\ \frac{-q^3q^1\sigma^{(M)}}{R^2} & \frac{-q^3q^2\sigma^{(M)}}{R^2} & -\sigma^{(M)}(1+\frac{(q^3)^2}{R^2}) \end{pmatrix}.$$

$$(8.72)$$

The equation of Coulomb law (8.71) can be compactly rewritten as:

$$-\left(\nabla_x^T \mathfrak{B}^{(M)} \nabla_x\right)\phi_B = -\eta\mathfrak{J}, \qquad (8.73)$$

where tensor $\mathfrak{B}^{(M)} := \{(B^{(M)})^{ij}\}$, operator $\nabla := \{\partial_i\}$, $\eta = \frac{4\pi e}{\sqrt{\det(B_{\mu\nu}^{(M)}(Q))}}$ and $\mathfrak{J} := \delta^{(3)}(\mathbf{x})$. Symmetric matrix $\mathfrak{B}^{(M)}$ can be diagonalized by similarity transformation via matrix P:

$$P^T\mathfrak{B}^{(M)}P = \Lambda_B = \begin{pmatrix} \lambda_1 & 0 & 0 \\ 0 & \lambda_2 & 0 \\ 0 & 0 & \lambda_3 \end{pmatrix}. \qquad (8.74)$$

Here, λ_i with $i = 1, 2, 3$ and matrix P can be found in $\det(\mathfrak{B}^{(M)} - \lambda\mathbf{I}) = 0$. From Eq.(8.72), the results are:

$$\lambda_1 = -\sigma^{(M)}\frac{(\mathbf{q})^2 + R^2}{R^2}, \quad \lambda_2 = \lambda_3 = -\sigma^{(M)}, \qquad (8.75)$$

$$P = \begin{pmatrix} \frac{q^1\sqrt{(q^3)^2}}{q^3\sqrt{\mathbf{q}^2}} & -\frac{q^1q^2}{\sqrt{\mathbf{q}^2[(q^1)^2+(q^3)^2]}} & -\frac{q^3}{\sqrt{(q^1)^2+(q^3)^2}} \\ \frac{q^2\sqrt{(q^3)^2}}{q^3\sqrt{\mathbf{q}^2}} & \frac{\sqrt{(q^1)^2+(q^3)^2}}{\sqrt{\mathbf{q}^2}} & 0 \\ \frac{\sqrt{(q^3)^2}}{\sqrt{\mathbf{q}^2}} & -\frac{q^3q^2}{\sqrt{\mathbf{q}^2[(q^1)^2+(q^3)^2]}} & \frac{\sqrt{(q^1)^2}}{\sqrt{(q^1)^2+(q^3)^2}} \end{pmatrix} \qquad (8.76)$$

where $q^i = Q^i - M^i$ with $i = \{1, 2, 3\}$ (see Eq.(8.34)), and $\mathbf{q}^2 = (q^1)^2 + (q^2)^2 + (q^3)^2$. It can be checked that

$$P^T P = P P^T = I, \tag{8.77}$$

where I is 3×3-unit matrix. So that Eq.(8.73) can be rewritten as follows

$$- \left(\nabla_x^T P P^T \mathfrak{B}^{(M)} P P^T \nabla_x \right) \phi_B = - \left((\nabla_x^T P)(P^T \mathfrak{B}^{(M)} P)(P^T \nabla_x) \right) \phi_B$$

$$\equiv - \left(\nabla_y^T \Lambda_B \nabla_y \right) \phi_B = -\eta \mathfrak{I} = \frac{-4\pi e}{\sqrt{-\det(B_{\mu\nu}^{(M)}(\mathcal{Q}))}} \delta^{(3)}(\mathbf{x}), \tag{8.78}$$

where

$$y \equiv Px, \quad \text{or} \quad y^i = P_{ij}x^j, \quad x^i = P_{ij}^T y^j. \tag{8.79}$$

$$\nabla_y = P^T \nabla_x, \quad \text{or} \quad \partial/\partial y^i = P_{ij}^T \partial/\partial x^j, \quad \partial/\partial x^i = P_{ij}\partial/\partial y^j. \tag{8.80}$$

Substituting Eqs.(8.74) and (8.79) into Eq.(8.78) gives

$$\left[\frac{\partial^2}{(\partial y^1/\sqrt{-\lambda_1})^2} + \frac{\partial^2}{(\partial y^2/\sqrt{-\lambda_2})^2} + \frac{\partial^2}{(\partial y^3/\sqrt{-\lambda_3})^2} \right] \phi_B$$

$$= -\eta \delta(P_{1j}^T y^j) \delta(P_{2j}^T y^j) \delta(P_{3j}^T y^j) \tag{8.81}$$

$$= -\eta \frac{\delta(y^1)\delta(y^2)\delta(y^3)}{|\det(P_{ij}^T)|} = \frac{-4\pi e}{\sqrt{-\det(B_{\mu\nu}^{(M)}(\mathcal{Q}))}} \frac{\delta(\frac{y^1}{\sqrt{-\lambda_1}})\delta(\frac{y^2}{\sqrt{-\lambda_2}})\delta(\frac{y^3}{\sqrt{-\lambda_3}})}{\sqrt{-\lambda_1\lambda_2\lambda_3}}, \tag{8.82}$$

where $\left| \det(P_{ij}^T) \right| = 1$ has been used. Setting

$$\tilde{x}^i \equiv \frac{y^i}{\sqrt{-\lambda^i}} \quad \text{with} \quad i = 1, 2, 3, \tag{8.83}$$

Eq.(8.82) becomes

$$\left[\frac{\partial^2}{\partial(\tilde{x}^1)^2} + \frac{\partial^2}{\partial(\tilde{x}^2)^2} + \frac{\partial^2}{\partial(\tilde{x}^3)^2} \right] \phi_B \equiv \nabla_{\tilde{x}}^2 \phi_B$$

$$= \frac{-4\pi e}{\sqrt{-\det(B_{\mu\nu}^{(M)}(\mathcal{Q}))}} \frac{\delta(\tilde{x}^1)\delta(\tilde{x}^2)\delta(\tilde{x}^3)}{\sqrt{-\lambda_1\lambda_2\lambda_3}}. \tag{8.84}$$

Then, we get the Coulomb potential:

$$\phi_B = \phi_B(\mathcal{Q}) = \frac{1}{\sqrt{-\det(B_{\mu\nu}^{(M)}(\mathcal{Q}))}} \frac{1}{\sqrt{-\lambda_1\lambda_2\lambda_3}} \frac{e}{r_B}, \tag{8.85}$$

where $r_B = \sqrt{(\tilde{x}^1)^2 + (\tilde{x}^2)^2 + (\tilde{x}^3)^2}$.

Exercise

Problem 1: Derive the Coulomb potential ϕ_B^M's expression (8.85) from the Maxwell equation in $\mathcal{B}^\mathcal{M}$.

8.7 Fine structure constant variation in $\{Q^0, Q^1, 0, 0\}$

At this stage, in order to get the expression of fine-structure constant $\alpha(\mathcal{Q}) \equiv \alpha_z(\Omega)$ at point $\mathcal{Q} \equiv \{Q^0, Q^1, Q^2, Q^3\}$, we should substitute Equations (8.55), (8.61), (8.70), (8.85) and (8.18) for $e_a{}^\mu$, $\omega^{ab}{}_\mu$, A^i, $A^0 \equiv \phi_B$ and $B_{\mu\nu}^{(M)}$ respectively, into the Dirac equation (8.20) of Hydrogen atom located in the local inertial coordinate system of the light-cone in RFW Universe with Λ (see Fig.8.4). Such an $\alpha(\mathcal{Q})$ will characterize the temporal and spatial variations of fine-structure constant α. When we assume $M = 0$, $\alpha(\mathcal{Q})|_{(M=0)}$ has been calculated in the Ref.[Yan (2012)] [Yan (2014)]. In the following sections of this chapter we study the observation results of Keck and VLT reported by [Webb *et al.* (2011)] [King *et al.* (2012)] recently by means of $M = \{M^0, M^1, 0, 0\}$-model.

To calculate $\alpha(\mathcal{Q})$ analytically, we build 3-dimension spatial Cartesian coordinate system $\{Q^1, Q^2, Q^3\}$ in the equatorial coordinate figure showing the Keck and VLT best-fit dipole structure of $\Delta\alpha/\alpha$ in Refs. [Webb *et al.* (2011)] [King *et al.* (2012)]. Denoting $\hat{Q}^i \equiv Q^i/|\mathbf{Q}|$ and Q^1 as the best-fit dipole position, the directions of the three axis $\{\hat{Q}^1, \hat{Q}^2, \hat{Q}^3\}$ in this figure are $\hat{Q}^1(\varphi_1^{[RA]}, \vartheta_1^{[D]}) = \{17.4 \text{ h}, -59°\}$, $\hat{Q}^2(\varphi_2^{[RA]}, \vartheta_2^{[D]}) = \{17.4 \text{ h}, 31°\}$, $\hat{Q}^3(\varphi_3^{[RA]}, \vartheta_3^{[D]}) = \{11.4 \text{ h}, 0°\}$ (see Fig.8.9 and its caption). Angle (Θ) between a quasar sight line $(\{\varphi_q^{[RA]}, \vartheta_q^{[D]}\})$ and axis Q^1 is determined by following

$$\cos\Theta = \cos[\vartheta_1^{[D]}] \cos[\vartheta_q^{[D]}] \cos[\frac{\varphi_1^{[RA]} - \varphi_q^{[RA]}}{12}\pi] + \sin[\vartheta_1^{[D]}] \sin[\vartheta_q^{[D]}].$$

(8.86)

In this section we calculate $\alpha(\mathcal{Q})$ for $\Theta = \pi$ and $\Theta = 0$ following the method described in Chapter 7 [Yan (2012)] [Yan (2014)].

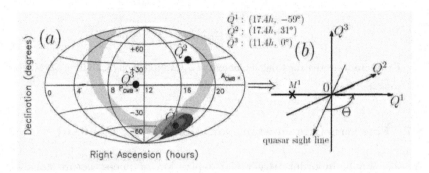

Fig. 8.9 (color online) The 3-dimension spatial Cartesian coordinate frame $\{Q^1,\ Q^2,\ Q^3\}$ on the equatorial coordinates. In left figure (a), the background is all-sky plot in equatorial coordinates showing the independent Keck (green, leftmost) and VLT (blue, rightmost) best-fit dipole s, and the combined samples (red, center), for the dipole model $\Delta\alpha/\alpha = A\cos\Theta$ copied from [Webb *et al.* (2011)] [King *et al.* (2012)]. The locations of axis $\{\hat{Q}^1,\ \hat{Q}^2,\ \hat{Q}^3\}$ in this figure are marked by "•". In [Webb *et al.* (2011)] [King *et al.* (2012)] it has been measured that the best-fit dipole is at right ascension $\varphi^{[RA]} = 17.4 \pm 0.9$ h, declination $\vartheta^{[D]} = -58 \pm 9$ dec. The cosmic microwave background dipole antipole are illustrated for comparison. The directions of 3-dimension spatial Cartesian coordinate system $\{Q^1,\ Q^2,\ Q^3\}$ that we take are: $\hat{Q}^1(\varphi_1^{[RA]}, \vartheta_1^{[D]}) = \{17.4\text{ h},\ -59°\}$, $\hat{Q}^2(\varphi_2^{[RA]}, \vartheta_2^{[D]}) = \{17.4\text{ h},\ 31°\}$, $\hat{Q}^3(\varphi_3^{[RA]}, \vartheta_3^{[D]}) = \{11.4\text{ h},\ 0°\}$. The gray shadow band is the Milky Way. In right figure (b), the 3-dimension spatial Cartesian coordinate system $\{Q^1,\ Q^2,\ Q^3\}$ with a non-zero space component M^1 of Minkowski point is drawn. Θ is angle between a quasar sight line $(\{\varphi_q^{[RA]}, \vartheta_q^{[D]}\})$ and axis Q^1. Formula for computing it is $\cos\Theta = \cos[\vartheta_1^{[D]}]\cos[\vartheta_q^{[D]}]\cos[\frac{\varphi_1^{[RA]}-\varphi_q^{[RA]}}{12}\pi] + \sin[\vartheta_1^{[D]}]\sin[\vartheta_q^{[D]}]$.

8.7.1 *Formulation of alpha-variation for case of* $\Theta = \pi$, *and* 0

When quasar sight line is anti-parallel (or parallel) to direction of $\hat{Q}^1(\varphi_1^{[RA]}, \vartheta_1^{[D]}) = \{17.4\text{ h},\ -59°\}$, the angle Θ between them is equal to π (or 0), and the locations of distant atoms on the light-cone in the RFW Universe are at $\{Q^0,\ Q^1 > 0\ (\text{or}\ < 0),\ Q^2 = 0,\ Q^3 = 0\}$ (see Fig.8.4). Then we have $q^0 = Q^0 - M^0$, $q^1 = Q^1 - M^1$, $q^2 = 0$, $q^3 = 0$, and

$$
B_{\mu\nu}^{(M)}(Q) = \begin{pmatrix} \frac{1}{\sigma^{(M)}} + \frac{(Q^0-M^0)^2}{R^2(\sigma^{(M)})^2} & -\frac{(Q^0-M^0)(Q^1-M^1)}{R^2(\sigma^{(M)})^2} & 0 & 0 \\ -\frac{(Q^0-M^0)(Q^1-M^1)}{R^2(\sigma^{(M)})^2} & \frac{-1}{\sigma^{(M)}} + \frac{(Q^1-M^1)^2}{R^2(\sigma^{(M)})^2} & 0 & 0 \\ 0 & 0 & \frac{-1}{\sigma^{(M)}} & 0 \\ 0 & 0 & 0 & \frac{-1}{\sigma^{(M)}} \end{pmatrix},\ (8.87)
$$

$$(B^{(M)})^{\mu\nu}(Q) = \begin{pmatrix} \sigma^{(M)}(1 - \frac{(Q^0 - M^0)^2}{R^2}) & -\frac{(Q^0 - M^0)(Q^1 - M^1)\sigma^{(M)}}{R^2} & 0 & 0 \\ -\frac{(Q^0 - M^0)(Q^1 - M^1)\sigma^{(M)}}{R^2} & -\sigma^{(M)}(1 + \frac{(Q^1 - M^1)^2}{R^2}) & 0 & 0 \\ 0 & 0 & -\sigma^{(M)} & 0 \\ 0 & 0 & 0 & -\sigma^{(M)} \end{pmatrix},$$

$$(8.88)$$

with

$$\sigma^{(M)} = 1 - \frac{(Q^0 - M^0)^2 - (Q^1 - M^1)^2}{R^2}. \tag{8.89}$$

Here, for a given red-shift z, $Q^0 \equiv ct(z)$ and $Q^1 = \pm\sqrt{Q(z)^2 - (Q^2)^2 - (Q^3)^2} = \pm Q(z)$ are determined from Eqs.(7.13) and (8.27) respectively (see also Figs.8.5 and 8.6). Then Eqs.(8.75) and (8.76) become

$$\lambda_1 = -\sigma^{(M)}\left(1 + \frac{(Q^1 - M^1)^2}{R^2}\right), \quad \lambda_2 = \lambda_3 = -\sigma^{(M)}, \tag{8.90}$$

$$P = \begin{pmatrix} 1 & 0 & 0 \\ 0 & 1 & 0 \\ 0 & 0 & 1 \end{pmatrix}, \tag{8.91}$$

Substituting Eqs.(8.55), (8.61), (8.70), (8.85) and (8.18) into Eq.(8.20) gives dS-SR Dirac equation for the electron in the distant Hydrogen located at the light-cone of RFW Universe:

$$\hbar c\beta \left[i\sqrt{\sigma^{(M)}}\gamma^\mu \mathcal{D}_\mu^L + i\frac{\sigma^{(M)} - \sqrt{\sigma^{(M)}}}{(1 - \sigma^{(M)})R^2}\eta_{ab}\delta_\lambda^a (Q^\lambda - M^\lambda)\gamma^b (Q^\mu - M^\mu)\mathcal{D}_\mu^L \right.$$

$$\left. -\frac{m_e c}{\hbar} \right]\psi = 0, \tag{8.92}$$

where factor $\hbar c\beta$ in the front of the equation is only for convenience, $L^\mu \simeq Q^\mu$ has been used (see Fig.8.4), and

$$\mathcal{D}_\mu^L \equiv \frac{\partial}{\partial L^\mu} - \frac{i}{4}\omega_\mu^{ab}\sigma_{ab} - i\frac{e}{c\hbar}B_{\mu\nu}^{(M)}A^\nu$$

$$= \frac{\partial}{\partial L^\mu} - \frac{i}{4}\frac{1}{R^2\left(1 + \sqrt{\sigma^{(M)}}\right)\sqrt{\sigma^{(M)}}}(\delta_\mu^a\delta_\lambda^b - \delta_\mu^b\delta_\lambda^a)(Q^\lambda - M^\lambda)\sigma_{ab}$$

$$-i\frac{e}{c\hbar}\frac{\delta_{\mu 0}B_{00}^{(M)}(Q) + \delta_{\mu i}B_{i0}^{(M)}(Q)}{\sqrt{-\det(B_{\mu\nu}^{(M)}(Q))}}\frac{1}{\sqrt{-\lambda_1\lambda_2\lambda_3}}\frac{e}{r_B}. \tag{8.93}$$

We use the method described in the Chapter 7 (see also [Yan (2012)] [Yan (2014)]) and expand each term of Eq.(8.92) to the order as follows:

(1) Since observed distant Hydrogen atom must be on the light cone and the location is $\{Q^0,\ Q^1,\ Q^2 = 0,\ Q^3 = 0\}$, then $\eta_{ab}L^a L^b \simeq \eta_{ab}Q^a Q^b = (Q^0)^2 - (Q^1)^2$, and the first term of Eq.(8.92) reads

$$\hbar c\beta i\sqrt{\sigma^{(M)}}\gamma^\mu \mathcal{D}^L_\mu = \sqrt{1 - \frac{(Q^0 - M^0)^2 - (Q^1 - M^1)^2}{R^2}}\, \hbar c\beta i\gamma^\mu \mathcal{D}^L_\mu \psi$$

(8.94)

with $\ \beta\gamma^\mu = \{\beta\gamma^0 = \beta^2 = 1,\ \beta\gamma^i = \alpha^i\}$ (8.95)

$$\hbar c\beta i\gamma^\mu \mathcal{D}^L_\mu \psi = (i\hbar\frac{\partial}{\partial t_L} + i\hbar c\vec{\alpha}\cdot\nabla + \frac{\hbar c\beta}{4}\omega^{ab}_\mu \gamma^\mu \sigma_{ab} + e\beta\gamma^\mu B_{\mu\nu}A^\nu)\psi.$$

(8.96)

In follows, we use variables $\{\tilde{x}^1, \tilde{x}^2, \tilde{x}^3\}$ (where since Eqs.(8.90) and (8.91), $\tilde{x}^i \equiv x^i/\sqrt{-\lambda^i}$, we have $\tilde{x}^1 \equiv \dfrac{1}{\sqrt{\sigma^{(M)}(1+(Q^1-M^1)^2/R^2)}}x^1$, $\tilde{x}^2 \equiv \dfrac{1}{\sqrt{\sigma^{(M)}}}x^2$ and $\tilde{x}^3 \equiv \dfrac{1}{\sqrt{\sigma^{(M)}}}x^3$) to replace $\{x^1, x^2, x^3\}$. The following notations are used hereafter:

$$\mathbf{r}_B = \mathbf{i}\tilde{x}^1 + \mathbf{j}\tilde{x}^2 + \mathbf{k}\tilde{x}^3, \quad |\mathbf{r}_B| = r_B,$$

(8.97)

$$\nabla_B = \mathbf{i}\frac{\partial}{\partial\tilde{x}^1} + \mathbf{j}\frac{\partial}{\partial\tilde{x}^2} + \mathbf{k}\frac{\partial}{\partial\tilde{x}^3}, \quad \tilde{x}^i \in \{\tilde{x}^1,\ \tilde{x}^2,\ \tilde{x}^3\}.$$

(8.98)

Then, noting $\frac{\partial}{\partial x^1} = \frac{\partial\tilde{x}^1}{\partial x^1}\frac{\partial}{\partial\tilde{x}^1} = \frac{1}{\sqrt{\sigma^{(M)}(1+(Q^1-M^1)^2/R^2)}}\frac{\partial}{\partial\tilde{x}^1}$, $\frac{\partial}{\partial x^i} = \frac{\partial\tilde{x}^i}{\partial x^i}\frac{\partial}{\partial\tilde{x}^i} = \frac{1}{\sqrt{\sigma^{(M)}}}\frac{\partial}{\partial\tilde{x}^i}$ with $i = \{2, 3\}$, the Eq.(8.96) becomes

$$\hbar c\beta i\gamma^\mu \mathcal{D}^L_\mu \psi = \left(i\hbar\frac{\partial}{\partial t_L} + i\frac{\hbar c}{\sqrt{\sigma^{(M)}}}\vec{\alpha}\cdot\nabla_B + i\frac{\hbar c}{\sqrt{\sigma^{(M)}}}\left[\frac{1}{\sqrt{1+(Q^1-M^1)^2/R^2}} - 1\right]\right.$$

$$\times\alpha^1\frac{\partial}{\partial\tilde{x}^1} + \frac{\hbar c\beta}{4}\omega^{ab}_\mu \gamma^\mu \sigma_{ab} + eB^{(M)}_{00}\phi_B + e\alpha^1 B^{(M)}_{10}\phi_B \Bigg)\psi$$

$$= \left(i\hbar\frac{\partial}{\partial t_L} + i\frac{\hbar c}{\sqrt{\sigma^{(M)}}}\vec{\alpha}\cdot\nabla_B + i\frac{\hbar c}{\sqrt{\sigma^{(M)}}}\left[\frac{1}{\sqrt{1+(Q^1-M^1)^2/R^2}} - 1\right]\alpha^1\frac{\partial}{\partial\tilde{x}^1}\right.$$

$$+ \frac{\hbar c\beta}{4}\omega^{ab}_\mu \gamma^\mu \sigma_{ab} + \left[\frac{1}{\sigma^{(M)}} + \frac{(Q^0-M^0)^2}{R^2(\sigma^{(M)})^2}\right]e\phi_B$$

$$\left. - \frac{(Q^1-M^1)(Q^0-M^0)}{R^2(\sigma^{(M)})^2}\alpha^1 e\phi_B \right)\psi.$$

(8.99)

(2) Estimation of the contributions of the fourth term in RSH of (8.99) (the spin-connection contributions) is as follows: From Eq.(8.61), the ratio of the fourth term to the first term of Eq.(8.99) is:

$$\left|\frac{\frac{\hbar c\beta}{4}\omega^{ab}_\mu \gamma^\mu \sigma_{ab}\psi}{i\hbar\partial_t\psi}\right| \sim \frac{\hbar c}{4}\frac{ct}{2R^2}\frac{1}{m_e c^2} = \frac{ct}{8R^2}\frac{\hbar}{m_e c} = \frac{1}{8}\frac{cta_c}{R^2} \sim 0,$$

(8.100)

where $a_c = \hbar/(m_e c) \simeq 0.3 \times 10^{-12}$m is the Compton wave length of electron. $\mathcal{O}(cta_c/R^2)$-term is neglectable. Therefore the 3-rd term in RSH of (8.96) has no contribution to our approximation calculations.

(3) Substituting Eqs.(8.100), (8.99) and (8.95) into Eq.(8.94) and noting $L^a \simeq Q^a$, we get the first term in LHS of Eq.(8.92)

$$\hbar c \beta i \sqrt{\sigma^{(M)}} \gamma^\mu \mathcal{D}_\mu^L \psi = \sqrt{\sigma^{(M)}} \left(i\hbar \frac{\partial}{\partial t_L} + i\frac{\hbar c}{\sqrt{\sigma^{(M)}}} \vec{\alpha} \cdot \nabla_B \right.$$

$$+i\frac{\hbar c}{\sqrt{\sigma^{(M)}}} \left[\frac{1}{\sqrt{1+(Q^1-M^1)^2/R^2}} - 1 \right] \alpha^1 \frac{\partial}{\partial \tilde{x}^1}$$

$$+ \left[\frac{1}{\sigma^{(M)}} + \frac{(Q^0-M^0)^2}{R^2(\sigma^{(M)})^2} \right] e\phi_B - \frac{(Q^1-M^1)(Q^0-M^0)}{R^2(\sigma^{(M)})^2} \alpha^1 e\phi_B \right) \psi.$$

$$(8.101)$$

(4) The second term of Eq.(8.92) is

$$\hbar c \beta i \frac{\sigma^{(M)} - \sqrt{\sigma^{(M)}}}{(1-\sigma^{(M)})R^2} \eta_{ab}(Q^a - M^a)\gamma^b(Q^\mu - M^\mu)\mathcal{D}_\mu^L \psi$$

$$= \hbar c \beta i \frac{\sigma^{(M)} - \sqrt{\sigma^{(M)}}}{(1-\sigma^{(M)})R^2} \left[\gamma^0(Q^0 - M^0) - \vec{\gamma} \cdot (\vec{Q} - \vec{M}) \right]$$

$$\times \left[(Q^0 - M^0)\mathcal{D}_0 + (Q^i - M^i)\mathcal{D}_i \right] \psi$$

$$= \hbar c i \frac{\sigma^{(M)} - \sqrt{\sigma^{(M)}}}{(1-\sigma^{(M)})R^2} \left[(Q^0 - M^0) - \alpha^1(Q^1 - M^1) \right]$$

$$\times \left[(Q^0 - M^0) \left(\partial_0^L - \frac{\gamma^0\gamma^1(Q^1 - M^1)}{2R^2(1+\sqrt{\sigma^{(M)}})\sqrt{\sigma^{(M)}}} \right. \right.$$

$$-\frac{ie}{c\hbar}(\frac{1}{\sigma^{(M)}} + \frac{(Q^0-M^0)^2}{R^2(\sigma^{(M)})^2})\phi_B \bigg) + (Q^1 - M^1) \left(\partial_1^L + \frac{\gamma^0\gamma^1(Q^0 - M^0)}{2R^2(1+\sqrt{\sigma^{(M)}})\sqrt{\sigma^{(M)}}} \right.$$

$$\left. \left. +\frac{ie}{c\hbar} \frac{(Q^0-M^0)(Q^1-M^1)}{R^2(\sigma^{(M)})^2}\phi_B \right) \right] \psi$$

$$= \frac{\sigma^{(M)} - \sqrt{\sigma^{(M)}}}{(1-\sigma^{(M)})R^2} \left[(Q^0 - M^0) - \alpha^1(Q^1 - M^1) \right]$$

$$\times \left\{ i\hbar c \left[(Q^0 - M^0)\frac{\partial}{\partial L^0} + (Q^1 - M^1)\frac{\partial}{\partial L^1} \right] + \frac{Q^0 - M^0}{(\sigma^{(M)})^2} e\phi_B \right\} \psi, \quad (8.102)$$

where $Q^2 = Q^3 = M^2 = M^3 = 0$, $\gamma^0 = \beta$, $\beta\gamma^1 = \alpha^1$ and Eqs.(8.93) and (8.89) were used. Noting $x^\mu \equiv L^\mu - Q^\mu$, $\frac{\partial}{\partial L^0} = \frac{\partial}{\partial x^0} = \frac{\partial}{c\partial t_L}$, $\frac{\partial}{\partial L^1} = \frac{\partial}{\partial x^1} = \frac{\partial \tilde{x}^1}{\partial x^1}\frac{\partial}{\partial \tilde{x}^1} = \frac{1}{\sqrt{\sigma^{(M)}}(1+(Q^1-M^1)^2/R^2)}\frac{\partial}{\partial \tilde{x}^1}$, (8.102) becomes

$$\hbar c \beta i \frac{\sigma^{(M)} - \sqrt{\sigma^{(M)}}}{(1 - \sigma^{(M)})R^2} \eta_{ab} \delta_\lambda^a (Q^\lambda - M^\lambda) \gamma^b (Q^\mu - M^\mu) \mathcal{D}_\mu^L \psi$$

$$= \frac{\sigma^{(M)} - \sqrt{\sigma^{(M)}}}{(1 - \sigma^{(M)})R^2} \left[(Q^0 - M^0) - \alpha^1 (Q^1 - M^1) \right]$$

$$\times \left\{ \left[(Q^0 - M^0) i\hbar \frac{\partial}{\partial t_L} + \frac{i\hbar c(Q^1 - M^1)}{\sqrt{\sigma^{(M)}(1 + (Q^1 - M^1)^2/R^2)}} \frac{\partial}{\partial \tilde{x}^1} \right] \right.$$

$$\left. + \frac{Q^0 - M^0}{(\sigma^{(M)})^2} e\phi_B \right\} \psi. \tag{8.103}$$

(5) Therefore, substituting Eqs.(8.101) and (8.103) into Eq.(8.92), we have

$$i\hbar \left(\sqrt{\sigma^{(M)}} + \frac{\sigma^{(M)} - \sqrt{\sigma^{(M)}}}{(1 - \sigma^{(M)})R^2} [(Q^0 - M^0)^2 - (Q^0 - M^0)(Q^1 - M^1)\alpha^1] \right) \frac{\partial}{\partial t_L} \psi$$

$$= \left\{ -i\hbar c \vec{\alpha} \cdot \nabla_B - \left[\frac{1}{\sqrt{\sigma^{(M)}}} + \frac{(Q^0 - M^0)^2}{R^2 \sigma^{(M)} \sqrt{\sigma^{(M)}}} \right. \right.$$

$$\left. + \frac{\sigma^{(M)} - \sqrt{\sigma^{(M)}}}{(1 - \sigma^{(M)})R^2} \frac{(Q^0 - M^0)^2}{(\sigma^{(M)})^2} \right] e\phi_B + m_e c^2 \beta \right\} \psi$$

$$+ \left\{ -i\hbar c \left(\frac{1}{\sqrt{1 + (Q^1 - M^1)^2/R^2}} - 1 \right) \alpha^1 \frac{\partial}{\partial \tilde{x}^1} + \frac{(Q^1 - M^1)(Q^0 - M^0)}{R^2 \sigma^{(M)} \sqrt{\sigma^{(M)}}} \alpha^1 e\phi_B \right.$$

$$\left. - \frac{\sigma^{(M)} - \sqrt{\sigma^{(M)}}}{(1 - \sqrt{\sigma^{(M)}})R^2} \left[\frac{i\hbar c(Q^1 Q^0 - (Q^1 - M^1)^2 \alpha^1)}{\sqrt{\sigma^{(M)}(1 + (Q^1 - M^1)^2/R^2)}} \frac{\partial}{\partial \tilde{x}^1} \right] \right\} \psi. \tag{8.104}$$

In order to discuss the spectra of Hydrogen atom in the dS-SR Dirac equation, we need to find its solutions with certain energy E for electron in the atom. From Eq.(6.10) in Chapter 6, we have

$$p^0 \equiv \frac{E}{c} = i\hbar \left[\left(\eta^{0\nu} - \frac{(Q^0 - M^0)(Q^\nu - M^\nu)}{R^2} \right) \partial_\nu + \frac{5(Q^0 - M^0)}{2R^2} \right],$$

$$E = i\hbar \left[\partial_{t_L} - \frac{(Q^0 - M^0)^2}{R^2} \partial_{t_L} - \frac{c(Q^0 - M^0)(Q^1 - M^1)}{R^2} \partial_{L^1} + \frac{5c(Q^0 - M^0)}{2R^2} \right],$$

$$E\psi \simeq i\hbar \left(1 - \frac{(Q^0 - M^0)^2}{R^2} \right) \partial_{t_L} \psi - i\hbar \frac{c(Q^0 - M^0)(Q^1 - M^1)}{R^2 \sqrt{\sigma^{(M)}(1 + (Q^1 - M^1)^2/R^2)}} \frac{\partial}{\partial \tilde{x}^1} \psi,$$

$$\tag{8.105}$$

where an estimation for the ratio of the 4-th term to the 2-nd of $E\psi$ were used:

$$\frac{|i\hbar \frac{5c^2 t}{2R^2} \psi|}{|\frac{-c^2 t^2}{R^2} i\hbar \partial_{t_L} \psi|} \sim \frac{|i\hbar \frac{5c^2 t}{2R^2}|}{|\frac{-2c^2 t^2}{R^2} E|} \sim \frac{5\hbar}{2tm_e c^2} \equiv \frac{5}{2} \frac{a_c}{ct},$$

where $a_c \equiv \hbar/(m_e c) \simeq 0.3 \times 10^{-12}$m is the Compton wave length of electron and ct is about the distance between earth and a distant atom near quasar. Obviously, $a_c/(ct)$ is ignorable. For instance, for an atom with $ct \sim 10^9$ly, $a_c/(ct) \sim 10^{-38} << (ct)^2/R^2 \sim 10^{-5}$. Hence the 4-th term of $E\psi$ was ignored. Equation(8.105) means

$$i\hbar \frac{\partial}{\partial t_L}\psi = \frac{E}{1-(Q^0-M^0)^2/R^2}\psi$$
$$+\frac{i\hbar c(Q^1-M^1)(Q^0-M^0)}{R^2(1-(Q^0-M^0)^2/R^2)}\frac{1}{\sqrt{\sigma^{(M)}(1+(Q^1-M^1)^2/R^2)}}\frac{\partial}{\partial \tilde{x}^1}\psi.$$

$$(8.106)$$

Then substituting Eq.(8.106) into Eq.(8.104) gives

$$E\psi = H_0\psi + H'\psi, \qquad (8.107)$$

where

$$H_0 = \left(1-\frac{(Q^0-M^0)^2}{R^2}\right)\left[\sqrt{\sigma^{(M)}}+\frac{(\sigma^{(M)}-\sqrt{\sigma^{(M)}})(Q^0-M^0)^2}{(1-\sigma^{(M)})R^2}\right]^{-1}$$
$$\times\left\{-i\hbar c\vec{\alpha}\cdot\nabla_B-\left[\frac{1}{\sqrt{\sigma^{(M)}}}+\frac{(Q^0-M^0)^2}{R^2\sigma^{(M)}}\left(\frac{1}{1+\sqrt{\sigma^{(M)}}}\right)\right]e\phi_B+m_ec^2\beta\right\}$$
$$\equiv -i\hbar_z(\Omega)c\vec{\alpha}\cdot\nabla_B-\frac{e_z(\Omega)^2}{r_B}+m_{e,\,z}(\Omega)c^2\beta, \qquad (8.108)$$

$$H' = \left(1-\frac{(Q^0-M^0)^2}{R^2}\right)\left[\sqrt{\sigma^{(M)}}+\frac{(\sigma^{(M)}-\sqrt{\sigma^{(M)}})(Q^0-M^0)^2}{(1-\sigma^{(M)})R^2}\right]^{-1}$$
$$\times\left\{-i\hbar c\left(\frac{1}{\sqrt{1+(Q^1-M^1)^2/R^2}}-1\right)\alpha^1\frac{\partial}{\partial\tilde{x}^1}+\frac{(Q^1-M^1)(Q^0-M^0)}{R^2\sigma^{(M)}\sqrt{\sigma^{(M)}}}\alpha^1 e\phi_B\right.$$
$$-\frac{\sigma^{(M)}-\sqrt{\sigma^{(M)}}}{(1-\sigma^{(M)})R^2}\left(i\hbar c((Q^1-M^1)(Q^0-M^0)-(Q^1-M^1)^2\alpha^1)\frac{\partial}{\partial\tilde{x}^1}\right.$$
$$\left.-\frac{(Q^1-M^1)(Q^0-M^0)^2}{(\sigma^{(M)})^2}\alpha^1 e\phi_B\right)$$
$$+\frac{(\sigma^{(M)}-\sqrt{\sigma^{(M)}})(Q^0-M^0)(Q^1-M^1)\alpha^1}{(1-(Q^0-M^0)^2/R^2)(1-\sigma^{(M)})R^2}$$
$$\left.\times\left(E+i\hbar c\frac{(Q^0-M^0)(Q^1-M^1)}{R^2\sqrt{1+(Q^1-M^1)^2/R^2}}\frac{\partial}{\partial\tilde{x}^1}\right)\right\}$$
$$-\frac{i\hbar c(Q^1-M^1)(Q^0-M^0)}{R^2\sqrt{\sigma^{(M)}(1+(Q^1-M^1)^2/R^2)}}\frac{\partial}{\partial\tilde{x}^1}$$
$$\equiv \sum_{i=1}^{3}C_i\hat{O}^i, \qquad (8.109)$$

where

$$\hat{Q}^1 = \alpha^1, \quad C_1 \propto \mathcal{O}(1/R^2); \qquad \hat{Q}^2 = \frac{\partial}{\partial \tilde{x}^1}, \quad C_2 \propto \mathcal{O}(1/R^2);$$

$$\hat{Q}^3 = \alpha^1 \frac{\partial}{\partial \tilde{x}^1}, \quad C_3 \propto \mathcal{O}(1/R^4), \tag{8.110}$$

and so we see that

$$H' \propto \mathcal{O}(1/R^2) << H_0 \propto \mathcal{O}(1). \tag{8.111}$$

We have mentioned in the Introduction section that the operator-structure of H_0 of (8.108) makes the corresponding eigenequation $E_0\psi = H_0\psi$ integrable. Hence Eq.(8.111) means that H_0 (Eq.(8.108)) and H' (Eq.(8.109)) can be legally treated as unperturbation Hamiltonian and perturbation Hamiltonian respectively in QM-problem with $H = H_0 + H'$.

(6) From Eq.(8.108), we have

$$\hbar_z(\Omega)c = \hbar c \left(1 - \frac{(Q^0 - M^0)^2}{R^2}\right) \left[\sqrt{\sigma^{(M)}} + \frac{(\sigma^{(M)} - \sqrt{\sigma^{(M)}})(Q^0 - M^0)^2}{(1 - \sigma^{(M)})R^2}\right]^{-1}, \tag{8.112}$$

$$e_z^2(\Omega) = e^2 \left(1 - \frac{(Q^0 - M^0)^2}{R^2}\right) \left[\sqrt{\sigma^{(M)}} + \frac{(\sigma^{(M)} - \sqrt{\sigma^{(M)}})(Q^0 - M^0)^2}{(1 - \sigma^{(M)})R^2}\right]^{-1}$$

$$\times \left\{ \left[\frac{1}{\sqrt{\sigma^{(M)}}} + \frac{(Q^0 - M^0)^2}{R^2 \sigma^{(M)}}\left(\frac{1}{1 + \sqrt{\sigma^{(M)}}}\right)\right] \frac{1}{\sqrt{-\det(B_{\mu\nu}^{(M)}(\mathcal{Q}))}} \frac{1}{\sqrt{-\lambda_1\lambda_2\lambda_3}} \right\}, \tag{8.113}$$

where the expression (8.85) for ϕ_B were used, and hence

$$\alpha_z(\Omega) = \alpha \frac{\frac{1}{\sqrt{\sigma^{(M)}}} + \frac{(Q^0 - M^0)^2}{R^2\sigma^{(M)}}\left(\frac{1}{1 + \sqrt{\sigma^{(M)}}}\right)}{\sqrt{-\det(B_{\mu\nu}^{(M)}(\mathcal{Q}))}\sqrt{-\lambda_1\lambda_2\lambda_3}}, \quad \text{with} \quad \alpha = \frac{e^2}{\hbar c}. \tag{8.114}$$

For case of $\Theta = \pi$, (or 0), $B_{\mu\nu}^{(M)}(\mathcal{Q})$ and λ_1, λ_2, λ_3 have been given in Eqs.(8.87) and (8.90) respectively, and hence we have

$$\sqrt{-\det(B_{\mu\nu}^{(M)}(\mathcal{Q}))}$$

$$= \frac{1}{\sigma^{(M)}}\left[\left(\frac{1}{\sigma^{(M)}} - \frac{(Q^1 - M^1)^2}{R^2(\sigma^{(M)})^2}\right)\left(\frac{1}{\sigma^{(M)}} + \frac{(Q^0 - M^0)^2}{R^2(\sigma^{(M)})^2}\right)\right.$$

$$\left. + \frac{(Q^0 - M^0)^2(Q^1 - M^1)^2}{R^4(\sigma^{(M)})^4}\right]^{1/2}, \tag{8.115}$$

$$\sqrt{-\lambda_1\lambda_2\lambda_3} = (\sigma^{(M)})^{3/2}\sqrt{1 + \frac{(Q^1 - M^1)^2}{R^2}}. \tag{8.116}$$

Substituting Eqs.(8.115) and (8.116) into Eq.(8.114) gives

$$\alpha_z(\Omega)$$
$$= \frac{\alpha \left[\frac{1}{\sigma^{(M)}} + \frac{(Q^0 - M^0)^2}{R^2 \sigma(M) \sqrt{\sigma^{(M)}}} \left(\frac{1}{1 + \sqrt{\sigma^{(M)}}} \right) \right] \left(1 + \frac{(Q^1 - M^1)^2}{R^2} \right)^{-1/2}}{\sqrt{\left[\left(\frac{1}{\sigma^{(M)}} - \frac{(Q^1 - M^1)^2}{R^2 (\sigma^{(M)})^2} \right) \left(\frac{1}{\sigma^{(M)}} + \frac{(Q^0 - M^0)^2}{R^2 (\sigma^{(M)})^2} \right) + \frac{(Q^0 - M^0)^2 (Q^1 - M^1)^2}{R^4 (\sigma^{(M)})^4} \right]}}.$$

(8.117)

When $z = 0$ (or $Q^0 \to 0$, and $Q^1 \to 0$), $\alpha_z(\Omega)$ should be normalized so that $\alpha_z(\Omega)|_{z=0} = \alpha_0$ which is the α-value measured in the Earth laboratories. So from Eq.(8.117) we have

$$\alpha_0 = \frac{\alpha \left[\frac{1}{\sigma_0^{(M)}} + \frac{(M^0)^2}{R^2 \sigma_0^{(M)} \sqrt{\sigma_0^{(M)}}} \left(\frac{1}{1 + \sqrt{\sigma_0^{(M)}}} \right) \right] \left(1 + \frac{(M^1)^2}{R^2} \right)^{-1/2}}{\sqrt{\left[\left(\frac{1}{\sigma_0^{(M)}} - \frac{(M^1)^2}{R^2 (\sigma_0^{(M)})^2} \right) \left(\frac{1}{\sigma_0^{(M)}} + \frac{(M^0)^2}{R^2 (\sigma_0^{(M)})^2} \right) + \frac{(M^0)^2 (M^1)^2}{R^4 (\sigma_0^{(M)})^4} \right]}}$$

$$\equiv \alpha N_0$$

(8.118)

where

$$N_0 = \frac{\left[\frac{1}{\sigma_0^{(M)}} + \frac{(M^0)^2}{R^2 \sigma_0^{(M)} \sqrt{\sigma_0^{(M)}}} \left(\frac{1}{1 + \sqrt{\sigma_0^{(M)}}} \right) \right] \left(1 + \frac{(M^1)^2}{R^2} \right)^{-1/2}}{\sqrt{\left[\left(\frac{1}{\sigma_0^{(M)}} - \frac{(M^1)^2}{R^2 (\sigma_0^{(M)})^2} \right) \left(\frac{1}{\sigma_0^{(M)}} + \frac{(M^0)^2}{R^2 (\sigma_0^{(M)})^2} \right) + \frac{(M^0)^2 (M^1)^2}{R^4 (\sigma_0^{(M)})^4} \right]}}$$

(8.119)

$$\sigma_0^{(M)} \equiv \sigma^{(M)} \Big|_{Q^0 = Q^1 = 0} = 1 - \frac{(M^0)^2 - (M^1)^2}{R^2}.$$

(8.120)

Therefore, from Eqs.(8.117) and (8.118), we obtain

$$\frac{\Delta \alpha}{\alpha_0} \equiv \frac{\alpha_z(\Omega) - \alpha_0}{\alpha_0}$$
$$= \frac{\left[\frac{1}{\sigma^{(M)}} + \frac{(Q^0 - M^0)^2}{R^2 \sigma(M) \sqrt{\sigma^{(M)}}} \left(\frac{1}{1 + \sqrt{\sigma^{(M)}}} \right) \right] \left(1 + \frac{(Q^1 - M^1)^2}{R^2} \right)^{-1/2}}{N_0 \sqrt{\left[\left(\frac{1}{\sigma^{(M)}} - \frac{(Q^1 - M^1)^2}{R^2 (\sigma^{(M)})^2} \right) \left(\frac{1}{\sigma^{(M)}} + \frac{(Q^0 - M^0)^2}{R^2 (\sigma^{(M)})^2} \right) + \frac{(Q^0 - M^0)^2 (Q^1 - M^1)^2}{R^4 (\sigma^{(M)})^4} \right]}}$$
$$- 1,$$

(8.121)

where
$Q^0(z) \equiv ct(z)$ and $Q^1(z) = \pm \sqrt{Q(z)^2 - (Q^2)^2 - (Q^3)^2}\Big|_{(Q^2 = Q^3 = 0)} = \pm Q(z)$ (i.e., $|Q^1(z)| = Q(z)$) have been given in Eqs.(8.24) and (8.27) respectively (see also Figs.8.5 and 8.6).

8.7.2 Comparisons with observations of alpha-varying for cases of $\Theta = 0, \pi$

Equation (8.121) is the prediction of α-varying derived from dS-SR Dirac equation of distant Hydrogen atom located on Q^1-axis. When the location $Q^1 > 0$, the direction of the corresponding quasar sight line is opposite to the direction of Q^1-axis (see Fig.8.9). For this case the corresponding angle (Θ) between the quasar sight line and the Q^1-axis is equal to π (i.e., $\Theta = \pi$ for this case). Oppositely, when the location $Q^1 < 0$, we have $\Theta = 0$, i.e., the direction of the quasar sight line is same as the direction of Q^1-axis. For both of cases, we can generally write Q^1 in Eq.(8.121) as

$$Q^1 = -|Q^1| \cos \Theta = -Q(z) \cos \Theta. \tag{8.122}$$

Looking back cosmic time variable is $Q^0 \equiv ct < 0$, and the coordinates of Minkowski point of Beltrami metric $B_{\mu\nu}^{(M)}$ are $M^0 < 0$, $M^1 > 0$, and $M^2 = M^3 = 0$. Substituting Eq.(8.121) into Eq.(8.121) gives

$$
\frac{\Delta\alpha}{\alpha_0}
= \frac{\left[\frac{1}{\sigma^{(M)}} + \frac{(Q^0(z)-M^0)^2}{R^2\sigma^{(M)}\sqrt{\sigma^{(M)}}}\left(\frac{1}{1+\sqrt{\sigma^{(M)}}}\right)\right]\left[N_0\sqrt{\left(1 + \frac{(-Q(z)\cos\Theta-M^1)^2}{R^2}\right)}\right]^{-1}}{\sqrt{\left[\left(\frac{1}{\sigma^{(M)}} - \frac{(-Q(z)\cos\Theta-M^1)^2}{R^2(\sigma^{(M)})^2}\right)\left(\frac{1}{\sigma^{(M)}} + \frac{(Q^0(z)-M^0)^2}{R^2(\sigma^{(M)})^2}\right) + \frac{(Q^0(z)-M^0)^2(-Q(z)\cos\Theta-M^1)^2}{R^4(\sigma^{(M)})^4}\right]}}
-1,
$$

$$\tag{8.123}$$

where N_0 is the same as (8.119) and

$$\sigma^{(M)} = 1 - \frac{(Q^0(z) - M^0)^2 - (-\cos\Theta Q(z) - M^1)^2}{R^2}. \tag{8.124}$$

Equation (8.123) is the theoretic prediction of dS-SR, whose variables are z (red shift) and $\Theta(= 0$, or $\pi)$, and the adjustable parameters are R, M^0, M^1. The discussions on it are follows:

(1) Since $|R|$ is the maximal length scale parameter in the theory (say $|R| \sim 10^{12}$lyr [Chen, Xiao, Yan (2008)]), we could deduce the $\Delta\alpha/\alpha_0$'s Taylor-power series of $1/R^2$ from (8.123) (for practical calculations, "*Mathematica*" is helpful):

$$
\frac{\Delta\alpha}{\alpha_0} \simeq \frac{1}{8R^4}\left[(M^1)^2 Q^0(z)(-2M^0 + Q^0(z)) + 2M^1(M^0 - Q^0(z))^2\right.
$$
$$
\left. \times Q(z)\cos\Theta + (M^0 - Q^0(z))^2(Q(z))^2\cos^2\Theta\right] + \mathcal{O}(1/R^6),
$$

$$\tag{8.125}$$

where the leading term $\propto \mathcal{O}(1/R^4)$ is dominating in the expansion of Eq.(8.123). Suppose the parameters and the variables are chosen such that

$$|M^0| >> |M^1|, \quad Q^0(z) \sim Q(z) \sim \epsilon, \qquad (8.126)$$

we have

$$\frac{\Delta\alpha}{\alpha_0} \sim \frac{1}{8R^4} 2\cos\Theta \, M^1(M^0)^2\epsilon, \qquad (8.127)$$

where the lesser terms $\propto \mathcal{O}((M^1)^2 M^0\epsilon/R^4)$ and $\propto \mathcal{O}(\epsilon^2/R^4)$ have been ignored, and Θ is only to be 0 or π (and noting $\cos(0) = 1$, $\cos(\pi) = -1$). Thus, when $\Theta = 0$, we have $\Delta\alpha/\alpha_0 > 0$, and when $\Theta = \pi$, oppositely, we have $\Delta\alpha/\alpha_0 < 0$. This is interesting since Eq.(8.127) indicates that the scenario reported by [Webb *et al.* (2011)] [King *et al.* (2012)] could be interpreted by Eq.(8.123) with a particular parameter setting in proper region of variables (8.126) in the theory that satisfies the requirement of Cosmological Principle.

(2) In this scenario we need to determine the parameters R, M^0, M^1 by comparing the theoretical predictions with the observation data. By using Keck +VLT data, the relations between $\Delta\alpha/\alpha_0$ and $r \equiv ct(z)$ along $\hat{Q}^1$-axis have been described in Section 8.2 (see Fig.8.1, Fig.8.2 and Fig.8.3) [Webb *et al.* (2011)] [King *et al.* (2012)]. Let's use Eqs.(8.123) and (8.124) with $\Theta = \pi$ and 0 to fit Fig.8.3 (i.e., Fig.3 in [Webb *et al.* (2011)]) which is based on the combined Keck and VLT data and $t(z)$ expression (8.24). The best fitting gives

$$R = 500 \text{ Glyr} = 0.5 \times 10^{12} \text{ Lyr},$$
$$M^0 \simeq -100 \text{ Glyr} = -1.0 \times 10^{11} \text{ Lyr}, \qquad (8.128)$$
$$M^1 \simeq -22 \text{ Glyr} = -2.2 \times 10^{10} \text{ Lyr},$$

which are consistent with requirement of Eq.(8.126). The fitted curve of Eq.(8.123) with Eq.(8.124) is shown in Fig.8.10.

(3) After determination of R, M^0, M^1 of Eq.(8.128), the metric $B_{\mu\nu}^{(M)}(Q)$ of local inertial coordinate system and then $\Delta\alpha/\alpha_0(z)$ along Q^1-axis are fully known. In Fig.8.11, the curves of $\Delta\alpha/\alpha_0(z)$ are plotted. The Keck's data in 2004 reported by [Murphy *et al.* (2004)] [Dent (2008)] are illustrated for comparison. Since the 2004-data were obtained by observations in all directions in Keck at that time, the deviations between the

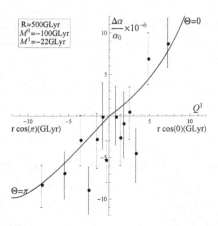

Fig. 8.10   Determination of parameters R, M^0, M^1 via fitting Keck+VLT's α-varying data reported by [Webb *et al.* (2011)] [King *et al.* (2012)] and described in Section 8.2. $\Delta\alpha/\alpha_0$ vs $A\ r(z)cos\Theta$ with $\Theta = \{0$, and $\pi\}$ shows an apparent gradient in α along the best-fit dipole. The best-fit direction is at $\hat{Q}^1(\varphi_1^{[RA]}, \vartheta_1^{[D]}) = \{17.4$ h, $-59°\}$ (see Fig.8.9). The data reported by [Webb *et al.* (2011)] [King *et al.* (2012)] and described in Section 8.2 are shown with error bars. A spatial gradient is statistically preferred over a monopole-only model at the 4.2σ level. A cosmology with parameters $(H_0,\ \Omega_M,\ \Omega_\Lambda)$ were given in Eq.(8.25). The fitted model's parameters are $R \simeq 500$GLyr, $M^0 \simeq -100$GLyr, $M^1 \simeq -22$GLyr. The resulting curve of $\Delta\alpha/\alpha_0(r(z))$ of Eq.(8.123) is shown.

data and the prediction curves of $\Delta\alpha/\alpha_0(z)$ are understandable. The point here is that the curves remarkably show a nontrivial behavior described in Section 8.2 and reported by Ref.[Webb *et al.* (2011)] [King *et al.* (2012)]. That is, in one direction in the sky α was smaller at the time of absorption, while in the opposite direction it was larger. More explicitly, we illustrate 3 pairs of $\Delta\alpha/\alpha_0(z)$-predictions in Table 1 as examples. In the table, $\Theta = 0$ (or π) means the quasar sight line is parallel (or anti-parallel) to direction of $\hat{Q}^1(\varphi_1^{[RA]}, \vartheta_1^{[D]}) = \{17.4$ h, $-59°\}$. For each z, there are two values of $(\Delta\alpha/\alpha_0)_{\text{th}}$ with opposite signs, which just matches the expectations of observations. Such a theoretical picture is subtle. In addition, it was also reported as a dipole form in [Webb *et al.* (2011)] [King *et al.* (2012)]

$$\frac{\Delta\alpha}{\alpha_0} \simeq \bar{A}_{\text{obs}} \cos\Theta, \quad \text{with } \bar{A}_{\text{obs}} = (1.02 \pm 0.21) \times 10^{-5}, \quad (8.129)$$

where $\bar{A}_{obs}$ means the observation value of amplitude $\bar{A}$. Theoretically,

Fig.8.11 indicates that when $z \sim 2$ to 4, $\bar{A}_{\text{th}} \simeq 1. \times 10^{-5}$ which coincides with $\bar{A}_{\text{obs}}$.

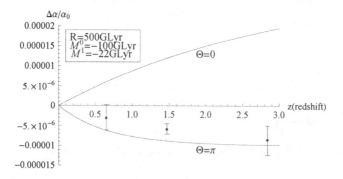

Fig. 8.11 α-varying in the region of $0 < z < 3$. Θ is angle between quasar sight line and axis Q^1. $\Theta = 0$ (or π) means the quasar sight line is parallel (or anti-parallel) to direction of $\hat{Q}^1(\varphi_1^{[RA]}, \vartheta_1^{[D]}) = \{17.4 \text{ h}, -59°\}$. When z fixes, there are two values of $\Delta\alpha/\alpha_0(z)$ with opposite signs. $\Delta\alpha/\alpha_0(z)$ is given by Eq.(8.123) with parameters Eq.(8.128). Three Keck's data in 2004 reported and discussed by [Murphy *et al.* (2004)] [Dent (2008)] are shown for comparison.

Table 8.1 Examples of predictions of $\Delta\alpha/\alpha_0$: $\Theta = \{0, \pi\}$ is angle between quasar sight line and axis Q^1. For each redshift z, there is a pair of $(\Delta\alpha/\alpha_0)_{\text{th}}$-predictions from Eq.(8.123) with parameters in Eq.(8.128).

z	0.65		1.47		2.84	
Θ	0	π	0	π	0	π
$(\Delta\alpha/\alpha_0)_{th}$	0.60×10^{-5}	-0.62×10^{-5}	1.20×10^{-5}	-0.88×10^{-5}	1.88×10^{-5}	-0.98×10^{-6}

(4) We further plot the curves of $\Delta\alpha/\alpha_0$ of Eq.(8.123) with Eq.(8.128) in a more wide z-range including radiation epoch of the Universe in Fig.8.12 (that epoch roughly corresponds to $z \geq 3 \times 10^3$). For $\Theta = 0$, the radiation epoch limit of $\Delta\alpha/\alpha$ is 4.7×10^{-5}, while for $\Theta = \pi$, this limit is about -5×10^{-6}. Therefore, we find that in that epoch the dipole form (8.129) is no longer true.

8.8 α-Varying in whole sky

In the last section, the α-varying for $\Theta = \{0, \pi\}$ (or for case that both distant atom and quasar lie at Q^1-axis) was studied and the model's parameters R, M^0, M^1 have been determined. Now we discuss general case

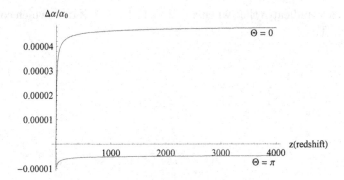

Fig. 8.12 α-varying in the range of $0 < z < 4000$.

of $\Theta \in \{0, \pi\}$, i.e., the case of $Q^1 \neq 0$, $Q^2 \neq 0$ and $Q^3 \neq 0$ (see Fig.8.9). The corresponding M-Beltrami metrics (8.35) reads

$$\{B_{\mu\nu}^{(M)}\} = \begin{pmatrix} \frac{(Q^0-M^0)^2}{R^2(\sigma^{(M)})^2} + \frac{1}{\sigma^{(M)}} & \frac{-(Q^0-M^0)(Q^1-M^1)}{R^2(\sigma^{(M)})^2} & \frac{-(Q^0-M^0)Q^2}{R^2(\sigma^{(M)})^2} & \frac{-(Q^0-M^0)Q^3}{R^2(\sigma^{(M)})^2} \\ \frac{-(Q^0-M^0)(Q^1-M^1)}{R^2(\sigma^{(M)})^2} & \frac{(Q^1-M^1)^2}{R^2(\sigma^{(M)})^2} - \frac{1}{\sigma^{(M)}} & \frac{(Q^1-M^1)Q^2}{R^2(\sigma^{(M)})^2} & \frac{(Q^1-M^1)Q^3}{R^2(\sigma^{(M)})^2} \\ \frac{-(Q^0-M^0)Q^2}{R^2(\sigma^{(M)})^2} & \frac{(Q^1-M^1)Q^2}{R^2(\sigma^{(M)})^2} & \frac{(Q^2)^2}{R^2(\sigma^{(M)})^2} - \frac{1}{\sigma^{(M)}} & \frac{Q^3Q^2}{R^2(\sigma^{(M)})^2} \\ \frac{-(Q^0-M^0)Q^3}{R^2(\sigma^{(M)})^2} & \frac{(Q^1-M^1)Q^3}{R^2(\sigma^{(M)})^2} & \frac{Q^3Q^2}{R^2(\sigma^{(M)})^2} & \frac{(Q^3)^2}{R^2(\sigma^{(M)})^2} - \frac{1}{\sigma^{(M)}} \end{pmatrix}$$

$$(8.130)$$

where

$$\sigma^{(M)} = 1 - \frac{(Q^0(z) - M^0)^2 - (Q^1(z) - M^1)^2 - (Q^2(z))^2 - (Q^3(z))^2}{R^2}$$

$$= 1 - \frac{1}{R^2}[(Q^0(z) - M^0)^2 - (Q(z)\cos\Theta - M^1)^2 - Q(z)^2\sin^2\Theta],$$

$$(8.131)$$

with $Q^1(z) = Q(z)\cos\Theta$ and $(Q^2(z))^2 + (Q^3(z))^2 = Q(z)^2 - Q^1(z)^2 = Q(z)^2\sin^2\Theta$ used. From Eqs.(8.130) and (8.131), we have

$$\det B_{\mu\nu}^{(M)}(\mathcal{Q}) = -(\sigma^{(M)})^{-5}. \qquad (8.132)$$

From Eq.(8.75), we have

$$\lambda_1 = -\sigma^{(M)}\left(1 + \frac{1}{R^2}[(Q(z)\cos\Theta - M^1)^2 + Q(z)^2\sin^2\Theta]\right),$$

$$\lambda_2 = \lambda_3 = -\sigma^{(M)}. \qquad (8.133)$$

We now focus on the derivations of α in this case. In Section 8.7, we presented the procedure for calculating α step by step in detail based on

$B_{\mu\nu}^{(M)}(\mathcal{Q})|_{(Q^2=Q^3=0)}$. Though the full $B_{\mu\nu}^{(M)}(\mathcal{Q})$ (8.130) is more complex than $B_{\mu\nu}^{(M)}(\mathcal{Q})|_{(Q^2=Q^3=0)}$ (8.87), the calculations in Section 8.7 can be repeated straightforwardly. The resulting $\alpha_z(\Omega)$-expression (8.114) keeps invariant except the $\{B_{\mu\nu}^{(M)}(\mathcal{Q}), \sigma^{(M)}, \lambda_i\}|_{(Q^2=Q^3=0)}$ in the formula should be replaced by $\{B_{\mu\nu}^{(M)}(\mathcal{Q}), \sigma^{(M)}, \lambda_i\}|_{(Q^i\neq0)}$ with $(i=1,2,3)$. This gives

$$\alpha_z(\Omega) = \alpha \frac{\frac{1}{\sqrt{\sigma^{(M)}}} + \frac{(Q^0-M^0)^2}{R^2\sigma^{(M)}}\left(\frac{1}{1+\sqrt{\sigma^{(M)}}}\right)}{\sqrt{-\det(B_{\mu\nu}^{(M)}(\mathcal{Q}))}\sqrt{-\lambda_1\lambda_2\lambda_3}}, \quad \text{with} \quad \alpha = \frac{e^2}{\hbar c}, \quad (8.134)$$

where $\sigma^{(M)}$, $\det B_{\mu\nu}^{(M)}(\mathcal{Q})$, λ_i are given in Eqs.(8.131), (8.132) and (8.133) respectively. The corresponding α-varying formula reads

$$\frac{\Delta\alpha}{\alpha_0} = \frac{\frac{1}{\sqrt{\sigma^{(M)}}} + \frac{(Q^0-M^0)^2}{R^2\sigma^{(M)}}\left(\frac{1}{1+\sqrt{\sigma^{(M)}}}\right)}{N_0\sqrt{-\det(B_{\mu\nu}^{(M)}(\mathcal{Q}))}\sqrt{-\lambda_1\lambda_2\lambda_3}} - 1, \quad (8.135)$$

where N_0 has been given in Eq.(8.119).

The α-varyings $\Delta\alpha/\alpha_0(z,\Theta)$ shown in Fig.8.13 is the plot of numerical result of Eq.(8.135) in which the curves correspond to z from 0 to 4.5 and $\Theta = \{0, \pi/4, \pi/3, 0.4\pi, \pi/2, 0.6\pi\ 2\pi/3, 3\pi/4, \pi\}$ respectively. We can see from the figure that: (i) When z is fixed, $\Delta\alpha/\alpha_0$ decreases along with Θ increases from 0 to π; (ii) In ranges of $\{0 \leq \Theta < 0.4\pi\}$ and $\{0.6\pi < \Theta \leq \pi\}$,α vary spatially. That is, α could be smaller in one direction in the sky yet larger in the opposite direction at the time of absorption. This feature is consistent to the observations in Keck and VLT reported by [Webb *et al.* (2011)] and [King *et al.* (2012)], described in Section 8.2; (iii) When $\Theta \sim \{0.5\pi, 0.6\pi, 0.7\pi\}$, the observation results of α-variations $\Delta\alpha/\alpha_0$ are nearly null.

In order to show $\Delta\alpha/\alpha_0$'s dipolar behavior more explicitly, we further plot figure of $\Delta\alpha/\alpha_0$ vs $A\, r\cos\Theta$ in Fig.8.14. Three theoretical prediction curves corresponding to $\Theta = \{0, \pi/4, 0.4\pi\}$ and the experiment observation data reported by [Webb *et al.* (2011)] [King *et al.* (2012)] described in Section 8.2 are shown in Fig.8.14 for comparison. It can been seen that the three curves are approximately close to each other in the region of $r\cos\Theta = \{-2.5\text{GLyr}, 2.5\text{GLyr}\}$, and their average gradient is about $A \simeq 1.0 \times 10^{-6}\text{GLyr}^{-1}$. This theoretical prediction value is consistent with the observation's $(1.1 \pm 0.25) \times 10^{-6}\text{GLyr}$ reported in [Webb *et al.* (2011)]

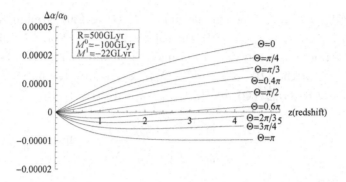

Fig. 8.13 α-varying $\Delta\alpha/\alpha_0(z, \Theta)$ in the range of $0 < z < 4.5$ and $0 \leq \Theta \leq \pi$. The parameters $\{R, M^0, M^1\}$ are shown in the figure.

[King *et al.* (2012)]. However, for absorbing systems with $\Theta \simeq 0.4\pi$ and $|r\cos\Theta| \geq 3.5\mathrm{GLyr}$, the curve with $\Theta = 0.4\pi$ in Fig.8.14 indicates that $\Delta\alpha/\alpha_0(\Theta \sim 0.4\pi) \neq -\Delta\alpha/\alpha_0(\Theta \sim (\pi - 0.4\pi))$. This means that the dipole term (i.e., the term $\propto \cos\Theta$) in the expansion of $\Delta\alpha/\alpha_0$ is no longer dominating. For the absorbing systems with $0.4\pi < \Theta < 0.5\pi$ the situations is also similar. Observational experiments to check such predictions is called for.

8.9   Summary and comments

The spacetime variations of fine-structure constant $\alpha \equiv e^2/\hbar c$ in cosmology is a new phenomenon beyond the Standard Model of physics. To reveal the meaning of such new physics is of utmost importance to a complete understanding of fundamental physics. The main conclusion to interpret such new physics presented in this chapter is that the phenomenon of α-varying cosmologically is due to the de Sitter (or anti de Sitter) spacetime symmetry in the de Sitter invariant special relativity (dS-SR). In order to understand this point, we address two issues here: (i) We have shown that the time-variation of α is caused by the quantum adiabatic evolutions in cosmology with dS-SR local inertial systems; (ii) In the framework of standard cosmology with RFW metric and ΛCDM model, the theory of non-isotropic spatial variation of α is achieved via considerations of dS-

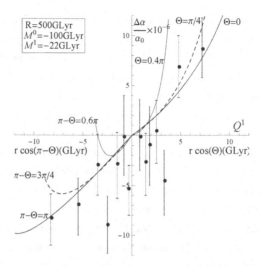

Fig. 8.14 Curves of $\Delta\alpha/\alpha_0(z,\Theta)$ vs $A\ r\cos\Theta$. Two solidline curves and one dotted curve are shown. One of the solidline curves corresponds to $\Theta = 0$ and the other is for $\Theta = 0.4\pi$. The dotted curve corresponds to $\Theta = \pi/4$. The horizontal axis shows projection of atom's "distance" $r(z) \equiv ct(z)$ onto Q^1 axis. $t(z)$ and cosmology parameters $(H_0,\ \Omega_M,\ \Omega_\Lambda)$ were given in Eqs.(8.24) and (8.25) respectively. The data reported by [Webb *et al.* (2011)] [King *et al.* (2012)] and described in Section 8.2 are plotted with error bars. The parameters $\{R,\ M^0,\ M^1\}$ are shown in the figure.

SR local inertial systems and the prediction of the dipole mode coinciding with the observations is obtained. This means that the observed dipole distributions of $\Delta\alpha/\alpha_0$ in the sky do not lead to a violation of the cosmology principle. This is an interesting result. Since the theory achieved in this Chapter has convincingly interpreted the aspects of the spacetime variations of fine-structure constant in cosmology, this theory could be considered a Geometry Model on Alpha-Variation (or shortly GM-AV).

More specifically, the logic that leads to GM-AV is recalled and summarized as follows:

(1) The Keck +VLT data that imply varying α are results of measuring spectra of atoms (or ions) in distant absorption clouds. So it is legitimate to use the electron wave equation of atoms (typically, the Hydrogen atom) to discuss this issue.

(2) As usual, QM equations for spectra in atoms are defined in inertial

coordinate systems to avoid ambiguities caused by inertial forces. So, it is necessary to take the local inertial coordinate systems in RFW Universe for discussing both laboratory atoms and the distant.

(3) When Einstein's cosmology constant $\Lambda \equiv 3/R^2 \neq 0$, the metric in the local inertial coordinate system s in RFW Universe has to be Beltrami metric $B_{\mu\nu}(x)$ or M-Beltrami metric $B_{\mu\nu}^{(M)}(x)$, but cannot be Minkowski metric $\eta_{\mu\nu}$.

(4) Since there exist both temporal and spatial variations for α in cosmology, $B_{\mu\nu}^{(M)}(x)$ is suitable. The de Sitter pseudo-radius parameter R and the Minkowski point parameters $\{M^0, M^1\}$ are expected to be determined by fitting to the observations.

(5) As usual, dS-Dirac equation for Hydrogen can always be reduced to spectrum equation of Hydrogen. In this way, both coefficient of Dirac-kinetic energy operator term $-i\vec{\alpha} \cdot \nabla$ (i.e., $\hbar_z(\Omega)c$) and coefficient of Coulomb potential term $-1/r$ (i.e., $e_z^2(\Omega)$) are derived explicitly in Section 8.6 and Section 8.7 (see also [Yan (2012)] [Yan (2014)]). Then $\alpha_z(\Omega) = e_z^2(\Omega)/\hbar_z(\Omega)c$ and $\Delta\alpha/\alpha_0 = (\alpha_z(\Omega) - \alpha_0)/\alpha_0$ have been calculated.

(6) According to Section 8.2, we focus on the best-fit direction about right ascension $\sim 17.5\ h$, declination $\sim -61\deg$, and calculate $\Delta\alpha/\alpha_0$ in this region. Comparing the theoretical prediction with the observation results in Section 8.2, the model's parameters are determined: $R \simeq 500\mathrm{GLyr}$, $M^0 \simeq -100\mathrm{GLyr}$ and $M^1 \simeq -22\mathrm{GLyr}$. Surprisingly but not strangely, the amazing observation about $\Delta\alpha/\alpha_0$-dipole in Section 8.2 is reproduced by this dS-SR theoretical model GM-AV. This is a remarkable result, which could be considered a highly nontrivial evidence to support dS-SR.

(7) The α-varying in the whole sky has also been studied in this model with the same parameters. The results are generally in agreement with the descriptions in Section 8.2 and the estimations in [Webb *et al.* (2011)] [King *et al.* (2012)]. Hence, GM-AV is logical.

To get things more straight, let's go back temporarily to the beginning of Section 8.3. If the theory of SR is exactly Einstein's SR (E-SR) with metric $\eta_{\mu\nu}$, the α will not vary over cosmic time, i.e., $\Delta\alpha/\alpha = 0$. However, the observations at cosmological distances go directly against it, i.e.,

$(\Delta\alpha/\alpha)_{obs} \neq 0$. Even though one would be open-minded to face this big challenge [Uzan (2003)], the simplest answer to the puzzle seems to be that E-SR might not be exact at cosmological spacetime scale, or E-SR needs to be extended. The most natural extension of E-SR in the framework of SR was attempted in works of Dirac(1935)-Inönü and Wigner(1953)-Gürsey and T.D. Lee (1968)-Lu, Zuo and Guo (1974) [Dirac (1935); Inönü, Wigner (1953); Gürsey, Lee (1963); Lu (1970); Lu, Zou and Guo (1974)]. Those works proposed the theory of dS-SR. Using the dS-SR in this chapter, we correctly work out the prediction of $\Delta\alpha/\alpha_0$ which is consistent with the data reported by [Webb *et al.* (2011); King *et al.* (2012)]. Hence the puzzle of α-varying over cosmological time could be considered solved. This approach could be considered a very simple answer to the problem, if not the simplest answer.

The relativity principle problem in curved spacetime with constant curvature has been solved in dS-SR in [Lu (1970); Lu, Zou and Guo (1974)] via introducing Beltrami metric (see Sections 3.1, 3.2 and 3.3, or References [Yan, Xiao, Huang and Li (2005)] and the Appendix of [Sun, Yan, Deng, Huang, Hu (2013)]), and hence E-SR is the limit of dS-SR with $|R| \to \infty$. Since $|R|$-value determined in Eq.(8.128) is fortunately cosmology huge (~ 500GLyr), we argue that such dS-SR would not contradict the experiments verified E-SR within the error bands. Furthermore, the Many-Multiplet (MM) method [Webb *et al.* (1999); Dzuba, Flambaum, Webb (1999a)] used in [Webb *et al.* (2011); King *et al.* (2012)] is itself R^{-1}-free. The huge (or almost infinite) R means that the estimates to $\Delta\alpha/\alpha$ in [Webb *et al.* (2011); King *et al.* (2012)] are reliable approximately.

Besides GM-AV based the geometric and symmetric considerations described above, there are some other prescriptions for understanding this puzzle. One of them is to treat α as a space-time function of $\alpha(x)$, and consider it as a scalar field $\varphi(x)$ or a function of $\varphi(x)$. The $\varphi(x)$ is a matter field satisfied some suitable dynamics, and fills the Universe everywhere (see, e.g., [Bekenstein (1982)] [Olive,Peloso,Uzan (2011)], and the references therein). Sometimes it is called dilaton-like scalar field. Nevertheless, there are still a number of unclear problems in this prescription. The introduction of this new scalar field with an unknown dynamics creates more questions than it might answer. Specifically, it will fundamentally change a number of aspects of modern physics such as the structure of our universe, the cosmological dynamics, the vacuum *et al.*, not the least of which being the implications to particle physics. In our view, introduction of the scalar field $\varphi(x)$ based on phenomenological considerations is an *ad hoc* approach. Comparatively, our spacetime symmetry approach is more natural and cost-effective, since our theory is based on the least assumption, i.e., de Sitter spacetime and de Sitter invariance and no fundamental and well-established physical laws need to be changed.

Chapter 9

De Sitter Invariance of Generally Covariant Dirac Equation

As has been proselytized in the earlier chapters, the fundamental spacetime at both cosmological scale and laboratory should be de Sitter spacetime if the effect of cosmological constant is not negligible. Since the standard Dirac equation in Minkowski spacetime enjoys the covariance under Lorentz transformation, which is the symmetry transformation of the Minkowski spacetime, it should be required that the Dirac equation in de Sitter spacetime, which is curved, should be covariant under de Sitter group. The first attempt was pioneered by Dirac using rotation generators of the 5-dimensional Minkowski spacetime [Dirac (1935)]. An equivalent version was derived in [Gürsey, Lee (1963)]. On the other hand, as has been widely recognized, the generalization of Dirac equation in Minkowski spacetime to curved spacetime is Eq.(6.38), which is covariant under *local* Lorentz transformation by construction. It remains to show that these two versions are equivalent and it is covariant under de Sitter transformation if the background spacetime is de Sitter. We here follow the work of Gürsey-Lee.

9.1 Gürsey-Lee metric of de Sitter spacetime

As explained in Chapter 3, de Sitter spacetime, denoted as dS, can be deemed as a hyperboloid embedded in a 5-dimensional flat Minkowski spacetime [Roberson (1928)]. Instead of using the Beltrami coordinates system which is a stereographic projection coordinate system, Gürsey-Lee utilized another coordinates system $(y^0, y^1, y^2, y^3, y^5)$ which mimic the spherical coordinates in 3-dimensions. The relation between coordinates ξ^A and the $(y^0, y^1, y^2, y^3, y^5)$ is

$$\xi^A = y^5 n^A(y^0, y^1, y^2, y^3), \tag{9.1}$$

where $y^5 = \sqrt{-\xi^A \xi_A}$ is the radial coordinates (like r in 3-dim polar system) and $n^A = \xi^A/\sqrt{-\xi^A \xi_A}$ is the angular part. Note that $\partial y^5/\partial \xi^A = -n_A, \partial \xi^A/\partial y^5 = n^A$. Like the local frame e_r, e_θ, e_φ in 3-dim Euclidean space, one can build local orthonormal frame $(e_a^A, e_5^A = n^A)$ such that $(a = 0, 1, 2, 3)$ $e_a^A e_b^B \eta_{AB} = \eta_{ab}, e_a^A n_A = 0$. We can express e_5 in terms of the rest as follows

$$n_A = \frac{1}{4!} \varepsilon^{a_1 a_2 a_3 a_4 5} \varepsilon_{A_1 A_2 A_3 A_4 A} e_{a_1}^{A_1} e_{a_2}^{A_2} e_{a_3}^{A_3} e_{a_4}^{A_4}. \tag{9.2}$$

There are infinite number of choices for the 4 vectors $e_a = e_a^A \partial/\partial \xi^A$. Since $n_A dn^A = 0$, we have

$$ds^2 = g_{\mu\nu} dy^\mu dy^\nu + (dy^5)^2, \tag{9.3}$$

where

$$g_{\mu\nu} = \frac{\partial \xi^A}{\partial y^\mu} \frac{\partial \xi_A}{\partial y^\nu} = (y^5)^2 \frac{\partial n^A}{\partial y^\mu} \frac{\partial n_A}{\partial y^\nu}. \tag{9.4}$$

Since

$$e_a = e_a^A \frac{\partial}{\partial \xi^A} = e_a^A \left(\frac{\partial y^\mu}{\partial \xi^A} \frac{\partial}{\partial y^\mu} + \frac{\partial y^5}{\partial \xi^A} \frac{\partial}{\partial y^5} \right) = e_a^A \left(\frac{\partial y^\mu}{\partial \xi^A} \frac{\partial}{\partial y^\mu} - n_A \frac{\partial}{\partial y^5} \right) = e_a^A \frac{\partial y^\mu}{\partial \xi^A} \frac{\partial}{\partial y^\mu}, \tag{9.5}$$

we have

$$e_a^\mu = e_a^A \frac{\partial y^\mu}{\partial \xi^A}. \tag{9.6}$$

From this relation, we have

$$\frac{\partial \xi^B}{\partial y^\mu} e_a^\mu = e_a^A \frac{\partial y^\mu}{\partial \xi^A} \frac{\partial \xi^B}{\partial y^\mu} = e_a^A (\delta_A^B + n_A n^B) = e_a^B. \tag{9.7}$$

So

$$\frac{\partial \xi^A}{\partial y^\mu} = e_a^A e_\mu^a, \tag{9.8}$$

from which it follows that

$$g_{\mu\nu} = \eta_{ab} e_\mu^a e_\nu^b. \tag{9.9}$$

Also from Eq.(9.6), we have(note that $\frac{\partial y^\mu}{\partial \xi^A} n^A = \frac{\partial y^\mu}{\partial \xi^A}\frac{\partial \xi^A}{\partial y^5} = \delta^\mu_5 = 0$)

$$e^a_B e^\mu_a = e^a_B e^A_a \frac{\partial y^\mu}{\partial \xi^A} = (\delta^A_B + n^A n_B)\frac{\partial y^\mu}{\partial \xi^A} = \frac{\partial y^\mu}{\partial \xi^B}. \tag{9.10}$$

Defining

$$\beta^\mu = \frac{\partial y^\mu}{\partial \xi^B}\gamma^B, \tag{9.11}$$

$$\beta^5 = n^A \gamma_A = \gamma(n), \tag{9.12}$$

we have $(\mu = 0, 1, 2, 3)$

$$\{\beta^\mu, \beta^5\} = 0, \tag{9.13}$$

$$\{\beta^\mu, \beta^\nu\} = g^{\mu\nu}, \tag{9.14}$$

where

$$g^{\mu\nu} = \frac{\partial y^\mu}{\partial \xi^A}\frac{\partial y^\nu}{\partial \xi_A} \tag{9.15}$$

Gürsey-Lee derived the Dirac equation in de Sitter spacetime from the 5-dimensional Lagrangian

$$L = \int_D [i\frac{1}{2}(\overline{\psi}\gamma^A \partial_{\xi_A}\psi - \partial_{\xi_A}\overline{\psi}\gamma^A\psi) - m\overline{\psi}\psi]d^5\xi, \tag{9.16}$$

where $\overline{\psi} = \psi^\dagger \gamma^0$. The integration region D is a shell bounded by $\sqrt{-\xi^A\xi_A} = R$ and $\sqrt{-\xi^A\xi_A} = R + \varepsilon$ where $\varepsilon > 0$ is an infinitesimal. The equation for ψ is given by $\delta L = 0$ for arbitrary $\delta\psi$ provided ψ is independent of y^5. Using the metric relation (9.3), we have

$$d^5\xi = \sqrt{-\det(g_{\mu\nu})}d^4y dy^5. \tag{9.17}$$

In terms of y^μ, β^μ, the Lagrangian becomes

$$L = \int_D [i\frac{1}{2}(\overline{\psi}\beta^\mu \partial_{y^\mu}\psi - \partial_{y^\mu}\overline{\psi}\beta^\mu\psi) - m\overline{\psi}\psi]\sqrt{-g}d^4y dy^5, \tag{9.18}$$

where $g = \det(g_{\mu\nu})$. Let $\overline{\psi}$ undergo an arbitrary y^5-independent variation $\delta\overline{\psi}$, we have

$$\delta L = \int_D [i\frac{1}{2}(\delta\overline{\psi}\beta^\mu \partial_{y^\mu}\psi - \partial_{y^\mu}\delta\overline{\psi}\beta^\mu\psi) - m\delta\overline{\psi}\psi]\sqrt{-g}d^4y dy^5 \tag{9.19}$$

$$= \int_D [i(\delta\overline{\psi}\beta^\mu \partial_{y^\mu}\psi - m\delta\overline{\psi}\psi)\sqrt{-g} - i\frac{1}{2}\delta\overline{\psi}\partial_{y^\mu}(\beta^\mu\sqrt{-g})\psi]d^4y dy^5. \tag{9.20}$$

Using the relation

$$\partial_{y^\mu}\left(\sqrt{-g}\frac{\partial y^\mu}{\partial \xi^B}\right) = \frac{4\xi_B}{-\xi^A\xi_A}\sqrt{-g}, \tag{9.21}$$

which will be proved later, we have,

$$\delta L = \int_D [i(\delta\overline{\psi}\beta^\mu\partial_{y^\mu}\psi - m\delta\overline{\psi}\psi) - i\delta\overline{\psi}\frac{2\beta^5}{y^5}\psi]\sqrt{-g}d^4ydy^5. \tag{9.22}$$

Here we need to take limit $\varepsilon \to 0$ and require that

$$\lim_{\varepsilon\to 0}\frac{\delta L}{\varepsilon} = 0, \tag{9.23}$$

which leads to

$$i\beta^\mu\partial_{y^\mu}\psi + \frac{2i}{R}\beta^5\psi - m\psi = 0. \tag{9.24}$$

Otherwise, we shall have

$$\frac{\delta L}{\delta\overline{\psi}(y^\mu)} = \int_0^\varepsilon [i(\beta^\mu\partial_{y^\mu}\psi - m\psi) - i\frac{2\beta^5}{y^5}\psi]\sqrt{-g}dy^5. \tag{9.25}$$

We can check with the Dirac operator derived from spinorial Gauss-Codazzi formula. Using

$$\frac{\partial}{\partial\xi^B}n^A = \frac{\delta^A_B + n^A n_B}{\sqrt{-\xi^C\xi_C}}, \tag{9.26}$$

we have the second fundamental form $K = \overline{\nabla}_A n^A = \frac{4}{R}$. Since $\frac{\partial\psi}{\partial y^5} = 0$ and

$$i\overline{D} = i\gamma^A\partial_A = i\gamma^A\left(\frac{\partial y^\mu}{\partial\xi^A}\partial_{y^\mu} + \frac{\partial y^5}{\partial\xi^A}\partial_{y^5}\right), \tag{9.27}$$

we have from the spinorial Gauss-Codazzi formula [Hijazi, Hijazi, Xiao (2001)] that

$$iD\psi = i\gamma^A\frac{\partial y^\mu}{\partial\xi^A}\partial_{y^\mu}\psi + i\frac{2}{R}\gamma(n)\psi, \tag{9.28}$$

which agrees with Eq.(9.24). Note that all five γ-matrices enter the first term.

Now we show that Eq.(9.24) is equivalent to Eq.(6.38). Defining $\alpha^\mu = e^\mu_a\gamma^a$ (here γ^a are the same as γ^A for $A = 0, 1, 2, 3$), then we have $\{\alpha^\mu, \alpha^\nu\} = 2g^{\mu\nu}$. Hence there must exits a matrix (A Lorentz transform on dS) such that

$$\alpha^\mu = S\beta^\mu S^{-1}, \tag{9.29}$$

and

$$S^{-1}\gamma^a S = \beta^a := e^a_\mu \beta^\mu, \tag{9.30}$$

here := means "defined as". Since $\{\alpha^\mu, \gamma^5\} = 0$, we have

$$S\beta^5 S^{-1} = \gamma^5. \tag{9.31}$$

This can also be proved rigorously as as follows.

Proof. From Eq.(9.29) we have

$$\beta^\mu = e^\mu_a S^{-1}\gamma^a S, \tag{9.32}$$

so

$$S^{-1}\gamma^a S = e^a_A \gamma^A. \tag{9.33}$$

Since $\gamma^5 = \frac{1}{4!}\varepsilon_{a_1 a_2 a_3 a_4 5}\gamma^{a_1}\gamma^{a_2}\gamma^{a_3}\gamma^{a_4}$, we have

$$S^{-1}\gamma^5 S = \frac{1}{4!}\varepsilon_{a_1 a_2 a_3 a_4 5} e^{a_1}_{A_1} e^{a_2}_{A_2} e^{a_3}_{A_3} e^{a_4}_{A_4}\gamma^{A_1}\gamma^{A_2}\gamma^{A_3}\gamma^{A_4}. \tag{9.34}$$

Using Eq.(9.2), we have then

$$S^{-1}\gamma^5 S = \varepsilon_{A_1 A_2 A_3 A_4} A n^A \gamma^{A_1}\gamma^{A_2}\gamma^{A_3}\gamma^{A_4}. \tag{9.35}$$

But

$$\gamma_A = \varepsilon_{A_1 A_2 A_3 A_4 A}\gamma^{A_1}\gamma^{A_2}\gamma^{A_3}\gamma^{A_4}. \tag{9.36}$$

$\square$

Using the relation

$$\alpha^\mu = S\frac{\partial y^\mu}{\partial \xi^A}\gamma^A S^{-1}, \tag{9.37}$$

we have

$$\frac{\partial \xi^B}{\partial y^\mu}\alpha^\mu = S\frac{\partial y^\mu}{\partial \xi^A}\gamma^A S^{-1}\frac{\partial \xi^B}{\partial y^\mu} = S\gamma^B S^{-1} - S\frac{\partial y^5}{\partial \xi^A}\gamma^A S^{-1}\frac{\partial \xi^B}{\partial y^5} \tag{9.38}$$

$$= S\gamma^B S^{-1} + Sn_A\gamma^A S^{-1}n^B.$$

So

$$\frac{\partial \xi^B}{\partial y^\mu}\alpha^\mu = S\gamma^B S^{-1} + \gamma^5 n^B. \tag{9.39}$$

Differentiating $S^{-1}\alpha^\mu S = \beta^\mu$, we have

$$S^{-1}[\Lambda_\lambda, \alpha^\mu]S + S^{-1}\partial_\lambda\alpha^\mu S = \partial_\lambda(\frac{\partial y^\mu}{\partial \xi^A})\gamma^A \tag{9.40}$$

where

$$\Lambda_\mu = S\partial_\mu S^{-1}. \tag{9.41}$$

So

$$[\Lambda_\lambda, \alpha^\mu] + \partial_\lambda \alpha^\mu = \partial_\lambda(\frac{\partial y^\mu}{\partial \xi^A})S\gamma^A S^{-1}. \tag{9.42}$$

Although we do not know the specific expression of S, we have from Eq.(9.39)

$$[\Lambda_\lambda, \alpha^\mu] + \partial_\lambda \alpha^\mu = \partial_\lambda(\frac{\partial y^\mu}{\partial \xi^A})\left[\frac{\partial \xi^A}{\partial y^\nu}\alpha^\nu - \gamma^5 n^A\right], \tag{9.43}$$

i.e.,

$$[\alpha^\mu, \Lambda_\lambda] = \partial_\lambda \alpha^\mu - \alpha^\nu \frac{\partial \xi^A}{\partial y^\nu}\partial_\lambda(\frac{\partial y^\mu}{\partial \xi^A}) + \gamma^5 n^A \partial_\lambda(\frac{\partial y^\mu}{\partial \xi^A}). \tag{9.44}$$

Defining

$$\Delta_\lambda := \frac{1}{2}e_\mu^a n^A \partial_\lambda(\frac{\partial y^\mu}{\partial \xi^A})\gamma^5\gamma_a, \tag{9.45}$$

we have

$$[\alpha^\mu, \Delta_\lambda] = -\gamma^5 n^A \partial_\lambda(\frac{\partial y^\mu}{\partial \xi^A}), \tag{9.46}$$

and

$$\begin{aligned}
\alpha^\lambda \Delta_\lambda &= \frac{1}{2}e_b^\lambda \gamma^b e_\mu^a n^A \partial_\lambda(\frac{\partial y^\mu}{\partial \xi^A})\gamma^5\gamma_a \\
&= -\frac{1}{2}e_b^\lambda \gamma^b e_\mu^a \frac{\partial y^\mu}{\partial \xi^A}\partial_\lambda n^A \gamma^5\gamma_a \\
&= -\frac{1}{2}e_b^\lambda \gamma^b e_\mu^a e_c^\mu e_A^c \partial_\lambda n^A \gamma^5\gamma_a \\
&= -\frac{1}{2}e_b^\lambda \gamma^b e_A^a \partial_\lambda n^A \gamma^5\gamma_a \\
&= -\frac{1}{2}e_b^B \gamma^b e_A^a(\frac{\partial y^\lambda}{\partial \xi^B}\partial_\lambda n^A)\gamma^5\gamma_a \\
&= -\frac{1}{2}e_b^B \gamma^b e_A^a(\frac{\partial}{\partial \xi^B}n^A - \frac{\partial y^5}{\partial \xi^B}\partial_5 n^A)\gamma^5\gamma_a. \tag{9.47}
\end{aligned}$$

Using $\frac{\partial y^5}{\partial \xi^B} = -n_B$, we have

$$\alpha^\lambda \Delta_\lambda = -\frac{1}{2\sqrt{-\xi^C \xi_C}}e_b^A \gamma^b e_A^a \gamma^5\gamma_a = \frac{2}{\sqrt{-\xi^C \xi_C}}\gamma^5. \tag{9.48}$$

So we can write

$$Si\beta^\mu S^{-1}\Lambda_\mu + \frac{2i}{R}\gamma^5 = i\alpha^\mu(\Lambda_\mu + \Delta_\mu). \tag{9.49}$$

Denoting

$$\omega_\lambda := \Lambda_\lambda - \Delta_\lambda, \tag{9.50}$$

we will show that ω_λ is just the spin connection in the generally covariant Dirac equation in curved spacetime. First we have

$$[\alpha^\mu, \omega_\lambda] = \partial_\lambda\alpha^\mu - \alpha^\nu\frac{\partial\xi^A}{\partial y^\nu}\partial_\lambda(\frac{\partial y^\mu}{\partial\xi^A}) = [\partial_\lambda e_a^\mu - e_a^\nu\frac{\partial\xi^A}{\partial y^\nu}\partial_\lambda(\frac{\partial y^\mu}{\partial\xi^A})]\gamma^a. \tag{9.51}$$

Hence

$$\{[\alpha^\mu, \omega_\lambda], \gamma_a\} = \partial_\lambda e_a^\mu - e_a^\nu\frac{\partial\xi^A}{\partial y^\nu}\partial_\lambda(\frac{\partial y^\mu}{\partial\xi^A}), \tag{9.52}$$

$$\{[\gamma_b, \omega_\lambda], \gamma_a\} = e_{b\mu}[\partial_\lambda e_a^\mu - e_a^\nu\frac{\partial\xi^A}{\partial y^\nu}\partial_\lambda(\frac{\partial y^\mu}{\partial\xi^A})]. \tag{9.53}$$

Let

$$\omega_\lambda = \frac{i}{4}\omega_{\lambda mn}\Sigma^{mn}. \tag{9.54}$$

Using

$$[\gamma^b, \Sigma_{mn}] = 2i(\delta_m^b\gamma_n - \delta_n^b\gamma_m). \tag{9.55}$$

We have

$$\omega_{\lambda ab} = e_{b\mu}[\partial_\lambda e_a^\mu - e_a^\nu\frac{\partial\xi^A}{\partial y^\nu}\partial_\lambda(\frac{\partial y^\mu}{\partial\xi^A})]. \tag{9.56}$$

Using Eqs.(9.8) and (9.10),

$$\begin{aligned}
\omega_{\lambda ab} &= e_{b\mu}[\partial_\lambda e_a^\mu - e_a^\nu e_c^A e_\nu^c\partial_\lambda(e_d^\mu e_A^d)] \\
&= e_{b\mu}\partial_\lambda e_a^\mu - e_{b\mu}e_a^A\partial_\lambda(e_d^\mu e_A^d) \\
&= -e_a^A\partial_\lambda e_{Ab} = -e_a^A\frac{\partial\xi^B}{\partial y^\lambda}\partial_B e_{Ab} = -e_a^A\frac{\partial\xi^B}{\partial y^\lambda}\overline{\nabla}_B e_{Ab}.
\end{aligned} \tag{9.57}$$

Here $\overline{\nabla}_B = \partial_{\xi^B}$ is the covariant differential operator in the 5-dimensional Minkowski spacetime. We use a bar over ∇, ω to represent covariant derivative, connection in the 5-dimensional Minkowski spacetime which is essentially flat. Recall that for the frame (e_a^A, n^A), the Levi-Civita connection components $\overline{\omega}_{Aab}$ are given by

$$\mathcal{D}_A e_{aB} := \overline{\nabla}_A e_{aB} - \overline{\omega}_{Aab}e_B^b - \overline{\omega}_{Aa5}e_B^5 = 0 \tag{9.58}$$

So

$$\overline{\omega}_{Aab} = e_b^B \overline{\nabla}_A e_{aB}, \tag{9.59}$$

we have

$$\omega_{\lambda ab} = \frac{\partial \xi^A}{\partial y^\lambda} \overline{\omega}_{Aab}. \tag{9.60}$$

Hence on dS, we have

$$\omega_{\mu ab} dy^\mu = \overline{\omega}_{Aab} d\xi^A_{|dS}. \tag{9.61}$$

This together with the fact that both $e_a = e_a^A \frac{\partial}{\partial \xi^A} = e_a^\mu \frac{\partial}{\partial y^\mu}$ and its dual 1-form $e_{aA} d\xi^A = e_{a\mu} dy^\mu$ (not only on dS), we can use coordinates-independent description. The first Maurer-Cartan equation is

$$de^a = \overline{\omega}^a{}_b \wedge e^b + \overline{\omega}^a{}_5 \wedge e^5. \tag{9.62}$$

Since

$$e^5 = n_A d\xi^A = -\frac{\partial y^5}{\partial \xi^A} d\xi^A = -\frac{\partial y^5}{\partial \xi^A} [\frac{\partial \xi^A}{\partial y^\mu} dy^\mu + \frac{\partial \xi^A}{\partial y^5} dy^5] = -\frac{\partial y^5}{\partial \xi^A} \frac{\partial \xi^A}{\partial y^5} dy^5, \tag{9.63}$$

we have $n_A d\xi^A_{|dS} = 0$. Hence,

$$de^a_{|dS} = \overline{\omega}^a{}_b \wedge e^b_{|dS}. \tag{9.64}$$

Therefore

$$de^a = \omega^a{}_b \wedge e^b, \tag{9.65}$$

which is exactly the first Maurer-Cartan equation on dS. So Eq.(9.24) becomes

$$\left[\alpha^\mu(i\partial_\mu - \frac{1}{4}\omega_{\mu ab}\Sigma^{ab}) - m\right]S\psi = 0. \tag{9.66}$$

Remark: One must have $\overline{\omega}^a{}_5 \wedge e^5 = 0$ since its components have the form $dy^\mu \wedge dy^5$ while other parts in Eq.(9.62) have only $dy^\mu \wedge dy^\nu$.

Combining $S^{-1}\gamma^5 S = e_A^5 \gamma^A, S^{-1}\gamma^a S = e_A^a \gamma^A$, we have

$$S^{-1}\gamma^A S = L^A{}_B \gamma^B, \tag{9.67}$$

where $L^A{}_B = (e^a{}_B, n_B)$. S is a function of L: $S = S(L)$

9.2 Proof of Eq.(9.21)

Differentiating

$$\frac{\partial \xi^A}{\partial y^\nu} \frac{\partial y^\mu}{\partial \xi^A} = \delta^\mu_\nu, \tag{9.68}$$

with respect to y^μ, we have

$$\frac{\partial y^\mu}{\partial \xi^A} \frac{\partial}{\partial y^\mu}\left(\frac{\partial \xi^A}{\partial y^\nu}\right) + \frac{\partial \xi^A}{\partial y^\nu} \frac{\partial}{\partial y^\mu}\left(\frac{\partial y^\mu}{\partial \xi^A}\right) = 0. \tag{9.69}$$

So

$$\frac{\partial y^\nu}{\partial \xi^B} \frac{\partial y^\mu}{\partial \xi^A} \frac{\partial}{\partial y^\mu}\left(\frac{\partial \xi^A}{\partial y^\nu}\right) + \frac{\partial y^\nu}{\partial \xi^B} \frac{\partial \xi^A}{\partial y^\nu} \frac{\partial}{\partial y^\mu}\left(\frac{\partial y^\mu}{\partial \xi^A}\right) = 0. \tag{9.70}$$

The second term is

$$\frac{\partial y^\nu}{\partial \xi^B} \frac{\partial \xi^A}{\partial y^\nu} \frac{\partial}{\partial y^\mu}\left(\frac{\partial y^\mu}{\partial \xi^A}\right) = \frac{\partial}{\partial y^\mu}\left(\frac{\partial y^\mu}{\partial \xi^B}\right) - \frac{\partial y^5}{\partial \xi^B} \frac{\partial \xi^A}{\partial y^5} \frac{\partial}{\partial y^\mu}\left(\frac{\partial y^\mu}{\partial \xi^A}\right). \tag{9.71}$$

Therefore

$$\begin{aligned}
\frac{\partial}{\partial y^\mu}\left(\frac{\partial y^\mu}{\partial \xi^B}\right) &= \frac{\partial y^5}{\partial \xi^B} \frac{\partial \xi^A}{\partial y^5} \frac{\partial}{\partial y^\mu}\left(\frac{\partial y^\mu}{\partial \xi^A}\right) + \frac{\partial y^\nu}{\partial \xi^B} \frac{\partial \xi^A}{\partial y^\nu} \frac{\partial}{\partial y^\mu}\left(\frac{\partial y^\mu}{\partial \xi^A}\right) \\
&= -\frac{\partial y^5}{\partial \xi^B} \frac{\partial y^\mu}{\partial \xi^A} \frac{\partial}{\partial y^\mu}\left(\frac{\partial \xi^A}{\partial y^5}\right) - \frac{\partial y^\nu}{\partial \xi^B} \frac{\partial y^\mu}{\partial \xi^A} \frac{\partial}{\partial y^\mu}\left(\frac{\partial \xi^A}{\partial y^\nu}\right).
\end{aligned} \tag{9.72}$$

But since $\frac{\partial \xi^A}{\partial y^5} = n^A$ which is independent of y^5, so

$$\frac{\partial y^\mu}{\partial \xi^A} \frac{\partial}{\partial y^\mu}\left(\frac{\partial \xi^A}{\partial y^5}\right) = \frac{\partial y^\mu}{\partial \xi^A} \frac{\partial}{\partial y^\mu}\left(\frac{\partial \xi^A}{\partial y^5}\right) + \frac{\partial y^5}{\partial \xi^A} \frac{\partial}{\partial y^5}\left(\frac{\partial \xi^A}{\partial y^5}\right) \tag{9.73}$$

$$= \frac{\partial}{\partial \xi^A}\left(\frac{\partial \xi^A}{\partial y^5}\right) = \frac{\partial}{\partial \xi^A} n^A = \frac{\partial}{\partial \xi^A}\left(\frac{\xi^A}{\sqrt{-\xi^B \xi_B}}\right) \tag{9.74}$$

$$= \frac{4}{\sqrt{-\xi^B \xi_B}}. \tag{9.75}$$

So

$$\frac{\partial}{\partial y^\mu}\left(\frac{\partial y^\mu}{\partial \xi^B}\right) = -\frac{\partial y^5}{\partial \xi^B} \frac{4}{\sqrt{-\xi^A \xi_A}} - \frac{\partial y^\nu}{\partial \xi^B} \frac{\partial y^\mu}{\partial \xi^A} \frac{\partial}{\partial y^\mu}\left(\frac{\partial \xi^A}{\partial y^\nu}\right) \tag{9.76}$$

$$= -\frac{4\xi_B}{\xi^A \xi_A} - \frac{\partial y^\nu}{\partial \xi^B} \frac{\partial y^\mu}{\partial \xi^A} \frac{\partial}{\partial y^\mu}\left(\frac{\partial \xi^A}{\partial y^\nu}\right). \tag{9.77}$$

From

$$\frac{\partial g}{\partial y^\mu} = g g^{\alpha\beta} \frac{\partial g_{\alpha\beta}}{\partial y^\mu} \qquad (9.78)$$

$$\frac{\partial y^\mu}{\partial \xi^A} \frac{\partial \xi^B}{\partial y^\mu} = \delta_A^B + n_A n^B \qquad (9.79)$$

$$\xi^A \frac{\partial y^\mu}{\partial \xi^A} = 0, \qquad (9.80)$$

we have

$$\frac{\partial}{\partial y^\mu} \sqrt{-g} = \sqrt{-g} g^{\alpha\beta} \frac{\partial \xi^A}{\partial y^\alpha} \frac{\partial}{\partial y^\mu} \left(\frac{\partial \xi_A}{\partial y^\beta} \right) \qquad (9.81)$$

$$= \sqrt{-g} \frac{\partial y^\alpha}{\partial \xi_C} \frac{\partial y^\beta}{\partial \xi^C} \frac{\partial \xi^A}{\partial y^\alpha} \frac{\partial}{\partial y^\mu} \left(\frac{\partial \xi_A}{\partial y^\beta} \right) \qquad (9.82)$$

$$= \sqrt{-g} \frac{\partial y^\beta}{\partial \xi^C} (\delta_C^A + n^A n_C) \frac{\partial}{\partial y^\mu} \left(\frac{\partial \xi_A}{\partial y^\beta} \right) \qquad (9.83)$$

$$= \sqrt{-g} \frac{\partial y^\nu}{\partial \xi^A} \frac{\partial}{\partial y^\mu} \left(\frac{\partial \xi^A}{\partial y^\nu} \right). \qquad (9.84)$$

By (9.77) and (9.84), we have

$$\frac{\partial}{\partial y^\mu} \left(\sqrt{-g} \frac{\partial y^\mu}{\partial \xi^B} \right) = -\frac{4\xi_B}{\xi^A \xi_A} \sqrt{-g}. \qquad (9.85)$$

9.3   de Sitter invariance

The invariance of the Eq.(9.66) under de Sitter transformation is hidden [Gürsey (1963)]. We show the invariance of Eq.(9.24). The 10 generators of de Sitter group are

$$J_{AB} = L_{AB} + \Sigma_{AB}, \qquad (9.86)$$

where $L_{AB} = i(\xi_A \partial_{\xi^B} - \xi_B \partial_{\xi^A})$, $\Sigma^{AB} = \frac{i}{4}[\gamma^A, \gamma^B]$. The de Sitter transformation is represented as $U = \exp(i\theta^{AB} J_{AB})$ with constant θ_{AB} and

$$\psi \to \psi' = U\psi = \exp(i\theta^{AB}\Sigma_{AB})\psi\left[\exp(-i\theta^{AB}L_{AB})\xi\right] =: U_S\psi(U_L^{-1}\xi), \qquad (9.87)$$

where $U_S := \exp(i\theta^{AB}\Sigma_{AB})$ and $U_L := \exp(i\theta^{AB}L_{AB})$. Since

$$U_S\gamma^A U_S^{-1} = (U_L^{-1})^A{}_B \gamma^B, \qquad (9.88)$$

and

$$U_L \frac{\partial y^\mu}{\partial \xi^A} U_L^{-1} = (U_L)_A{}^B \frac{\partial y^\mu}{\partial \xi^B}, \qquad (9.89)$$

where $\mathbf{U}_L$ is the vector representation of U_L. So

$$U\beta^\mu U^\dagger = \beta^\mu. \tag{9.90}$$

Similarly

$$U\beta^5 U^\dagger = U_L n_A U_L^{-1} U_S \gamma^A U_S^{-1} = n_A \gamma^A = \beta^5. \tag{9.91}$$

Note that under de Sitter transform, a point ξ on dS is transformed to another point still on dS. So the de Sitter invariance is proved.

Appendix A

Solutions of Dirac Equation of Hydrogen in Minkowski Spacetime

In order to clarify the notations used in the text of this book and the standard solutions of Dirac equation of Hydrogen in Minkowski spacetime, we provide this Appendix. The descriptions follow [Strange (2008)][1].

A.1 Pauli equation in a central potential

It is well known that spin, which is hidden in in relativistic quantum theory, does not appear in the Schrödinger equation. However, spin can be included within a non-relativistic framework, and this was done initially by Pauli [Pauli (1927)]. The Pauli equation can be written in matrix form (see, e.g., [Strange (2008)] [Baym (1967)]):

$$i\hbar\frac{\partial}{\partial t}\begin{pmatrix} \psi_\uparrow(\mathbf{r},t) \\ \psi_\downarrow(\mathbf{r},t) \end{pmatrix} =$$

$$\left(\frac{1}{2m}(\hat{\mathbf{p}} - e\mathbf{A}(\mathbf{r}))^2 + V(r) + \frac{\hbar}{4m^2c^2}\frac{1}{r}\frac{dV(r)}{dr}\boldsymbol{\sigma}\cdot\hat{\mathbf{L}}\right)\begin{pmatrix} \psi_\uparrow(\mathbf{r},t) \\ \psi_\downarrow(\mathbf{r},t) \end{pmatrix}, \quad (A.1)$$

where $\hat{\mathbf{L}} \equiv \mathbf{r} \times \hat{\mathbf{p}}$ represents the orbital angular momentum, and $\boldsymbol{\sigma} \equiv \sigma^1\mathbf{i} + \sigma^2\mathbf{j} + \sigma^3\mathbf{k}$ is (2×2)-Pauli matrices with

$$\sigma^1 = \begin{pmatrix} 0 & 1 \\ 1 & 0 \end{pmatrix}, \quad \sigma^2 = \begin{pmatrix} 0 & -i \\ i & 0 \end{pmatrix}, \quad \sigma^3 = \begin{pmatrix} 1 & 0 \\ 0 & -1 \end{pmatrix}.$$

The spin operator is

$$\hat{\mathbf{S}} = \frac{\hbar}{2}\boldsymbol{\sigma}, \quad with \quad \hat{S}_x = \frac{\hbar}{2}\sigma^1, \quad \hat{S}_y = \frac{\hbar}{2}\sigma^2, \quad \hat{S}_z = \frac{\hbar}{2}\sigma^3. \quad (A.2)$$

---

[1] There are, of cause, variant descriptions on this subject in popular textbooks, but they are basically equivalent. The presentation of [Strange (2008)] is adopted in this book.

And then we have

$$[\hat{L}_x, \hat{L}_y] = i\hbar\hat{L}_z, \ [\hat{L}_y, \hat{L}_z] = i\hbar\hat{L}_x, \ [\hat{L}_z, \hat{L}_x] = i\hbar\hat{L}_y, \ [\hat{\mathbf{L}}^2, \hat{L}_i] = 0, \quad (A.3)$$

$$[\hat{S}_x, \hat{S}_y] = i\hbar\hat{S}_z, \ [\hat{S}_y, \hat{S}_z] = i\hbar\hat{S}_x, \ [\hat{S}_z, \hat{S}_x] = i\hbar\hat{S}_y, \ [\hat{\mathbf{S}}^2, \hat{S}_i] = 0, \quad (A.4)$$

with $i = x, \ y, \ z$. The total angular momentum operator is:

$$\hat{\mathbf{J}} \equiv \hat{J}_x \mathbf{i} + \hat{J}_y \mathbf{j} + \hat{J}_z \mathbf{k} = \hat{\mathbf{L}} + \hat{\mathbf{S}} = \hat{\mathbf{L}} + \frac{\hbar}{2}\boldsymbol{\sigma}. \qquad (A.5)$$

Equation (A.1) is an equation of form

$$i\hbar\frac{\partial}{\partial t}\psi(\mathbf{r}, t) = \hat{H}\psi(\mathbf{r}, t), \qquad (A.6)$$

and so the variables can be separated in the usual way, giving us

$$\psi(\mathbf{r}, t) = \begin{pmatrix} \psi_\uparrow(\mathbf{r}, t) \\ \psi_\downarrow(\mathbf{r}, t) \end{pmatrix} = \begin{pmatrix} \psi_\uparrow(\mathbf{r}) \\ \psi_\downarrow(\mathbf{r}) \end{pmatrix} e^{-iEt/\hbar} = \psi(\mathbf{r})e^{-iEt/\hbar}, \qquad (A.7)$$

where E is an eigenvalue of the Hamiltonian operator, and

$$\left(\frac{1}{2m}(\hat{\mathbf{p}} - e\mathbf{A}(\mathbf{r}))^2 + V(r) + \frac{\hbar}{4m^2c^2}\frac{1}{r}\frac{dV(r)}{dr}\boldsymbol{\sigma}\cdot\hat{\mathbf{L}} - E \right) \begin{pmatrix} \psi_\uparrow(\mathbf{r}, t) \\ \psi_\downarrow(\mathbf{r}, t) \end{pmatrix} = 0.$$
$$(A.8)$$

This is the eigenvalue form of the Pauli equation.

We know from the standard separation of the variables of Schrödinger equation that the solutions in the absence of the spin-orbit coupling term take the form

$$\psi_{nlm}(\mathbf{r}) = R_{nl}(r, E)Y_l^m(\hat{\mathbf{r}}), \qquad (A.9)$$

where $Y_l^m(\hat{\mathbf{r}})$ are the usual spherical harmonics and $R_{nl}(e, E)$ are radial functions. Clearly it would be desirable to have solutions to Eq.(A.8) that are close to this form and reduce to Eq.(A.9) when the spin-orbit term tend to zero (as $c \to \infty$). We know that if the spin-orbit term in Eq.(A.8) is omitted the Hamiltonian will commute with $\hat{\mathbf{S}}^2$, $\hat{S}_z$, $\hat{\mathbf{L}}^2$, and $\hat{L}_z$. It commutes with the spin operators because there would be no spin dependence in the Hamiltonian at all, and it commutes with the orbital angular momentum operators as we know it does from the non-relativistic theory (see, e.g., [Gasiorowicz (1974)]).

Making use of Eqs.(A.3) and (A.4), we have

$$[\hat{S}_z, \hat{\mathbf{L}}\cdot\hat{\mathbf{S}}] = -i\hbar(\hat{S}_x\hat{L}_y - \hat{S}_y\hat{L}_x), \qquad (A.10)$$

$$[\hat{L}_z, \hat{\mathbf{L}}\cdot\hat{\mathbf{S}}] = -i\hbar(\hat{L}_x\hat{S}_y - \hat{L}_y\hat{S}_x). \qquad (A.11)$$

Now look what happens if we add Eqs.(A.10) and (A.11) together

$$[\hat{S}_z, \hat{\mathbf{L}} \cdot \hat{\mathbf{S}}] + [\hat{L}_z, \hat{\mathbf{L}} \cdot \hat{\mathbf{S}}] = [\hat{L}_z + \hat{S}_z, \hat{\mathbf{L}} \cdot \hat{\mathbf{S}}] = [\hat{J}_z, \hat{\mathbf{L}} \cdot \hat{\mathbf{S}}]$$
$$= -i\hbar(\hat{S}_x\hat{L}_y - \hat{S}_y\hat{L}_x) - i\hbar(\hat{L}_x\hat{S}_y - \hat{L}_y\hat{S}_x) = 0. \tag{A.12}$$

Equations (A.10) and (A.11) show that neither $\hat{S}_z$ nor $\hat{L}_z$ commutes with the Pauli Hamiltonian. However, Eq.(A.12) shows that, despite this, $\hat{J}_z$ does commute with the Pauli Hamiltonian. Clearly we could have gone through the above mathematics for the x- and y-components of $\hat{\mathbf{J}}$ and found that they also commute with the Hamiltonian. We know that if a quantity commutes with Hamiltonian it is conserved and may have simultaneous eigenfunctions with the Hamiltonian. Hence we expect our solutions to Eq.(A.8) to be simultaneous eigenfunctions of $\hat{H}$, $\hat{\mathbf{J}}^2$ and $\hat{J}_z$. Furthermore, Eqs.(A.10) to (A.12) (and their cyclic permutations) are proofs that in relativistic quantum mechanics the spin angular momentum and orbital angular momentum are not conserved separately, but that total angular momentum is conserved.

Making use the matrix form of $\hat{J}_z$ we have

$$\hat{J}_z\psi(r) = \begin{pmatrix} \hat{L}_z + \hbar/2 & 0 \\ 0 & \hat{L}_z - \hbar/2 \end{pmatrix} \begin{pmatrix} \psi_\uparrow(\mathbf{r}) \\ \psi_\downarrow(\mathbf{r}) \end{pmatrix}$$
$$= \begin{pmatrix} \hat{L}_z + \hbar/2 & 0 \\ 0 & \hat{L}_z - \hbar/2 \end{pmatrix} \begin{pmatrix} \psi_\uparrow(\mathbf{r}) \\ \psi_\downarrow(\mathbf{r}) \end{pmatrix} = m_j\hbar \begin{pmatrix} \psi_\uparrow(\mathbf{r}) \\ \psi_\downarrow(\mathbf{r}) \end{pmatrix}, \tag{A.13}$$

and then

$$\begin{pmatrix} (\hat{L}_z - (m_j - 1/2)\hbar)\psi_\uparrow(\mathbf{r}) \\ (\hat{L}_z - (m_j + 1/2)\hbar)\psi_\downarrow(\mathbf{r}) \end{pmatrix} = 0. \tag{A.14}$$

A solution of this is that $\psi_\uparrow$ and $\psi_\downarrow$ both have an angular dependence which is defined by a spherical harmonic with the same value of the l quantum number and with values of the m_l quantum number differing by one. If this is the case then the eigenfunctions of the Hamiltonian are also eigenfunctions of $\hat{\mathbf{L}}^2$. So we have

$$\psi_\uparrow(\mathbf{r}) = g(r)Y_l^{m_l}(\hat{\mathbf{r}}), \qquad \psi_\downarrow(\mathbf{r}) = f(r)Y_l^{m_l+1}(\hat{\mathbf{r}}). \tag{A.15}$$

Next we are going to define the operator $\hat{K}_1$.

$$\hat{K}_1 = \hat{\mathbf{L}} \cdot \boldsymbol{\sigma} + \hbar. \tag{A.16}$$

This turns out to be a useful operator, although here it has been defined rather arbitrarily. The $\hbar$ in Eq.(A.16) is understood to be multiplied by unit matrix.

Let us see what the eigenvalues of $\hat{K}_1$ are. To discover this we will operate with it on ψ, and use the square of equation (A.5).

$$\hat{K}_1\psi = (\hat{\mathbf{L}} \cdot \boldsymbol{\sigma} + \hbar)\psi = \left(\frac{2}{\hbar}\hat{\mathbf{L}} \cdot \hat{\mathbf{S}} + \hbar\right)\psi$$

$$= \left(\frac{2}{\hbar}\frac{1}{2}(\hat{\mathbf{J}}^2 - \hat{\mathbf{L}}^2 - \hat{\mathbf{S}}^2) + \hbar\right)\psi$$

$$= \left(j(j+1) - l(l+1) + \frac{1}{4}\right) = -\hbar\kappa\psi, \qquad (A.17)$$

where we have explicitly used the fact that $s = 1/2$ for all the particles in which we are interested. In Eq.(A.17) we have defined the eigenvalues of $\hat{K}_1$ as $-\kappa\hbar$, but the expression for them can be simplified considerably. There are two possibilities, $j = l + 1/2$ and $j = l - 1/2$. In each case it is simple to use Eq.(A.17) to show that

$$j = l + 1/2 \quad \Rightarrow \quad \kappa = -l - 1, \qquad (A.18)$$

$$j = l - 1/2 \quad \Rightarrow \quad \kappa = l. \qquad (A.19)$$

So we see that the eigenvalues of this operator are integers and an equivalent representation of the total momentum to $\hat{\mathbf{J}}$. If $\kappa < 0$ we have $j = l + 1/2$ and if $\kappa > 0$ we have $j = l - 1/2$. There are two values of κ for every value of l. Finally note that any integer value of κ is permissible except $\kappa = 0$. Let us operate with $\hat{K}_1$ on $\psi(\mathbf{r})$

$$\hat{K}_1\psi(\mathbf{r}) = \begin{pmatrix} \hat{L}_z + \hbar & \hat{L}_x - i\hat{L}_y \\ \hat{L}_x + i\hat{L}_y & \hat{L}_z - \hbar \end{pmatrix} = -\hbar\kappa \begin{pmatrix} \psi_\uparrow(\mathbf{r}) \\ \psi_\downarrow(\mathbf{r}) \end{pmatrix}. \qquad (A.20)$$

The component form of (A.20) is:

$$(\hat{L}_z + \hbar)\psi_\uparrow + (\hat{L}_x - i\hat{L}_y)\psi_\downarrow = -\hbar\kappa\psi_\uparrow \qquad (A.21)$$

$$(-\hat{L}_z + \hbar)\psi_\downarrow + (\hat{L}_x + i\hat{L}_y)\psi_\uparrow = -\hbar\kappa\psi_\downarrow. \qquad (A.22)$$

It turns out that Eq.(A.22) is just a multiple of Eq.(A.21), so we only need to work with one of them. Noting the following formulas

$$\hat{L}_\pm \equiv \hat{L}_x \pm i\hat{L}_\pm, \qquad (A.23)$$

$$\hat{L}_+Y_l^{m_l} = (l - m_l)^{1/2}(l + m_l + 1)^{1/2}\hbar Y_l^{m_l+1}, \qquad (A.24)$$

$$\hat{L}_-Y_l^{m_l} = (l + m_l)^{1/2}(l - m_l + 1)^{1/2}\hbar Y_l^{m_l-1}, \qquad (A.25)$$

and substituting Eq.(A.15) into Eq.(A.21) we have

$$f(r)[(l + m_l + 1)(l - m_l)]^{1/2} Y_l^{m_l} = g(r)(-\kappa - m_l - 1) Y_l^{m_l}. \quad (A.26)$$

Here the spherical harmonics cancel, and this relation shows us that the radial parts of $\psi_\uparrow$ and $\psi_\downarrow$ only differ from one another by a constant. For $\kappa < 0$ Eq.(A.26) gives

$$(l + m_l + 1)^{1/2} f(r) = (l - m_l)^{1/2} g(r), \qquad (\kappa = -l - 1). \quad (A.27)$$

For $\kappa > 0$ we have

$$(l - m_l)^{1/2} f(r) = -(l + m_l + 1)^{1/2} g(r), \qquad (\kappa = l). \quad (A.28)$$

These equations show that for a particular set of quantum numbers, the radial parts of the eigenfunctions defined by Eq.(A.15) are the same except for a simple multiplicative factor. So we can write

$$\psi(\mathbf{r}) = f(r) \begin{pmatrix} c_1 Y_l^{m_l}(\hat{\mathbf{r}}) \\ c_2 Y_l^{m_l+1}(\hat{\mathbf{r}}) \end{pmatrix}. \quad (A.29)$$

We have shown the relative sizes of c_1 and c_2 in Eqs.(A.27) and (A.28), but their absolute values depend upon the normalization:

$$\int d^3\mathbf{r} \left(\psi_\uparrow^*(\mathbf{r})\psi_\uparrow(\mathbf{r}) + \psi_\downarrow^*(\mathbf{r})\psi_\downarrow(\mathbf{r}) \right) = 1. \quad (A.30)$$

There is still some arbitrariness in this integral because we can choose the values of the constants c and the radial part of the wavefunction in Eq.(A.29) in an infinite number of ways and still satisfy Eq.(A.30), but the usual choice is:

$$\int r^2 |f(r)|^2 dr = 1 \quad (A.31)$$

and hence

$$|c_1|^2 + |c_2|^2 = 1. \quad (A.32)$$

Comparison of Eq.(A.32) and Eqs.(A.27) and (A.28) gives

$$c_1 = \left(\frac{l - m_l}{2l + 1} \right)^{1/2}, \quad c_2 = \left(\frac{l + m_l + 1}{2l + 1} \right)^{1/2} \quad (\kappa < 0); \quad (A.33)$$

$$c_1 = -\left(\frac{l + m_l + 1}{2l + 1} \right)^{1/2}, \quad c_2 = \left(\frac{l - m_l}{2l + 1} \right)^{1/2} \quad (\kappa > 0). \quad (A.34)$$

Finally, since function of Eq.(A.29) is a row function with two components, and $\{Y_l^{m_l},\ Y_l^{m_l+1}\}$ are the common eigenfunctions of $\{\hat{\mathbf{L}}^2,\ \hat{L}_z\}$, when we introduce notations:

$$\chi_{1/2}^{1/2} = \begin{pmatrix} 1 \\ 0 \end{pmatrix}, \quad \chi_{1/2}^{-1/2} = \begin{pmatrix} 0 \\ 1 \end{pmatrix}, \tag{A.35}$$

the function of (A.29) can be written to be

$$\psi(\mathbf{r}) = f(r)\left(c_1 Y_l^{m_l}(\hat{\mathbf{r}})\chi_{1/2}^{1/2} + c_2 Y_l^{m_l+1}(\hat{\mathbf{r}})\chi_{1/2}^{-1/2}\right). \tag{A.36}$$

In this case c_1, c_2 could be understood as suitable Clebsch-Gordan coefficients. We will show this issue below.

A.2   Relativistic   quantum   numbers   and   spin-angular functions

In Eq.(A.17) we introduced the quantum number κ. Here we are going to introduce two more numbers which turn out to be useful, and we will also summarize all the relations between quantum numbers.

$$\kappa = -l - 1 = -(j + 1/2) \quad (j = l + 1/2), \tag{A.37}$$
$$\kappa = l = (j + 1/2) \quad (j = l - 1/2). \tag{A.38}$$

It will also be worthwhile to define the quantum number $\bar{l}$, which is the value of the l quantum number associated with $-\kappa$

$$\bar{l} = l + 1 = -\kappa \quad (\kappa < 0),$$
$$\bar{l} = l - 1 = \kappa - 1 \quad (\kappa > 0). \tag{A.39}$$

We also define

$$S_\kappa = \frac{\kappa}{|\kappa|}, \tag{A.40}$$

so that

$$S_\kappa = -1 \quad (j = l + 1/2), \tag{A.41}$$
$$S_\kappa = +1 \quad (j = l - 1/2), \tag{A.42}$$

and

$$S_\kappa = l - \bar{l} = 2(l - j). \tag{A.43}$$

We are now in a position to examine the properties of the angular part of our solution to the Pauli equation in a central field. The symbol $\chi_\kappa^{m_j}(\hat{\mathbf{r}})$ is usually used to describe these quantities and they are commonly known as spin-angular function. Using expressions of the spin eigenvectors

$$\chi^{1/2} = \begin{pmatrix} 1 \\ 0 \end{pmatrix}, \quad \chi^{-1/2} = \begin{pmatrix} 0 \\ 1 \end{pmatrix}, \tag{A.44}$$

and Eqs.(A.29) and (A.36) we have

$$\psi_j^{m_j}(\mathbf{r}) = f(r) \sum_{m_s} C(l\tfrac{1}{2}j; m_j - m_s, m_s) Y_l^{m_j - m_s}(\hat{\mathbf{r}}) \chi^{m_s}, \tag{A.45}$$

where we have replaced the subscripts 1 and 2 (in Eq.(A.29)) with the quantum numbers j and m_j, and $C(l\tfrac{1}{2}j; m_j - m_s, m_s)$ is the Clebsch-Gordan coefficients listed in Table A.1.

Table A.1 The Clebsch-Gordan coefficients $C(l\tfrac{1}{2}j; m_j - m_s, m_s)$.

j	$m_s = 1/2$	$m_s = -1/2$
$l - 1/2$	$\left(\frac{l+m_j+1/2}{2l+1}\right)^{1/2}$	$-\left(\frac{l-m_j+1/2}{2l+1}\right)^{1/2}$
$l + 1/2$	$\left(\frac{l-m_j+1/2}{2l+1}\right)^{1/2}$	$\left(\frac{l+m_j+1/2}{2l+1}\right)^{1/2}$

Equation (A.45) can also been written in form:

$$\psi_j^{m_j}(\mathbf{r}) \equiv f(r) \chi_\kappa^{m_j}(\hat{\mathbf{r}})$$

$$= f(r) \sum_{m_s=-1/2}^{1/2} C_{l,m_j-m_s;\,1/2,m_s}^{j\,m_j} Y_l^{m_j-m_s}(\hat{\mathbf{r}}) \chi^{m_s}, \tag{A.46}$$

where

$$\chi_\kappa^{m_j}(\hat{\mathbf{r}}) = \sum_{m_s=-1/2}^{1/2} C_{l,m_j-m_s;\,1/2,m_s}^{j\,m_j} Y_l^{m_j-m_s}(\hat{\mathbf{r}}) \chi^{m_s}, \tag{A.47}$$

$$\text{with} \quad \chi^{m_s=1/2} = \begin{pmatrix} 1 \\ 0 \end{pmatrix}, \quad \chi^{m_s=-1/2} = \begin{pmatrix} 0 \\ 1 \end{pmatrix}.$$

We further introduce notation:

$$\chi_{-\kappa}^{m_j}(\hat{\mathbf{r}}) = -\sigma_r \chi_\kappa^{m_j}(\hat{\mathbf{r}}),$$

$$\text{with} \quad \sigma_r = \hat{\mathbf{r}} \cdot \vec{\sigma} = \begin{pmatrix} \cos\theta & \sin\theta e^{-i\phi} \\ \sin\theta e^{i\phi} & -\cos\theta \end{pmatrix}. \tag{A.48}$$

A.3 Dirac equation for Hydrogen

The Dirac equation in Minkowski spacetime for a spherical potential $V(\mathbf{r}) = V(r)$ is

$$(c\boldsymbol{\alpha} \cdot \hat{\mathbf{p}} + \beta mc^2 + V(r))\psi(\mathbf{r}) = W\psi(\mathbf{r}). \qquad (A.49)$$

where

$$\boldsymbol{\alpha} = \begin{pmatrix} 0 & \boldsymbol{\sigma} \\ \boldsymbol{\sigma} & 0 \end{pmatrix}, \quad \beta = \begin{pmatrix} I & 0 \\ 0 & -I \end{pmatrix},$$

where $\boldsymbol{\sigma} \equiv \sigma^1 \mathbf{i} + \sigma^2 \mathbf{j} + \sigma^3 \mathbf{k}$ is (2×2)-Pauli matrix, and I is (2×2)-unit matrix. For Hydrogen atom, $m = m_e$ and $V(r) = e/r$. The origin is taken at the nucleus. For this problem it is most convenient to solve the Dirac equation in spherical polar coordinates. Obviously the βmc^2 term is independent of coordinate system and the potential is most easily written in radial form. So we only have to transform the kinetic energy term. To do this we use a vector identity

$$\nabla = \hat{\mathbf{r}}(\hat{\mathbf{r}} \cdot \nabla) - \hat{\mathbf{r}} \times (\hat{\mathbf{r}} \times \nabla), \qquad (A.50)$$

where $\hat{\mathbf{r}}$ is a unit radial vector: $\hat{\mathbf{r}} \equiv \mathbf{r}/|\mathbf{r}| = \mathbf{r}/r$. Equation (A.50) follows directly from the vector identity:

$$\mathbf{A} \times (\mathbf{B} \times \mathbf{C}) = \mathbf{B}(\mathbf{A} \cdot \mathbf{C}) - \mathbf{C}(\mathbf{A} \cdot \mathbf{B}).$$

The orbital angular momentum operator is given by $\hat{\mathbf{L}} = \hat{\mathbf{r}} \times \hat{\mathbf{p}} = -i\hbar\mathbf{r} \times \nabla$. Here $\hat{\mathbf{r}}$ is the vector operator defining the position of the electron relative to the source of the field. For our case of a spherically symmetric potential, Eq.(A.50) simplifies because $\partial/\partial\theta$ and $\partial/\partial\phi$ are both equal to zero so $\mathbf{r} \cdot \nabla = r\partial/\partial r$. Thus, with some trivial manipulation, Eq.(A.50) can be written

$$\nabla = \hat{\mathbf{r}}\frac{\partial}{\partial r} - \frac{i}{\hbar}\frac{\hat{\mathbf{r}}}{|\mathbf{r}|} \times \hat{\mathbf{L}}. \qquad (A.51)$$

We write using it the relativistic kinetic energy operator in radial form

$$\boldsymbol{\alpha} \cdot \hat{\mathbf{p}} = -i\hbar\boldsymbol{\alpha} \cdot \nabla = -i\hbar \cdot \hat{\mathbf{r}}\frac{\partial}{\partial r} - \frac{1}{|\mathbf{r}|}\boldsymbol{\alpha} \cdot \hat{\mathbf{r}} \times \hat{\mathbf{L}}. \qquad (A.52)$$

Noting the operator $\hat{\mathbf{L}}$ is perpendicular to $\mathbf{r}$, i.e., $\hat{L}^i\hat{r}^i = 0$, and

$$\alpha^i\alpha^j = i\epsilon^{ijk}\Sigma^k + \delta^{ij}, \quad with \quad \Sigma^k = \begin{pmatrix} \sigma^k & 0 \\ 0 & \sigma^k \end{pmatrix},$$

we can see that

$$\boldsymbol{\alpha} \cdot \hat{\boldsymbol{r}} \boldsymbol{\alpha} \cdot \hat{\mathbf{L}} = i\boldsymbol{\Sigma} \cdot \hat{\mathbf{r}} \times \hat{\mathbf{L}}. \tag{A.53}$$

We know that γ_5 commutes with the $\boldsymbol{\alpha}$- and the $\boldsymbol{\Sigma}$-matrices. Now, all these matrices are closely related to one another through $\gamma_5 \boldsymbol{\alpha} = -\boldsymbol{\Sigma}$ and $\gamma_5 \boldsymbol{\Sigma} = -\boldsymbol{\alpha}$. So we can postmultiply each side of Eq.(A.53) by γ_5 to find

$$i\boldsymbol{\alpha} \cdot \hat{\mathbf{r}} \boldsymbol{\Sigma} \cdot \hat{\mathbf{L}} = -\boldsymbol{\alpha} \cdot \hat{\mathbf{r}} \times \hat{\mathbf{L}}. \tag{A.54}$$

Now we can substitute Eq.(A.54) into the second term of Eq.(A.52)

$$\boldsymbol{\alpha} \cdot \hat{\mathbf{p}} = -i\hbar \boldsymbol{\alpha} \cdot \hat{\mathbf{p}} \frac{\partial}{\partial r} + \frac{i}{|\mathbf{r}|} \boldsymbol{\alpha} \cdot \hat{\mathbf{r}} \boldsymbol{\Sigma} \cdot \hat{\mathbf{L}}. \tag{A.55}$$

We define $\hat{K}$ operator:

$$\hat{K} \equiv \beta(\boldsymbol{\Sigma} \cdot \hat{\mathbf{L}} + \hbar). \tag{A.56}$$

(note it is different from $\hat{K}_1$ in Eq.(A.16)). This can be substituted into Eq.(A.55) to give

$$\boldsymbol{\alpha} \cdot \hat{\mathbf{p}} = -i\hbar \alpha_r + \frac{i}{|\mathbf{r}|} \alpha_r (\beta \hat{K} - \hbar) \tag{A.57}$$

with $\alpha_r \equiv \boldsymbol{\alpha} \cdot \hat{\mathbf{r}}$. This is form in which we want $\boldsymbol{\alpha} \cdot \mathbf{p}$ to substitute into the Dirac equation (A.49). Doing this and a little algebra leads to

$$\left[ic\gamma_5 \Sigma_r \left(\hbar \frac{\partial}{\partial r} + \frac{\hbar}{r} - \frac{\beta \hat{K}}{r} \right) + \beta mc^2 + V(r) \right] \psi(r) = W \psi(r), \tag{A.58}$$

with $\Sigma_r = \boldsymbol{\Sigma} \cdot \hat{\mathbf{r}}$. We make use of Eq.(A.45) and are going to insist that the four-component eigenfunctions take on a form as similar as possible to that found when we discussed two-component Pauli spinor, i.e.,

$$\psi_\kappa^{m_j}(\mathbf{r}) = \begin{pmatrix} g_\kappa(r) \chi_\kappa^{m_j}(\hat{\mathbf{r}}) \\ if_\kappa(r) \chi_{-\kappa}^{m_j}(\hat{\mathbf{r}}) \end{pmatrix}, \tag{A.59}$$

where the $\chi_\kappa^{m_j}(\hat{\mathbf{r}})$ and $\chi_{-\kappa}^{m_j}(\hat{\mathbf{r}})$ are the spin-angular functions of Eqs.(A.47) and (A.48). This is a reasonable form for the wavefunctions. The reason for the negative κ and the i in the lower part of the wavefunction will become clear shortly. This choice of eigenfunction means that the solutions are also eigenfunction of $\hat{\mathbf{J}}^2$, $\hat{J}_z$ and $\hat{K}$. If we substitute Eq.(A.59) into Eq.(A.58) we obtain four equations, of cause, but it turns out that there are really only two separate equations. Remembering that the eigenvalue of the $\hat{K}$ operator is $-\hbar\kappa$, these are

$$-c\hbar\left(\frac{\partial}{\partial r}+\frac{1}{r}+\frac{\kappa}{r}\right)g_\kappa(r)\chi_{-\kappa}^{m_j}+(W-V(r)+mc^2)f_\kappa(r)\chi_{-\kappa}^{m_j}=0, \quad (A.60)$$

$$-c\hbar\left(\frac{\partial}{\partial r}+\frac{1}{r}-\frac{\kappa}{r}\right)f_\kappa(r)\chi_{\kappa}^{m_j}+(W-V(r)-mc^2)g_\kappa(r)\chi_{\kappa}^{m_j}=0. \quad (A.61)$$

Now the $\chi_\kappa^{m_j}$ cancel nicely and the role of i in the solutions (A.59) is apparent. It was inserted to ensure that the radial part of the eigenfunction in manifestly real. We can write these equations in the form

$$\frac{\partial g_\kappa(r)}{\partial r}=-\frac{\kappa+1}{r}g_\kappa(r)+\frac{1}{c\hbar}(W-V(r)+mc^2)f_\kappa(r), \quad (A.62)$$

$$\frac{\partial f_\kappa(r)}{\partial r}=\frac{\kappa-1}{r}f_\kappa(r)+\frac{1}{c\hbar}(W-V(r)-mc^2)g_\kappa(r). \quad (A.63)$$

This is the usual form of the radial Dirac equation. It is sometimes useful to make the substitution $u_\kappa(r)=rg_\kappa(r)$ and $v_\kappa(r)=rf_\kappa(r)$. In that case Eqs.(A.62) and (A.63) become

$$\frac{\partial u_\kappa(r)}{\partial r}=-\frac{\kappa}{r}u_\kappa(r)+\frac{1}{c\hbar}(W-V(r)+mc^2)v_\kappa(r), \quad (A.64)$$

$$\frac{\partial v_\kappa(r)}{\partial r}=\frac{\kappa}{r}v_\kappa(r)-\frac{1}{c\hbar}(W-V(r)-mc^2)u_\kappa(r). \quad (A.65)$$

Let's consider one-electron atoms. Namely, we are going to study the motion of one electron of mass m moving under the influence of the spherically symmetric Coulomb potential

$$V(r)=-\frac{Ze^2}{r}. \quad (A.66)$$

For ease of writing we put

$$\xi\equiv\frac{Ze^2}{c\hbar}=Z\alpha. \quad (A.67)$$

Equations (A.64) and (A.65) become

$$\frac{\partial u_\kappa(r)}{\partial r}=-\frac{\kappa}{r}u_\kappa(r)+\left(W_C+\frac{\xi}{r}+k_C\right)v_\kappa(r), \quad (A.68)$$

$$\frac{\partial v_\kappa(r)}{\partial r}=\frac{\kappa}{r}v_\kappa(r)-\left(W_C+\frac{\xi}{r}-k_C\right)u_\kappa(r), \quad (A.69)$$

where $k_C = mc/\hbar$ is the Compton wavevector, and $W_C = W/(c\hbar)$. Introducing $\phi_1(r)$, $\phi_2(r)$ through following equations

$$u_\kappa(r) = (k_C + W_C)^{1/2}e^{-\lambda r}(\phi_1 + \phi_2), \tag{A.70}$$

$$v_\kappa(r) = (k_C + W_C)^{1/2}e^{-\lambda r}(\phi_1 - \phi_2), \tag{A.71}$$

where $\lambda \equiv (k_C - W_C)^{1/2}$, and substituting Eqs.(A.70) and (A.71) into Eqs.(A.68) and (A.69) we find:

$$\lambda\frac{\partial\phi_1}{\partial r} - 2\lambda^2\phi_1 + \frac{\kappa\lambda}{r}\phi_2 + \frac{\xi W_C}{r}\phi_1 + \frac{\xi k_C}{r}\phi_2 = 0, \tag{A.72}$$

$$\lambda\frac{\partial\phi_2}{\partial r} + \frac{\kappa\lambda}{r}\phi_1 - \frac{\xi W_C}{r}\phi_1 - \frac{\xi k_C}{r}\phi_1 = 0. \tag{A.73}$$

Let us change to a dimensionless variable $\rho \equiv 2\lambda r$ and then we have

$$\frac{\partial\phi_1}{\partial\rho} = \left(1 - \frac{\xi W_C}{\lambda\rho}\right)\phi_1 - \left(\frac{\kappa}{\rho} + \frac{\xi k_C}{\lambda\rho}\right)\phi_2, \tag{A.74}$$

$$\frac{\partial\phi_2}{\partial\rho} = \frac{\xi W_C}{\lambda\rho}\phi_2 + \left(-\frac{\kappa}{\rho} + \frac{\xi k_C}{\lambda\rho}\right)\phi_1. \tag{A.75}$$

We solve them in the following.

A power series in ρ for $\phi_1(\rho)$ and $\phi_2(\rho)$ is:

$$\phi_1(\rho) = \rho^s \sum_{m=0}^{\infty} \alpha_m\rho^m, \tag{A.76}$$

$$\phi_2(\rho) = \rho^s \sum_{m=0}^{\infty} \beta_m\rho^m. \tag{A.77}$$

(There should be no confusion between the coefficients α_m and α-matrices in Dirac equation (A.49) and etc.) Putting Eqs.(A.76) and (A.77) into Eqs.(A.74) and (A.75) is straightforward. Then if we compare coefficients of equal powers of ρ in each term we find the following recursion relation for the α_m and β_m coefficients:

$$\alpha_m(m + s) = \alpha_{m-1} - \frac{W_C\xi}{\lambda}\alpha_m - \left(\kappa + \frac{\xi k_C}{\lambda}\right)\beta_m, \tag{A.78}$$

$$\beta_m(m + s) = \left(-\kappa + \frac{\xi k_C}{\lambda}\right)\alpha_m + \frac{W_C\xi}{\lambda}\beta_m. \tag{A.79}$$

For $m > 0$ this determines all the α_m and β_m coefficients in terms of α_0 and β_0. However, as yet we do not know these, nor do we know the correct value of s. To find the latter we let $m = 0$ in Eqs.(A.78) and (A.79).

Then, by definition, from Eqs.(A.76) and (A.77), $\alpha_{-1} = 0$ and we have a pair of simultaneous equations for α_0 and β_0. These equations will have a non-trivial solution if the determinant of the coefficients is equal to zero.

$$\begin{vmatrix} s + W_C\xi/\lambda & \kappa + \xi k_C/\lambda \\ \kappa - \xi k_C/\lambda & s - W_C\xi/\lambda \end{vmatrix} = 0, \tag{A.80}$$

i.e.,

$$s^2 - \frac{W_C^2\xi^2}{\lambda^2} = \kappa^2 - \xi^2 k_C^2/\lambda^2. \tag{A.81}$$

But, by definition, $\lambda^2 = k_C^2 - W_C^2$ and so we are left with

$$s = +\sqrt{\kappa^2 - \xi^2}. \tag{A.82}$$

Here we give up $s < 0$-solution, because the positive square root in Eq.(A.82) ensures the wavefunctions are always square integrable, but the $s < 0$ solution cannot. The next stage in this analysis is to find α_m in terms of α_{m-1} and β_m in terms of β_{m-1}. This can be done using Eqs.(A.78) and (A.79). Rearranging Eq.(A.79) gives us

$$\frac{\beta_m}{\alpha_m} = \frac{\kappa - \xi k_C/\lambda}{W_C\xi/\lambda - m - s} = \frac{\kappa - \xi k_C/\lambda}{n' - m}, \tag{A.83}$$

where $n' \equiv (W_C\xi/\lambda) - s$. Combining Eqs.(A.83) and (A.78) gives

$$\alpha_m = -\frac{n' - m}{m(2s + m)}\alpha_{m-1}. \tag{A.84}$$

We now have α_m in terms of α_{m-1}, but m just represents an arbitrary term in the expansions (A.76). Using Eq.(A.84) but reducing m by one means we can find α_{m-1} in terms of α_{m-2}. This process can be continued down to α_0 to give us

$$\alpha_m = (-1)^m \frac{(n' - 1)(n' - 2)\cdots(n' - m)}{m!(2s + 1)(2s + 2)\cdots(2s + m)}\alpha_0. \tag{A.85}$$

We can also use Eq.(A.83) to eliminate α_m and α_{m-1} in Eq.(A.78). This gives us an expression for β_m in terms of β_{m-1}

$$\beta_m = -\frac{n' - m + 1}{m(2s + m)}\beta_{m-1}. \tag{A.86}$$

and hence

$$\beta_m = (-1)^m \frac{n'(n' - 1)\cdots(n' - m + 1)}{m!(2s + 1)(2s + 2)\cdots(2s + m)}\beta_0. \tag{A.87}$$

Now the only quantities we don't know are α_0 and β_0. However, these are known in terms of each other from Eq.(A.83):

$$\beta_0 = \frac{\kappa - \xi k_C/\lambda}{n'}\alpha_0. \tag{A.88}$$

This arbitrariness can be removed later by our choice of normalization. Substituting Eqs.(A.85) and (A.86) into Eq.(A.76) gives

$$
\begin{aligned}
\phi_1(\rho) &= \rho^s \alpha_0 \sum_{m=0}^{\infty} \frac{(1-n')(2-n')\cdots(m-n')}{m!(2s+1)(2s+2)\cdots(2s+m)}\rho^m \\
&= \rho^s \alpha_0 \left(\frac{(1-n')}{2s+1}\rho + \frac{(1-n')(2-n')}{(2s+1)(2s+2)}\frac{\rho^2}{1!} + \cdots\right) \\
&= \rho^s \alpha_0 M(1-n', 2s+1, \rho),
\end{aligned} \tag{A.89}
$$

$$
\begin{aligned}
\phi_2(\rho) &= \rho^s \beta_0 \sum_{m=0}^{\infty}(-1)^m \frac{n'(n'-1)\cdots(n'-m+1)}{m!(2s+1)(2s+2)\cdots(2s+m)}\rho^m \\
&= \rho^s \beta_0 \left(1 - \frac{n'}{2s+1}\rho + \frac{n'(n'-1)}{(2s+1)(2s+2)}\frac{\rho^2}{2!} + \cdots\right) \\
&= \rho^s \beta_0 M(-n', 2s+1, \rho) \\
&= \rho^s \frac{\kappa - \xi k_C/\lambda}{n'}\alpha_0 M(-n', 2s+1, \rho),
\end{aligned} \tag{A.90}
$$

where $M(a, b, \rho)$ is confluent hypergeometric functions which will be discussed in Appendix B, and Eq.(A.88) were used in the last step of Eq.(A.90).

Up to this moment we have not yet applied any boundary conditions. The boundary condition we use is the usual one for such problems: the wavefunction must remain finite as $r \to \infty$. To implement this we have to look at the asymptotic behavior of the hypergeometric of the hypergeometric functions. This is given by Eq.(B.4). At large z this expression is dominated by the exponential and so diverges. As stated in Appendix B, if $-n'$ is a negative integer and $2s + 1$ is not a negative integer it is easy to see from Eq.(B.2) that the confluent hypergeometric function becomes a simple polynomial of order n' in ρ. So

$$n' = 0, 1, 2, 3, \cdots. \tag{A.91}$$

We see here that n' can take on only integer values and it is closely related to the principal quantum number defined in non-relativistic quantum mechanics. There is a problem with the $n' = 0$: the confluent hypergeometric function in Eq.(A.89) diverges; the first parameter is not a negative integer. Therefore $\alpha_0 = 0$ and hence $\phi_1(\rho) = 0$. In this case α_0/n' in Eq.(A.90) is just a normalization constant. In general for $n' = 0$, $\phi_2(\rho)$ is non-zero unless

$$\kappa = \frac{\xi k_C}{\lambda}. \tag{A.92}$$

As all quantities on the right hand side of Eq.(A.92) are positive by definition it can only be true for $\kappa \geq 1$.

We have now solved the Dirac equation for the radial parts of the Hydrogen ic wavefunctions. Let us combine the definitions of $u(r)$, $v(r)$ and ρ with Eqs.(A.70), (A.71), (A.89) and (A.90) to give a final expressions for $g(r)$ and $f(r)$:

$$f_\kappa(r) = 2\lambda(k_C - W_C)^{1/2} e^{-\lambda r} (2sr)^{s-1} \alpha_0$$
$$\times \left(M(1 - n', 2s + 1, 2\lambda r) - \frac{\kappa - \xi k_C/\lambda}{n'} M(-n', 2s + 1, 2\lambda r) \right), \tag{A.93}$$

$$g_\kappa(r) = 2\lambda(k_C + W_C)^{1/2} e^{-\lambda r} (2sr)^{s-1} \alpha_0$$
$$\times \left(M(1 - n', 2s + 1, 2\lambda r) + \frac{\kappa - \xi k_C/\lambda}{n'} M(-n', 2s + 1, 2\lambda r) \right). \tag{A.94}$$

Now we have the eigenvectors, the next task is to find the eigenvalues. To do this it is useful to define the *principal quantum number*:

$$n = n' + |\kappa|. \tag{A.95}$$

When n' was defined earlier it was in the middle of finding the eigenvectors of Eqs.(A.93) and (A.94), and, so as not to divert attention of n', which gives us an expression for the energy eigenvalues. We remedy that omission here. Recall that

$$n' = \frac{W_C \xi}{\lambda} - s = \frac{W_C \xi}{(k_C^2 - W_C^2)^{1/2}} - s. \tag{A.96}$$

Rearranging this and substituting for n' from (A.95) gives

$$W_C^2 = k_C^2 \left(1 + \frac{\xi^2}{(n - |\kappa| + s)^2} \right)^{-1}, \tag{A.97}$$

and from the definitions of k_C and W_C we find

$$W = mc^2 \left(1 + \frac{\xi^2}{(n - |\kappa| + s)^2} \right)^{-1/2}, \tag{A.98}$$

where $\xi = Z\alpha$ (see Eq.(A.67)).

A.4 Explicit wave functions for one-electron atom with principal quantum number $n = 1$ and 2

Recalling Eqs.(A.59) with (A.47) and (A.48), we have known that the wave functions for one-electron atom are in the form:

$$\psi_\kappa^{m_j}(\mathbf{r}) = \begin{pmatrix} g_\kappa(r)\chi_\kappa^{m_j}(\hat{\mathbf{r}}) \\ if_\kappa(r)\chi_{-\kappa}^{m_j}(\hat{\mathbf{r}}) \end{pmatrix}, \tag{A.99}$$

where $f_\kappa(r)$ and $g_\kappa(r)$ have been given in Eqs.(A.93) and (A.94) respectively. Those two equations look exceedingly complicated, but they can be simplified to some extent. Incorporating a factor of $1/n'$ into α_0 enables us to rewrite Eqs.(A.93) and (A.94) in the following form

$$f_\kappa(r) = 2\lambda(k_C - W_C)^{1/2}e^{-\lambda r}(2sr)^{s-1}\alpha_0'$$
$$\times \left(n'M(1 - n', 2s + 1, 2\lambda r) - (\kappa - \frac{\xi k_C}{\lambda})M(-n', 2s + 1, 2\lambda r) \right), \tag{A.100}$$

$$g_\kappa(r) = 2\lambda(k_C + W_C)^{1/2}e^{-\lambda r}(2sr)^{s-1}\alpha_0'$$
$$\times \left(n'M(1 - n', 2s + 1, 2\lambda r) + (\kappa - \frac{\xi k_C}{\lambda})M(-n', 2s + 1, 2\lambda r) \right), \tag{A.101}$$

where $\alpha_0' = \alpha_0/n'$. We write down explicit expressions for the lowest few eigenfunctions of the one-electron atom here.

For the $1s$ state:

$$g(r) = \frac{(2\lambda)^{s+1/2}}{(2k_C\Gamma(2s + 1))^{1/2}}(k_C + W_C)^{1/2}r^{s-1}e^{-\lambda r}(b_0 + b_1 r) \tag{A.102}$$

$$f(r) = -\frac{(2\lambda)^{s+1/2}}{(2k_C\Gamma(2s+1))^{1/2}}(k_C - W_C)^{1/2}r^{s-1}e^{-\lambda r}(a_0 + a_1 r) \tag{A.103}$$

For the $2s$ state:

$$g(r) = \left(\frac{(2\lambda)^{2s+1}k_C(2s + 1)(k_C + W_C)}{2W_C(2W_C + k_C)\Gamma(2s + 1)} \right)^{1/2} r^{s-1}e^{-\lambda r}(b_0 + b_1 r) \tag{A.104}$$

$$f(r) = -\left(\frac{(2\lambda)^{2s+1}k_C(2s + 1)(k_C - W_C)}{2W_C(2W_C + k_C)\Gamma(2s + 1)} \right)^{1/2} r^{s-1}e^{-\lambda r}(a_0 + a_1 r) \tag{A.105}$$

Table A.2　Parameters defining the eigenfunctions of the one-electron atom for the principal numbers $n = 1$ and $n = 2$.

Orbital	$1s$	$2s$	$2p^{1/2}$	$2p^{3/2}$
n	1	2	2	2
n'	0	1	1	0
κ	-1	-1	1	-2
l	0	0	1	1
$\bar{l}$	1	1	0	2
s	$(1-\xi^2)^{1/2}$	$(1-\xi^2)^{1/2}$	$(1-\xi^2)^{1/2}$	$(4-\xi^2)^{1/2}$
W_C	$s\lambda/\xi$	$\xi k_C^2/2\lambda$	$\xi k_C^2/2\lambda$	$s\lambda/\xi$
a_0	1	$\frac{W_C}{k_C}+1$	$\frac{W_C}{k_C}$	1
a_1	0	$\frac{-\lambda}{2s+1}\left(1+\frac{2W_C}{k_C}\right)$	$\frac{\lambda}{2s+1}\left(1-\frac{2W_C}{k_C}\right)$	0
b_0	1	$\frac{W_C}{k_C}$	$\frac{W_C}{k_C}-1$	1
b_1	0	$\frac{-\lambda}{2s+1}\left(1+\frac{2W_C}{k_C}\right)$	$\frac{\lambda}{2s+1}\left(1-\frac{2W_C}{k_C}\right)$	0

For the $2p^{1/2}$ state:

$$g(r) = \left(\frac{(2\lambda)^{2s+1}k_C(2s+1)(k_C+W_C)}{2W_C(2W_C-k_C)\Gamma(2s+1)}\right)^{1/2} r^{s-1}e^{-\lambda r}(b_0+b_1 r) \quad \text{(A.106)}$$

$$f(r) = -\left(\frac{(2\lambda)^{2s+1}k_C(2s+1)(k_C-W_C)}{2W_C(2W_C-k_C)\Gamma(2s+1)}\right)^{1/2} r^{s-1}e^{-\lambda r}(a_0+a_1 r) \quad \text{(A.107)}$$

For the $2p^{3/2}$ state[2]:

$$g(r) = \left(\frac{(2\lambda)^{s+1/2}(k_C+W_C)}{2k_C\Gamma(2s+1)}\right)^{1/2} r^{s-1}e^{-\lambda r}(b_0+b_1 r), \quad \text{(A.108)}$$

$$f(r) = -\left(\frac{(2\lambda)^{s+1/2}(k_C-W_C)}{2k_C\Gamma(2s+1)}\right)^{1/2} r^{s-1}e^{-\lambda r}(a_0+a_1 r). \quad \text{(A.109)}$$

All parameters defining the eigenfunctions in the above are in Table A.2.

A.5　Mathematica calculation for practice

"Mathematica" is a very convenient tool for practical calculations to the

---

[2]Some mistakes in writing of Eqs.(8.53g) and (8.53h) in [Strange (2008)] have been corrected.

problems in relativistic quantum mechanics. Here we show a simple example: to calculate the energy levels for $n \leq 2$ of one-electron atom with $Z = 92$. The readers could understand others from it by analogy. The programs and results are as follows:

1, For $n = 1$, $\kappa = -1$ case:

In[1]:= Z=92;

    $\alpha = 1/137.003599911$;   (* fine-construct constant *)

    mcc = 510998.918;   (* mass of electron: mcc $\equiv m_e c^2 = 510998.918$eV *)

    $n = 1$;   (* principle quantum number *)

    $\kappa = -1$;   (* relativistic quantum number *)

    $s = (\kappa^2 - Z^2\alpha^2)^{1/2}$;

    levelenergy=mcc $* (1 + Z\hat{\ }2\alpha\hat{\ }2/(n - \text{Abs}[\kappa] + s)\hat{\ }2)\hat{\ }(-1/2) - $mcc

    (* see Eq.(A.98) *)

Out[1]= -132353   (* unit: eV *)

2, For $n = 2$, $\kappa = -1$ case:

In[2]:= Z=92;

    $\alpha = 1/137.003599911$;   (* fine-construct constant *)

    mcc = 510998.918;   (* mass of electron: mcc $\equiv m_e c^2 = 510998.918$eV *)

    $n = 2$;   (* principle quantum number *)

    $\kappa = -1$;   (* relativistic quantum number *)

    $s = (\kappa^2 - Z^2\alpha^2)^{1/2}$;

    levelenergy=mcc $* (1 + Z\hat{\ }2\alpha\hat{\ }2/(n - \text{Abs}[\kappa] + s)\hat{\ }2)\hat{\ }(-1/2) - $mcc

    (* see Eq.(A.98) *)

Out[2]= -34235.2   (* unit: eV *)

3, For $n = 2$, $\kappa = 1$ case:

In[3]:= Z=92;

$\alpha = 1/137.003599911;$ (* fine-construct constant *)

mcc = 510998.918;   (* mass of electron: mcc $\equiv m_e c^2 = 510998.918$eV *)

$n = 2;$ (* principle quantum number *)

$\kappa = 1;$ (* relativistic quantum number *)

$s = (\kappa^2 - Z^2\alpha^2)^{1/2};$

levelenergy=mcc $* (1 + Z^\wedge 2\alpha^\wedge 2/(n - \text{Abs}[\kappa] + s)^\wedge 2)^\wedge(-1/2) - $ mcc

(* see Eq.(A.98) *)

Out[3]= -34235.2   (* unit: eV *)

4, For $n = 2$, $\kappa = -2$ case:

In[4]:= Z=92;

$\alpha = 1/137.003599911;$ (* fine-construct constant *)

mcc = 510998.918;   (* mass of electron: mcc $\equiv m_e c^2 = 510998.918$eV *)

$n = 2;$ (* principle quantum number *)

$\kappa = -1;$ (* relativistic quantum number *)

$s = (\kappa^2 - Z^2\alpha^2)^{1/2};$

levelenergy=mcc $* (1 + Z^\wedge 2\alpha^\wedge 2/(n - \text{Abs}[\kappa] + s)^\wedge 2)^\wedge(-1/2) - $ mcc

(* see Eq.(A.98) *)

Out[4]= -29664.3   (* unit: eV *)

These results are shown in Fig.8.9 in text.

Appendix B

The Confluent Hypergeometric Function

The confluent hypergeometric function is defined as the solution of the differential equation

$$z\frac{d^2 M}{dz^2} + (b - z)\frac{dM}{dz} - aM = 0. \tag{B.1}$$

The solution of this is

$$M(a, b, z) = 1 + \frac{az}{b} + \frac{(a)_2 z^2}{2!(b)_2} + \cdots \frac{(a)_n z^n}{n!(b)_n} \cdots \tag{B.2}$$

where

$$(a)_0 = 1, \quad (a)_n = a(a + 1)(a + 2) \cdots (a + n - 1). \tag{B.3}$$

Note that Eq.(B.2) is an infinite series in general. However, it can be terminated if a is equal to a negative integer $-m$ and b is not equal to a negative integer. In that case the mth term in the series for $(a)_n$ and all subsequent terms are zero. Useful asymptotic forms of the confluent hypergeometric function are

$$\lim_{z \to \infty} M(a, b, z) = \frac{\Gamma(b)}{\Gamma(a)} e^z z^{a-b} (1 + \mathcal{O}(|z|^{-1})), \quad Re\,z > 0, \tag{B.4}$$

$$\lim_{z \to \infty} M(a, b, z) = \frac{\Gamma(b)}{\Gamma(b - a)} (-z)^{-a} (1 + \mathcal{O}(|z|^{-1})), \quad Re\,z < 0, \tag{B.5}$$

where $\Gamma(b)$ is the Gamma function and $\mathcal{O}(x)$ means terms of order x. Also note that

$$M(0, b, z) = 1, \quad M(a, 0, z) = \infty. \tag{B.6}$$

Appendix C

Spherical Harmonics

The spherical harmonics form a complete set of functions. The first few of them for low values of quantum numbers are

$$Y_0^0(\theta, \phi) = \frac{1}{\sqrt{4\pi}} \tag{C.1}$$

$$Y_1^0(\theta, \phi) = \sqrt{\frac{3}{4\pi}} \cos\theta \tag{C.2}$$

$$Y_1^1(\theta, \phi) = -\sqrt{\frac{3}{4\pi}} e^{i\phi} \sin\theta \tag{C.3}$$

$$Y_2^0(\theta, \phi) = \sqrt{\frac{5}{16\pi}} (4\cos^2\theta - 1) \tag{C.4}$$

$$Y_2^1(\theta, \phi) = -\sqrt{\frac{15}{8\pi}} e^{i\phi} \sin\theta \cos\theta \tag{C.5}$$

$$Y_2^2(\theta, \phi) = \sqrt{\frac{15}{32\pi}} e^{2i\phi} \sin^2\theta \tag{C.6}$$

$$Y_3^0(\theta, \phi) = \sqrt{\frac{7}{16\pi}} (5\cos^3\theta - \cos\theta) \tag{C.7}$$

$$Y_3^1(\theta, \phi) = -\frac{1}{4}\sqrt{\frac{21}{4\pi}} e^{i\phi} \sin\theta(5\cos^3\theta - 1) \tag{C.8}$$

$$Y_3^2(\theta, \phi) = \frac{1}{4}\sqrt{\frac{105}{2\pi}} e^{2i\phi} \sin^2\theta \cos\theta \tag{C.9}$$

$$Y_3^3(\theta, \phi) = -\frac{1}{4}\sqrt{\frac{35}{4\pi}} e^{3i\phi} \sin^3\theta \tag{C.10}$$

$$Y_4^0(\theta, \phi) = \frac{1}{8}\sqrt{\frac{9}{4\pi}} (35\cos\theta - 30\cos^2\theta + 3) \tag{C.11}$$

$$Y_4^1(\theta, \phi) = -\frac{3}{4}\sqrt{\frac{5}{4\pi}}e^{i\phi}\sin\theta(7\cos^3\theta - 3\cos\theta) \qquad (C.12)$$

$$Y_4^2(\theta, \phi) = \frac{3}{4}\sqrt{\frac{5}{8\pi}}e^{2i\phi}\sin^2\theta(7\cos^2\theta - 1) \qquad (C.13)$$

$$Y_4^3(\theta, \phi) = -\frac{3}{4}\sqrt{\frac{35}{4\pi}}e^{3i\phi}\sin^3\theta\cos\theta \qquad (C.14)$$

$$Y_4^4(\theta, \phi) = \frac{3}{8}\sqrt{\frac{35}{8\pi}}e^{4i\phi}\sin^4\theta \qquad (C.15)$$

Spherical harmonics with negative values of m are related to these by

$$Y_l^{-m}(\theta, \phi) = (-1)^m Y_l^{m*}(\theta, \phi). \qquad (C.16)$$

The spherical harmonics are written in terms of Legendre functions as

$$Y_l^m(\theta, \phi) = (-1)^m\left(\frac{2l+1}{4\pi}\frac{(l-m)!}{(l+m)!}\right)^{1/2}P_l^m(\cos\theta)e^{im\phi}. \qquad (C.17)$$

The normalization is such that

$$\int_0^{2\pi}d\phi\int_0^\pi\sin\theta Y_l^{m*}(\theta, \phi)Y_{l'}^{m'}(\theta, \phi)d\theta = \delta_{ll'}\delta_{mm'} \qquad (C.18)$$

and it is this integral that often determines selection rules in physical processes involving electronic transitions. The integral of three spherical harmonics is also a useful quality:

$$
\begin{aligned}
C_{lml'm'}^{l''m''} &= \int Y_{l''}^{m''*}(\Omega)Y_{l'}^{m'}(\Omega)Y_l^m(\Omega)d\Omega \\
&= \left(\frac{(2l+1)(2l'+1)}{4\pi(2l''+1)}\right)^{1/2}C(ll'l''; m, m')C(ll'l''; 0, 0)\delta_{m'', m'+m},
\end{aligned}
$$
$$(C.19)$$

where $C(ll'l''; m, m')$ and $C(ll'l''; 0, 0)$ are Clebsch-Gordan coefficients. The spherical harmonic with argument $\hat{\mathbf{r}}$ is related to the spherical harmonic with argument $-\hat{\mathbf{r}}$ via the equation

$$Y_l^m(-\hat{\mathbf{r}}) = (-1)^{l+m}Y_l^{-m*}(\hat{\mathbf{r}}). \qquad (C.20)$$

Finally, it is well known and extremely useful in scattering theory that the exponential function can be written as a sum of terms involving Bessel functions and spherical harmonics

$$e^{i\mathbf{k}\cdot\mathbf{r}} = 4\pi\sum_{l,m}i^l j_l(kr)Y_l^{m*}(\hat{\mathbf{k}})Y_l^m(\hat{\mathbf{r}}), \qquad (C.21)$$

where $j_l(kr)$ is Bessel functions.

Appendix D

Electric Coulomb Law in QSO-Light-Cone Space

Let's derive Eq.(7.55). The action for deriving electrostatic potential of proton located at $Q^\mu = \{Q^0 = ct,\ Q^1,\ Q^2 = 0,\ Q^3 = 0\}$ with background space-time metric $g_{\mu\nu} \equiv B_{\mu\nu}(Q)$ of Eq.(7.9) in the Gaussian system of units reads

$$S = -\frac{1}{16\pi c} \int d^4 L \sqrt{-g}\, F_{\mu\nu} F^{\mu\nu} - \frac{e}{c} \int d^4 L \sqrt{-g}\, j^\mu A_\mu, \qquad (D.1)$$

where $g = \det(B_{\mu\nu})$, $F_{\mu\nu} = \frac{\partial A_\nu}{\partial L^\mu} - \frac{\partial A_\mu}{\partial L^\nu}$ and $j^\mu \equiv \{j^0 = c\rho_{proton}/\sqrt{B_{00}},\ \mathbf{j}\}$ is 4-current density vector of proton (see, e.g, Ref.[Landau, Lifshitz (1987)]: *Chapter 4; Chapter 10, Eq.(90.3)*). The explicit matrix-expressions for $B_{\mu\nu}(Q)$ and $B^{\mu\nu}(Q)$ up to $\mathcal{O}(1/R^2)$ are follows:

$$B_{\mu\nu}(Q) = \begin{pmatrix} 1 + \frac{2(Q^0)^2 - (Q^1)^2}{R^2} & -\frac{Q^0 Q^1}{R^2} & 0 & 0 \\ -\frac{Q^1 Q^0}{R^2} & -1 + \frac{2(Q^1)^2 - (Q^0)^2}{R^2} & 0 & 0 \\ 0 & 0 & -1 - \frac{(Q^0)^2 - (Q^1)^2}{R^2} & 0 \\ 0 & 0 & 0 & -1 - \frac{(Q^0)^2 - (Q^1)^2}{R^2} \end{pmatrix},$$

$$B^{\mu\nu}(Q) = \begin{pmatrix} 1 - \frac{2(Q^0)^2 - (Q^1)^2}{R^2} & \frac{Q^0 Q^1}{R^2} & 0 & 0 \\ \frac{Q^1 Q^0}{R^2} & -1 - \frac{2(Q^1)^2 - (Q^0)^2}{R^2} & 0 & 0 \\ 0 & 0 & -1 + \frac{(Q^0)^2 - (Q^1)^2}{R^2} & 0 \\ 0 & 0 & 0 & -1 + \frac{(Q^0)^2 - (Q^1)^2}{R^2} \end{pmatrix}.$$

Making space-time variable change of $L^\mu \to L^\mu - Q^\mu \equiv x^\mu = \{x^0 = ct_L - ct,\ x^i = L^i - Q^i\}$, we have action S as

$$S = -\frac{1}{16\pi c}\int d^4x \sqrt{-\det(B_{\mu\nu}(Q))}\, F_{\mu\nu}F^{\mu\nu} - \frac{e}{c}\int d^4x \sqrt{-\det(B_{\mu\nu}(Q))}\, j^\mu A_\mu$$

$$= \left(-\frac{1}{16\pi c}B_{\mu\lambda}(Q)B_{\nu\rho}(Q)\int d^4x F^{\mu\nu}(x)F^{\lambda\rho}(x)\right.$$

$$\left. -\frac{e}{c}\int d^4x j^\mu(x)A_\mu(x)\right)\sqrt{-\det(B_{\mu\nu}(Q))}, \qquad (D.2)$$

and the equation of motion $\delta S/\delta A_\mu(x) = 0$ as follows (see, e.g, [Landau, Lifshitz (1987)], Eq.(90.6), pp.257)

$$\partial_\nu F^{\mu\nu} = B^{\nu\lambda}\partial_\nu F^\mu{}_\lambda = -\frac{4\pi}{c}j^\mu. \qquad (D.3)$$

In Beltrami space, $A^\mu = \{\phi_B, \mathbf{A}\}$ (see, e.g., [Landau, Lifshitz (1987)], Eq.(16.2) in *pp.45*) and 4-charge current $j^\mu = \{c\rho_{proton}/\sqrt{B_{00}}, \mathbf{j}\}$. According to the expression of charge density in curved space in Ref. [Landau, Lifshitz (1987)], (*pp.256, Eq.(90.4)*), $\rho_{proton} \equiv \rho_B = \frac{e}{\sqrt{\gamma}}\delta^{(3)}(\mathbf{x})$ and $\mathbf{j} = 0$, where

$$\gamma = \det(\gamma_{ij}), \qquad (D.4)$$

$$dl^2 = \gamma_{ij}dx^i dx^j = \left(-g_{ij} + \frac{g_{0i}g_{j0}}{g_{00}}\right)dx^i dx^j \quad (\textit{see Eq.(84.7) in Ref.[37]})$$

$$= \left(-B_{ij} + \frac{B_{0i}B_{j0}}{B_{00}}\right)dx^i dx^j \qquad (D.5)$$

Noting $B_{01} = B_{10} = -\frac{C^2 t^2}{R^2}$, $B_{00} \sim 1$, and $B_{01}B_{10} \simeq \mathcal{O}(1/R^4) \sim 0$, we have

$$\sqrt{\gamma} \equiv \sqrt{\det(\gamma_{ij})} \simeq \sqrt{-\det(B_{ij})}, \qquad (D.6)$$

and hence

$$\rho_{proton} \equiv \rho_B = \frac{e\delta^{(3)}(\mathbf{x})}{\sqrt{-\det(B_{ij})}}, \quad \mathbf{j} = 0. \qquad (D.7)$$

(1) When $\mu = 0$ in Eq.(D.3), we have the Coulomb's law (7.50), i.e.,

$$-B^{ij}(Q)\partial_i\partial_j\phi_B(x) = \left[\left(1 - \frac{(Q^0)^2 - (Q^1)^2}{R^2}\right)\nabla^2 + \frac{(Q^1)^2}{R^2}\frac{\partial^2}{\partial(x^1)^2}\right]\phi_B(x)$$

$$= -\frac{4\pi}{c}j^0 = \frac{-4\pi e}{\sqrt{-\det(B_{ij}(Q))B_{00}(Q)}}\delta^{(3)}(\mathbf{x}), \qquad (D.8)$$

where $B^{ij}(Q) = \eta^{ij} + \frac{(Q^0)^2 - 2(Q^1)^2}{R^2}\delta_{i1}\delta_{j1} + \frac{(Q^0)^2 - (Q^1)^2}{R^2}\delta_{i2}\delta_{j2} + \frac{(Q^0)^2 - (Q^1)^2}{R^2}\delta_{i3}\delta_{j3} + \mathcal{O}(R^{-4})$ has been used, and B_{ij} was given in

Eq.(7.9). Expanding (D.8), we have

$$\left[\frac{\partial^2}{\partial(x^1/[1+\frac{(Q^1)^2}{2R^2}])^2} + \frac{\partial^2}{\partial(x^2)^2} + \frac{\partial^2}{\partial(x^3)^2} \right] \phi_B(x)$$

$$= -4\pi \left(1 - \frac{3(Q^0)^2 - 4(Q^1)^2}{2R^2} \right)$$

$$\times \left(1 - \frac{2(Q^0)^2 - (Q^1)^2}{2R^2} \right) \left(1 + \frac{(Q^0)^2 - (Q^1)^2}{R^2} \right) e\delta(x^1)\delta(x^2)\delta(x^3)$$

$$= -4\pi e \left(1 - \frac{3[(Q^0)^2 - (Q^1)^2]}{2R^2} \right) \delta(x^1)\delta(x^2)\delta(x^3).$$

Noting $\delta(x^1) = \delta(x^1/[1+(Q^1)^2/2R^2])(1-(Q^1)^2/2R^2)$, we rewrite above equation as follows

$$\left[\frac{\partial^2}{\partial(x^1/[1+\frac{(Q^1)^2}{2R^2}])^2} + \frac{\partial^2}{\partial(x^2)^2} + \frac{\partial^2}{\partial(x^3)^2} \right] \phi_B(x)$$

$$= -4\pi e \left(1 - \frac{3[(Q^0)^2 - (Q^1)^2]}{2R^2} \right)$$

$$\times \left(1 - \frac{(Q^1)^2}{2R^2} \right) \delta(x^1/[1 + \frac{(Q^1)^2}{2R^2}])\delta(x^2)\delta(x^3)$$

$$= -4\pi e \left(1 - \frac{3(Q^0)^2 - 2(Q^1)^2}{2R^2} \right) \delta(x^1/[1 + \frac{(Q^1)^2}{2R^2}])\delta(x^2)\delta(x^3).$$

Setting

$$\tilde{x}^1 \equiv x^1/[1 + \frac{(Q^1)^2}{2R^2}], \tag{D.9}$$

the above equation becomes

$$\left[\frac{\partial^2}{\partial(\tilde{x}^1)^2} + \frac{\partial^2}{\partial(x^2)^2} + \frac{\partial^2}{\partial(x^3)^2} \right] \phi_B(x) \tag{D.10}$$

$$= -4\pi e \left(1 - \frac{3(Q^0)^2 - 2(Q^1)^2}{2R^2} \right) \delta(\tilde{x}^1)\delta(x^2)\delta(x^3). \tag{D.11}$$

Then the solution is $\phi_B(x) = \left(1 - \frac{3(Q^0)^2 - 2(Q^1)^2}{2R^2} \right) e/r_B$ (see, e.g., [Landau, Lifshitz (1987)], Eq.(36.9) in *pp.89*) with

$$r_B = \sqrt{(\tilde{x}^1)^2 + (x^2)^2 + (x^3)^2}$$

$$= \left((1 - \frac{2(Q^1)^2 - (Q^0)^2}{2R^2})^2 (x^1)^2 + (x^2)^2 + (x^3)^2 \right)^{1/2}.$$

Therefore, we have

$$\phi_B = \left(1 - \frac{3(Q^0)^2 - 2(Q^1)^2}{2R^2} \right) \frac{e}{r_B}, \tag{D.12}$$

which is the scalar potential in Eq.(7.55) in the text.

(2) When $\mu = i$ $(i = 1, 2, 3)$ in Eq.(D.3), we have

$$\partial^i \partial_\mu A^\mu - B^{\mu\nu} \partial_\mu \partial_\nu A^i = -\frac{4\pi}{c} j^i = 0. \tag{D.13}$$

By means of the gauge condition

$$\partial_\mu A^\mu = 0, \tag{D.14}$$

we have

$$B^{\mu\nu} \partial_\mu \partial_\nu A^i = 0. \tag{D.15}$$

Then

$$A^i = 0 \tag{D.16}$$

is a solution that satisfies the gauge condition (D.14) (noting $\partial_0 A^0 = \frac{\partial}{\partial x^0} \phi_B(r_B) = 0$ due to $\frac{\partial Q^0}{\partial x^0} = \frac{\partial Q^0}{\partial L^0} = 0$). Equation (D.16) is the vector potential in Eq.(7.55) in the text.

Appendix E

Adiabatic Approximative Wave Functions in vacde Sitter–Dirac Equation of Hydrogen

Now we derive the wave function of Eq.(7.76) in the text. We start with Eq.(7.73), i.e.

$$i\hbar\partial_t\psi = H(t)\psi = [H_0(r_B, \hbar, m_e, e) + H_0'(t)]\psi, \tag{E.1}$$

where

$$H(t) = H_0(r_B, \hbar, m_e, e) + H_0'(t), \tag{E.2}$$

$$H_0(r_B, \hbar, m_e, e) = -i\hbar c\vec{\alpha}\cdot\nabla_B + m_e c^2\beta - \frac{e^2}{r_B} \quad (see\ Eq.(7.26)) \tag{E.3}$$

$$H_0'(t) = -\left(\frac{c^2 t^2}{2R^2}\right) H_0(r_B, \hbar, 3m_e, e). \tag{E.4}$$

Suppose the modification of $H(t)$ along with the time change is sufficiently slow, the system could be quasi-stationary in any instant θ. Then, in the Schrödinger picture, the quasi-stationary equation of $H(\theta)$

$$H(\theta)U_n(\mathbf{x}, \theta) = W_n(\theta)U_n(\mathbf{x}, \theta) \tag{E.5}$$

can be solved. By Eqs.(E.2), (E.3) and (E.4) and $t \to \theta$, the solutions are as follows (similar to Eq.(8.11) in text)

$$W_n(\theta) \equiv W_{n,\kappa}(\theta) = m_e^{(\theta)}c^2\left(1 + \frac{\alpha^2}{(n - |\kappa| + s)^2}\right)^{-1/2} \tag{E.6}$$

$$\alpha \equiv \frac{e_\theta^2}{\hbar_\theta c} \simeq \frac{e^2}{\hbar c} + \mathcal{O}(\frac{c^4\theta^4}{R^4}), \quad |\kappa| = (j + 1/2) = 1, 2, 3 \cdots$$

$$s = \sqrt{\kappa^2 - \alpha^2}, \quad n = 1, 2, 3 \cdots.$$

where (see Eqs.(7.68), (7.69) and (7.70) in text)

$$\hbar_\theta = \left(1 - \frac{c^2\theta^2}{2R^2}\right)\hbar, \tag{E.7}$$

$$m_e^{(\theta)} = \left(1 - \frac{(Q^1(\theta))^2}{2R^2}\right)m_e, \tag{E.8}$$

$$e_\theta = \left(1 - \frac{c^2\theta^2}{4R^2}\right)e, \tag{E.9}$$

The complete set of commutative observables is $\{H, \ \kappa, \ \mathbf{j}^2, \ j_z\}$, so that we have

$$U_n(\mathbf{x}, \theta) = \psi_{n,\kappa,j,j_z}(\mathbf{r}_B, \hbar_\theta, m_e^{(\theta)}, e_\theta), \tag{E.10}$$

where $\mathbf{j} = \mathbf{L} + \frac{\hbar}{2}\mathbf{\Sigma}$, $\hbar\kappa = \beta(\mathbf{\Sigma} \cdot \mathbf{L} + \hbar)$. $[U_n(\mathbf{x}, \theta)]$ is a complete set and satisfies

$$\int d^3x U_n(\mathbf{x}, \theta)U_m^*(\mathbf{x}, \theta) = \delta_{mn}, \quad n = \{n_r, K, j, j_z\}. \tag{E.11}$$

Thus, the solution of time-dependent Schrödinger equation (or Dirac equation) (E.1) can be expanded as follows

$$\psi(\mathbf{x}, t) = \sum_n C_n(t)U_n(\mathbf{x}, t)\exp\left[-i\int_0^t \omega_n(\theta)d\theta\right], \quad \omega_n(\theta) = \frac{W_n(\theta)}{\hbar}. \tag{E.12}$$

Substituting Eq.(E.12) into Eq.(E.1), we have

$$i\hbar\sum_n(\dot{C}_nU_n + C_n\dot{U}_n)\exp\left[-i\int_0^t \omega_n(\theta)d\theta\right] = 0. \tag{E.13}$$

Multiplying both sides of Eq.(E.13) by $U_m^*\exp\left[i\int_0^t \omega_m(\theta)d\theta\right]$, and integrating $\mathbf{x}$ using Eq.(E.11), we have

$$\dot{C}_m + C_m\int d^3x U_m^*\dot{U}_m + \sum_n{}' C_n\int d^3x U_m^*\dot{U}_n\exp\left[-i\int_0^t (\omega_n - \omega_m)d\theta\right] = 0,$$

$$m = 1, 2, 3, \cdots \tag{E.14}$$

where $\sum_n'$ means that $n \neq m$ in the summation over n. Noting Eq.(E.11), we have

$$\int \dot{U}_m^*U_m d^3x + \int U_m^*\dot{U}_m d^3x = 0, \tag{E.15}$$

and hence

$$\int U_m^*\dot{U}_m d^3x = i\beta' \tag{E.16}$$

is a purely imaginary number. Denoting

$$\alpha_{mn} = \int U_m^* \dot{U}_n d^3x, \quad \text{and} \quad \omega_{nm} = \omega_n - \omega_m, \tag{E.17}$$

then Eq.(E.14) becomes

$$\dot{C}_m + i\beta' C_m + \sum_n{}' C_n \alpha_{mn} \exp\left[-i\int_0^t \omega_{nm} d\theta\right] = 0. \quad m = 1, 2, 3, \cdots$$

$$\tag{E.18}$$

To further simplify it, we set

$$V_n(\mathbf{x}, t) = U_n(\mathbf{x}, t) \exp\left[-i\int_0^t \beta_n'(\theta)d\theta\right], \tag{E.19}$$

then

$$\psi(\mathbf{x}, t) = \sum_n C_n'(t) V_n(\mathbf{x}, t) \exp\left[-i\int_0^t \omega_n(\theta)d\theta\right], \tag{E.20}$$

where $C_n'(t) = C_n(t) \exp\left[i\int_0^t \beta_n'(\theta)d\theta\right]$, and

$$\dot{C}_m'(t) = [\dot{C}_m + i\beta_m' C_m(t)] \exp\left(i\int_0^t \beta_n'(\theta)d\theta\right) \tag{E.21}$$

Substituting (E.21) into (E.18), we finally get

$$\dot{C}_m' + \sum_n{}' C_n' \alpha_{mn} \exp\left[-i\int_0^t \omega_{nm}' d\theta\right] = 0. \quad m = 1, 2, 3, \cdots \tag{E.22}$$

where

$$\omega_{mn}' = \omega_n' - \omega_m', \quad \omega_n' = \frac{1}{\hbar}W_n + \beta_n'. \tag{E.23}$$

Now let's solve Eq.(E.22). Firstly, we derive α_{mn}. By (E.5), we have

$$\frac{\partial H}{\partial t}U_n + H\dot{U}_n = \dot{W}_n U_n + W_n \dot{U}_n. \tag{E.24}$$

By multiplying U_m^* and integrating over $\mathbf{x}$, we have

$$\int U_m^* \dot{H} U_n d^3x + \int U_m^* H \dot{U}_n d^3x = W_n \int U_m^* \dot{U}_n d^3x,$$

i.e., $\quad \dot{H}_{mn} + W_m \alpha_{mn} = W_n \alpha_{mn}, \tag{E.25}$

so that

$$\alpha_{mn} = \int U_m^* \dot{U}_n d^3x = \frac{1}{W_n - W_m}\dot{H}_{mn}, \quad m \neq n. \tag{E.26}$$

Therefore Eq.(E.22) becomes

$$\dot{C}'_m + \sum_n{}' C'_n \frac{\dot{H}_{mn} \exp\left(-i\int_0^t \omega'_{nm} d\theta\right)}{\hbar\omega_{nm}} = 0. \quad m = 1, 2, 3, \cdots \quad (E.27)$$

Suppose at the initial time the system is in s-state, i.e., $C_n(0) = C'_n(0) = \delta_{ns}$. For adiabatic process, $\dot{H}(t) \to 0$, then the 0th-order approximate solution of Eq.(E.27) is

$$[C'_m(t)]_0 = \delta_{ms}. \tag{E.28}$$

Substituting Eq.(E.28) into Eq.(E.27), we get the first order correction to the approximation

$$[\dot{C}'_m]_1 = \frac{-\dot{H}_{ms}}{\hbar\omega_{ms}} \exp\left(-i\int_0^t \omega'_{ms} d\theta\right) = 0, \quad m \neq s. \tag{E.29}$$

Since the dependence on time t of $U_n(t)$ is weak for adiabatic process, Eq.(E.16) indicates β'_n is small, and by Eq.(E.23), we have $\omega'_{ms} \approx \omega_{ms}$. Then, from Eq.(E.29), the first order correction to the solution is

$$[C'_m]_1 = \frac{\dot{H}_{ms}}{i\hbar\omega_{ms}} \left(e^{i\omega_{ms}t} - 1\right), \quad m \neq s. \tag{E.30}$$

Substituting Eqs.(E.29) and (E.30) into Eq.(E.20) and neglecting β_n, we get the wave function as follows

$$\psi(\mathbf{x}, t) \simeq U_s(\mathbf{x}, t)e^{-i\frac{W_s t}{\hbar}} + \sum_{m \neq s} \frac{\dot{H}_{ms}}{i\hbar\omega_{ms}} \left(e^{i\omega_{ms}t} - 1\right) U_m(\mathbf{x}, t)e^{\left(-i\int_0^t \frac{W_m(\theta)}{\hbar} d\theta\right)}.$$

$$\tag{E.31}$$

By using Eqs.(E.10), (E.8), (E.7), (E.9), we finally obtain the desired results

$$\psi(t) \simeq \psi_s(\mathbf{r}_B, \hbar_t, m_e^{(t)}, e_t)e^{-i\frac{W_s}{\hbar}t}$$

$$+ \sum_{m \neq s} \frac{\dot{H}'(t)_{ms}}{i\hbar\omega_{ms}^2} \left(e^{i\omega_{ms}t} - 1\right) \psi_m(\mathbf{r}_B, \hbar_t, m_e^{(t)}, e_t) \times e^{\left(-i\int_0^t \frac{W_m(\theta)}{\hbar} d\theta\right)},$$

$$\tag{E.32}$$

where

$$\hbar_t = \left(1 - \frac{c^2 t^2}{2R^2}\right)\hbar, \tag{E.33}$$

$$m_e^{(t)} = \left(1 - \frac{(Q^1(t))^2}{2R^2}\right)m_e, \tag{E.34}$$

$$e_t = \left(1 - \frac{c^2 t^2}{4R^2}\right)e. \tag{E.35}$$

Equations (E.33), (E.34) and (E.35) are Eqs.(7.68), (7.69) and (7.70) in the text. Equation (E.32) is just Eq.(7.76) in the text.

Appendix F

Matrix Elements of H′ in $(2s^{1/2}\text{-}2p^{1/2})$-Hilbert Space

Now we derive Eq.(7.109) and Eq.(7.115). We start with the dS-SR Dirac spectrum equation of Hydrogen , which has been shown in Eqs.(7.89)–(7.93) in the text:

$$H_{(dS-SR)}\psi = (\ H_0(r, \hbar_t, \mu_t, e_t) + H'\ )\psi = E\psi, \qquad (\text{F.1})$$

where

$$H_0(r, \hbar_t, \mu_t, e_t) = -i\hbar_t c\vec{\alpha}\cdot\nabla + \mu_t c^2\beta - \frac{e_t^2}{r}, \qquad (\text{F.2})$$

$$H' = \frac{1}{2}(H'^\dagger + H') \equiv H_1' + H_2', \qquad (\text{F.3})$$

with

$$H_1' = -\frac{Q^1 Q^0}{4R^2}\left(\alpha^1 H_0(r, \hbar, \mu, e) + H_0(r, \hbar, \mu, e)\alpha^1\right), \qquad (\text{F.4})$$

$$H_2' = \frac{Q^1 Q^0}{4R^2}\left(i\hbar c\overrightarrow{\frac{\partial}{\partial x^1}} - i\hbar c\overleftarrow{\frac{\partial}{\partial x^1}}\right). \qquad (\text{F.5})$$

The definition of H'-elements in the H_0-eigenstate space, $\langle H'\rangle_{2L^{1/2},2L'^{1/2}}^{m_j,\,m_j'}$, has been given in Eqs.(7.106) and (7.107). The eigen values and eigen states of H_0 are given in the section IV.

(I) H_1'-matrix elements:

(1) $\langle H_1'\rangle^{1/2,\,-1/2}_{2s^{1/2},2p^{1/2}}$:

$$\langle H_1'\rangle^{1/2,\,-1/2}_{2s^{1/2},2p^{1/2}} = \int drr^2 \int d\Omega\, \psi^{1/2\dagger}_{(2s)j=1/2}(\mathbf{r})\, H_1'\, \psi^{-1/2}_{(2p)j=1/2}(\mathbf{r})$$

$$= -\frac{Q^1 Q^0}{4R^2} W \langle (2s^{1/2})^{1/2}|\alpha^1|(2p^{1/2})^{-1/2}\rangle$$

$$= -\frac{Q^1 Q^0}{4R^2} W \int drr^2 \int d\Omega \Big(g_{(2s^{1/2})}(r)\chi^{1/2\dagger}_{\kappa}(\hat{\mathbf{r}})_{(2s^{1/2})},\ -i f_{(2s^{1/2})}(r)\chi^{1/2\dagger}_{-\kappa}(\hat{\mathbf{r}})_{(2s^{1/2})}\Big)$$

$$\times \begin{pmatrix} 0 & \sigma^1 \\ \sigma^1 & 0 \end{pmatrix} \begin{pmatrix} g_{(2p^{1/2})}(r)\chi^{-1/2}_{\kappa}(\hat{\mathbf{r}})_{(2p^{1/2})} \\ i f_{(2p^{1/2})}(r)\chi^{-1/2}_{-\kappa}(\hat{\mathbf{r}})_{(2p^{1/2})} \end{pmatrix}$$

$$= -i\frac{Q^1 Q^0}{4R^2} W \int_0^\infty drr^2 \Big\{ g_{(2s^{1/2})}(r) f_{(2p^{1/2})}(r) \int d\Omega \chi^{1/2\dagger}_{\kappa}(\hat{\mathbf{r}})_{(2s^{1/2})}\sigma^1 \chi^{-1/2}_{-\kappa}(\hat{\mathbf{r}})_{(2p^{1/2})}$$

$$ - f_{(2s^{1/2})}(r) g_{(2p^{1/2})}(r) \int d\Omega \chi^{1/2\dagger}_{-\kappa}(\hat{\mathbf{r}})_{(2s^{1/2})}\sigma^1 \chi^{-1/2}_{\kappa}(\hat{\mathbf{r}})_{(2p^{1/2})} \Big\} \tag{F.6}$$

where $W = W_{(n=2,\kappa=\pm 1)}$, and the explicit expressions of $2s^{1/2}$- and $2p^{1/2}$-wave functions of $\{g_{(2s^{1/2})}(r),\ f_{(2s^{1/2})}(r),\ g_{(2p^{1/2})}(r),\ f_{(2p^{1/2})}(r),\ \chi^{\pm 1/2}_{\pm\kappa}(\hat{\mathbf{r}})_{(2s^{1/2})},\ \chi^{\pm 1/2}_{\pm\kappa}(\hat{\mathbf{r}})_{(2p^{1/2})}\}$ are given in Eqs.(7.36)−(7.45) in the text. From them, we have

$$\int d\Omega \chi^{1/2\dagger}_{\kappa}(\hat{\mathbf{r}})_{(2s^{1/2})}\sigma^1 \chi^{-1/2}_{-\kappa}(\hat{\mathbf{r}})_{(2p^{1/2})}$$

$$= \int_0^{2\pi} d\phi \int_0^{\pi} d\theta \sin\theta$$

$$\times (Y_0^0, 0) \begin{pmatrix} 0 & 1 \\ 1 & 0 \end{pmatrix} \begin{pmatrix} \sqrt{\frac{2}{3}}\cos\theta Y_1^{-1}(\theta\phi) - \sqrt{\frac{1}{3}}\sin\theta e^{-i\phi} Y_1^0(\theta\phi) \\ \sqrt{\frac{2}{3}}\sin\theta e^{i\phi} Y_1^{-1}(\theta\phi) - \sqrt{\frac{1}{3}}\cos\theta Y_1^0(\theta\phi) \end{pmatrix}$$

$$= \frac{1}{2}\int_0^{\pi} d\theta \sin\theta (\sin\theta\cos\theta + \cos^2\theta) = \frac{1}{3}; \tag{F.7}$$

$$\int d\Omega \chi^{1/2\dagger}_{-\kappa}(\hat{\mathbf{r}})_{(2s^{1/2})}\sigma^1 \chi^{-1/2}_{\kappa}(\hat{\mathbf{r}})_{(2p^{1/2})}$$

$$= \int_0^{2\pi} d\phi \int_0^{\pi} d\theta$$

$$\times \sin\theta \left(-\cos\theta Y_0^0, -\sin\theta e^{-i\phi} Y_0^0\right) \begin{pmatrix} 0 & 1 \\ 1 & 0 \end{pmatrix} \begin{pmatrix} -\sqrt{\frac{2}{3}} Y_1^{-1}(\theta\phi) \\ \sqrt{\frac{1}{3}} Y_1^0(\theta\phi) \end{pmatrix}$$

$$= -\frac{1}{2}\int_0^{\pi} d\theta \sin\theta \cos^2\theta = -\frac{1}{3}, \tag{F.8}$$

where $Y_1^0(\theta\phi) = \sqrt{\frac{3}{4\pi}}\cos\theta$, $Y_1^1(\theta\phi) = -\sqrt{\frac{3}{8\pi}}e^{i\phi}\sin\theta$, $Y_1^{-1}(\theta\phi) = \sqrt{\frac{3}{8\pi}}e^{-i\phi}\sin\theta$ and $Y_0^0(\theta\phi) = \sqrt{\frac{1}{4\pi}}$ have been used. Substituting Eqs.(F.7) (F.8) into (F.6), we get

$$\langle H_1'\rangle_{2s^{1/2},2p^{1/2}}^{1/2,\,-1/2}$$

$$= -i\frac{Q^1Q^0}{4R^2}\frac{W}{3}\int_0^\infty dr r^2 \left\{ g_{(2s^{1/2})}(r)f_{(2p^{1/2})}(r) + f_{(2s^{1/2})}(r)g_{(2p^{1/2})}(r) \right\}.$$

$$\text{(F.9)}$$

Inserting the explicit expressions of the radial wave functions $g_{(2s^{1/2})}(r)$, $f_{(2s^{1/2})}(r)$, and $g_{(2p^{1/2})}(r)$, $f_{(2p^{1/2})}(r)$ (i.e., Eqs.(7.37), (7.38), (7.42) and (7.43) in the text) into the integral in Eq.(F.9), and accomplishing the calculations, we have

$$\int_0^\infty dr r^2 \left\{ g_{(2s^{1/2})}(r)f_{(2p^{1/2})}(r) + f_{(2s^{1/2})}(r)g_{(2p^{1/2})}(r) \right\}$$

$$= \sqrt{\frac{k_C^2 - W_C^2}{4W_C^2 - k_C^2}}\left(\frac{W_C}{k_C} - \frac{k_C}{2W_C}(s+1) \right), \qquad \text{(F.10)}$$

where formula $\int_0^\infty dr r^{\nu-1}\exp(-\mu r) = \Gamma(\nu)/\mu^\nu$ was used. Consequently, we obtain upon substituting Eq.(F.10) into Eq.(F.9),

$$\langle H_1'\rangle_{2s^{1/2},2p^{1/2}}^{1/2,\,-1/2} \equiv -i\Theta_1$$

$$= -i\frac{Q^1Q^0}{4R^2}\frac{W}{3}\sqrt{\frac{k_C^2 - W_C^2}{4W_C^2 - k_C^2}}\left(\frac{W_C}{k_C} - \frac{k_C}{2W_C}(s+1) \right), \quad \text{(F.11)}$$

which is just the desired result Eq.(7.109). All above calculations have been checked by the *Mathematica*.

(2) By means of similar calculations we get also that

$$\langle H_1'\rangle_{2s^{1/2},2p^{1/2}}^{-1/2,\,1/2} = -i\Theta_1. \qquad \text{(F.12)}$$

Since $H_1' = H_1'^\dagger$, we have

$$\langle H_1'\rangle_{2p^{1/2},2s^{1/2}}^{-1/2,\,1/2} = (\langle H_1'\rangle_{2s^{1/2},2p^{1/2}}^{1/2,\,-1/2})^* = i\Theta_1, \qquad \text{(F.13)}$$

$$\langle H_1'\rangle_{2p^{1/2},2s^{1/2}}^{1/2,\,-1/2} = (\langle H_1'\rangle_{2s^{1/2},2p^{1/2}}^{-1/2,\,1/2})^* = i\Theta_1. \qquad \text{(F.14)}$$

Furthermore, for all other elements of H_1'-matrix, since $\int_0^{2\pi} d\phi\,\exp(\pm in\phi) = 0$ and $\int_0^\pi d\theta\sin\theta\cos^{2n+1}\theta = 0$ and etc, the explicit calculations show that all those H_1'-matrix elements vanish. Consequently, all elements of H_1' are calculated, and Eq.(7.109) is proved.

(II) H_2'-matrix elements:

H_2' has been given in Eqs.(7.113) and (7.114) in the text, which is as follows

$$H_2' = \frac{Q^1 Q^0}{4R^2}\left(i\hbar c\,\overrightarrow{\frac{\partial}{\partial x^1}} - i\hbar c\,\overleftarrow{\frac{\partial}{\partial x^1}}\right), \qquad (F.15)$$

$$\text{where}\ \ \frac{\partial}{\partial x^1} \equiv \partial_1 = \sin\theta\cos\phi\frac{\partial}{\partial r} + \cos\theta\cos\phi\frac{1}{r}\frac{\partial}{\partial\theta} - \frac{\sin\phi}{r\sin\theta}\frac{\partial}{\partial\phi}. \quad (F.16)$$

We derive Eq.(7.115) now.

(1) $\langle H_2'\rangle_{2s^{1/2},2p^{1/2}}^{1/2,\,-1/2}$:

$$\langle H_2'\rangle_{2s^{1/2},2p^{1/2}}^{1/2,\,-1/2} = \int dr\,r^2 \int d\Omega\ \psi_{(2s)j=1/2}^{1/2\dagger}(\mathbf{r})\,H_2'\,\psi_{(2p)j=1/2}^{-1/2}(\mathbf{r})$$

$$= -\frac{Q^1 Q^0}{4R^2}i\hbar c\!\int dr\,r^2 \int\! d\Omega\left(g_{(2s^{1/2})}(r)\chi_\kappa^{1/2\dagger}(\hat{\mathbf{r}})_{(2s^{1/2})},\, -if_{(2s^{1/2})}(r)\chi_{-\kappa}^{1/2\dagger}(\hat{\mathbf{r}})_{(2s^{1/2})}\right)$$

$$\times\left(\overrightarrow{\partial}_1 - \overleftarrow{\partial}_1\right)\begin{pmatrix} g_{(2p^{1/2})}(r)\chi_\kappa^{-1/2}(\hat{\mathbf{r}})_{(2p^{1/2})} \\ if_{(2p^{1/2})}(r)\chi_{-\kappa}^{-1/2}(\hat{\mathbf{r}})_{(2p^{1/2})} \end{pmatrix}$$

$$= -i\frac{Q^1 Q^0}{4R^2}\hbar c\int dr\,r^2\!\!\int\! d\Omega\left\{g_{(2s^{1/2})}(r)[\partial_1 g_{(2p^{1/2})}(r)]\chi_\kappa^{1/2\dagger}(\hat{\mathbf{r}})_{(2s^{1/2})}\chi_{-\kappa}^{-1/2}(\hat{\mathbf{r}})_{(2p^{1/2})}\right.$$

$$+ g_{(2s^{1/2})}(r)g_{(2p^{1/2})}(r)\chi_\kappa^{1/2\dagger}(\hat{\mathbf{r}})_{(2s^{1/2})}[\partial_1\chi_{-\kappa}^{-1/2}(\hat{\mathbf{r}})_{(2p^{1/2})}]$$

$$+ f_{(2s^{1/2})}(r)[\partial_1 f_{(2p^{1/2})}(r)]\chi_\kappa^{1/2\dagger}(\hat{\mathbf{r}})_{(2s^{1/2})}\chi_{-\kappa}^{-1/2}(\hat{\mathbf{r}})_{(2p^{1/2})}$$

$$+ f_{(2s^{1/2})}(r)f_{(2p^{1/2})}(r)\chi_\kappa^{1/2\dagger}(\hat{\mathbf{r}})_{(2s^{1/2})}[\partial_1\chi_{-\kappa}^{-1/2}(\hat{\mathbf{r}})_{(2p^{1/2})}]$$

$$- [\partial_1 g_{(2s^{1/2})}(r)]g_{(2p^{1/2})}(r)\chi_\kappa^{1/2\dagger}(\hat{\mathbf{r}})_{(2s^{1/2})}\chi_{-\kappa}^{-1/2}(\hat{\mathbf{r}})_{(2p^{1/2})}$$

$$- g_{(2s^{1/2})}(r)g_{(2p^{1/2})}(r)\chi_\kappa^{1/2\dagger}(\hat{\mathbf{r}})_{(2s^{1/2})}[\partial_1\chi_{-\kappa}^{-1/2}(\hat{\mathbf{r}})_{(2p^{1/2})}]$$

$$- [\partial_1 f_{(2s^{1/2})}(r)]f_{(2p^{1/2})}(r)\chi_\kappa^{1/2\dagger}(\hat{\mathbf{r}})_{(2s^{1/2})}\chi_{-\kappa}^{-1/2}(\hat{\mathbf{r}})_{(2p^{1/2})}$$

$$\left.- f_{(2s^{1/2})}(r)f_{(2p^{1/2})}(r)\chi_\kappa^{1/2\dagger}(\hat{\mathbf{r}})_{(2s^{1/2})}[\partial_1\chi_{-\kappa}^{-1/2}(\hat{\mathbf{r}})_{(2p^{1/2})}]\right\}$$

$$= -i\frac{Q^1 Q^0}{4R^2}\hbar c\left\{\frac{1}{3}\int_0^\infty dr\,r^2\left[\frac{\partial g_{(2s^{1/2})}(r)}{\partial r}g_{(2p^{1/2})}(r) - \frac{\partial g_{(2p^{1/2})}(r)}{\partial r}g_{(2s^{1/2})}(r)\right.\right.$$

$$\left.+ \frac{\partial f_{(2s^{1/2})}(r)}{\partial r}f_{(2p^{1/2})}(r) - \frac{\partial f_{(2p^{1/2})}(r)}{\partial r}f_{(2s^{1/2})}(r)\right]$$

$$\left.- \frac{2}{3}\int_0^\infty dr\left[g_{(2s^{1/2})}(r)g_{(2p^{1/2})}(r) - f_{(2s^{1/2})}(r)f_{(2p^{1/2})}(r)\right]\right\}, \qquad (F.17)$$

where the integrals for $d\Omega$ have been implemented in terms of the explicit expressions of $\chi_{\pm\kappa}^{1/2}(\hat{\mathbf{r}})_{(2s^{1/2})}$, $\chi_{\pm\kappa}^{-1/2}(\hat{\mathbf{r}})_{(2p^{1/2})}$ in Eqs.(7.39), (7.40), (7.44), (7.45) and Eq.(F.16). Substituting expressions (7.37), (7.38), (7.42) and (7.43) into Eq.(F.17), and finishing the integrals, we get

$$\langle H'_2 \rangle^{1/2,\,-1/2}_{2s^{1/2},2p^{1/2}} \equiv -i\Theta_2$$
$$= i\frac{Q^1 Q^0}{2R^2} \frac{\hbar c\lambda}{6\sqrt{4W_C^2 - k_C^2}} \left(\frac{k_C^2}{W_C} - 2(\frac{1}{s} + 1)k_C - W_C + \frac{2}{k_C s}W_C^2 \right),$$

$$(F.18)$$

which is just Eq.(7.115). And all above result can be checked by the *Mathematica* calculations.

(2) By means of similar calculations we get also that

$$\langle H'_2 \rangle^{-1/2,\,1/2}_{2s^{1/2},2p^{1/2}} = -i\Theta_2. \tag{F.19}$$

Since $H'_2 = H'^{\dagger}_2$, we have

$$\langle H'_2 \rangle^{-1/2,\,1/2}_{2p^{1/2},2s^{1/2}} = (\langle H'_2 \rangle^{1/2,\,-1/2}_{2s^{1/2},2p^{1/2}})^* = i\Theta_2, \tag{F.20}$$

$$\langle H'_2 \rangle^{1/2,\,-1/2}_{2p^{1/2},2s^{1/2}} = (\langle H'_2 \rangle^{-1/2,\,1/2}_{2s^{1/2},2p^{1/2}})^* = i\Theta_2. \tag{F.21}$$

Furthermore, for all other elements of H'_2-matrix, since $\int_0^{2\pi} \exp(\pm in\phi)d\phi = 0$ and $\int_0^{\pi} d\theta \sin\theta \cos^{2n+1}\theta = 0$ and etc, the explicit calculations show that all those H'_2-matrix elements vanish. Consequently, all elements of H'_2 are calculated, and Eq.(7.115) is proved.

Appendix G

Solutions of Exercises

Chapter 3

Section 3.1

(1) Problem 1:

Solution: From Eq.(3.6), we have

$$B_{00}(x) = \frac{1}{\sigma} + \frac{(x^0)^2}{R^2\sigma^2}, \quad B_{01}(x) = -\frac{x^0 x^1}{R^2\sigma^2}, \quad B_{11}(x) = \frac{-1}{\sigma} + \frac{(x^1)^2}{R^2\sigma^2},$$

where $\sigma = 1 - [(x^0)^2 - (x^1)^2]/R^2$ with $x^0 = ct$.

(2) Problem 2:

Solution: From Eq.(3.6), we have

$$
\begin{aligned}
ds^2_{Bel} &= B_{\mu\nu}(x)dx^\mu dx^\nu \\
&= \frac{R^2(R^2 + r^2)}{(R^2 + r^2 - c^2t^2)^2}c^2 dt^2 - \frac{2rR^2 ct}{(R^2 + r^2 - c^2t^2)^2}cdtdr \\
&\quad - \frac{R^2(R^2 - c^2t^2)}{(R^2 + r^2 - c^2t^2)^2}dr^2 - \frac{r^2 R^2}{R^2 + r^2 - c^2t^2}(d\theta^2 + \sin^2\theta d\phi^2).
\end{aligned}
$$
(G.1)

Section 3.2

(1) Problem 1:

Solution: From Eq.(3.32), we have

$$
\begin{aligned}
\partial_{x^1} L_{dS} &= \frac{m^2 c^4 R}{2L_{dS}} \left\{ \frac{-4x^1[R^2(c^2 - \dot{\mathbf{x}}^2) - \mathbf{x}^2\dot{\mathbf{x}}^2 + (\mathbf{x}\cdot\dot{\mathbf{x}})^2 + c^2(\mathbf{x} - \dot{\mathbf{x}}t)^2]}{(R^2 + \mathbf{x}^2 - c^2t^2)^3} \right. \\
&\quad \left. + \frac{-2x^1\dot{\mathbf{x}}^2 + 2(\mathbf{x}\cdot\dot{\mathbf{x}})\dot{x}^1 + 2c^2(x^1 - \dot{x}^1 t)}{c^2(R^2 + \mathbf{x}^2 - c^2t^2)^2} \right\}.
\end{aligned}
$$
(G.2)

(2) Problem 2:

Solution:

$$
\partial_{\dot{x}^1} L_{dS} = \frac{m^2 c^4 R^2}{2L_{dS}} \frac{-2R^2\dot{x}^1 - 2\mathbf{x}^2\dot{x}^1 + 2(\mathbf{x}\cdot\dot{\mathbf{x}})x^1 - 2c^2t(x^1 - \dot{x}^1 t)}{c^2(R^2 + \mathbf{x}^2 - c^2t^2)^2}.
$$
(G.3)

(3) Problem 3:

Solution:

$$\partial_t(\partial_{\dot{x}^1} L_{dS}) = -\frac{m^4 c^8 R^4}{4L_{dS}^3}\left\{\frac{4c^2 t[R^2(c^2 - \dot{\mathbf{x}}^2) - \mathbf{x}^2\dot{\mathbf{x}}^2 + (\mathbf{x}\cdot\dot{\mathbf{x}})^2 + c^2(\mathbf{x} - \dot{\mathbf{x}}t)^2]}{c^2(R^2 + \mathbf{x}^2 - c^2 t^2)^3}\right.$$

$$\left. + \frac{2(\mathbf{x} - \dot{\mathbf{x}}t)\cdot\dot{\mathbf{x}}}{(R^2 + \mathbf{x}^2 - c^2 t^2)^2}\right\}$$

$$+ \frac{m^2 c^4 R^2}{2L_{dS}}\left\{\frac{(-2)(-2c^2 t)[-2R^2\dot{x}^1 - 2\mathbf{x}^2\dot{x}^1 + 2(\mathbf{x}\cdot\dot{\mathbf{x}})x^1 - 2c^2 t(x^1 - \dot{x}^1 t)]}{c^2(R^2 + \mathbf{x}^2 - c^2 t^2)^3}\right.$$

$$\left. + \frac{-2c^2(x^1 - \dot{x}^1 t) + 2c^2 t^2\dot{x}^1}{c^2(R^2 + \mathbf{x}^2 - c^2 t^2)^2}\right\} \tag{G.4}$$

(4) Problem 4:

Solution:

$$\dot{x}^1\partial_{x^1}(\partial_{\dot{x}^1} L_{dS}) = \frac{-m^4 c^8 R^4}{4L_{dS}^3}\dot{x}^1\left\{\frac{-4x^1[R^2(c^2 - \dot{\mathbf{x}}^2) - \mathbf{x}^2\dot{\mathbf{x}}^2 + (\mathbf{x}\cdot\dot{\mathbf{x}})^2 + c^2(\mathbf{x} - \dot{\mathbf{x}}t)^2]}{(R^2 + \mathbf{x}^2 - c^2 t^2)^3}\right.$$

$$\left. + \frac{-2x^1\dot{\mathbf{x}}^2 + 2(\mathbf{x}\cdot\dot{\mathbf{x}})\dot{x}^1 + 2c^2(x^1 - \dot{x}^1 t)}{c^2(R^2 + \mathbf{x}^2 - c^2 t^2)^2}\right\}$$

$$+ \frac{m^2 c^4 R^2}{2L_{dS}}\dot{x}^1\left\{\frac{(-2)2x^1[-2R^2\dot{x}^1 - 2\mathbf{x}^2\dot{x}^1 + 2(\mathbf{x}\cdot\dot{\mathbf{x}})x^1 - 2c^2 t(x^1 - \dot{x}^1 t)]}{c^2(R^2 + \mathbf{x}^2 - c^2 t^2)^3}\right.$$

$$\left. + \frac{-2x^1\dot{x}^1 + 2(\mathbf{x}\cdot\dot{\mathbf{x}}) - 2c^2 t}{c^2(R^2 + \mathbf{x}^2 - c^2 t^2)^2}\right\} \tag{G.5}$$

$$\dot{x}^2\partial_{x^2}(\partial_{\dot{x}^1} L_{dS}) = \frac{-m^4 c^8 R^4}{4L_{dS}^3}\dot{x}^2\left\{\frac{-4x^2[R^2(c^2 - \dot{\mathbf{x}}^2) - \mathbf{x}^2\dot{\mathbf{x}}^2 + (\mathbf{x}\cdot\dot{\mathbf{x}})^2 + c^2(\mathbf{x} - \dot{\mathbf{x}}t)^2]}{(R^2 + \mathbf{x}^2 - c^2 t^2)^3}\right.$$

$$\left. + \frac{-2x^2\dot{\mathbf{x}}^2 + 2(\mathbf{x}\cdot\dot{\mathbf{x}})\dot{x}^2 + 2c^2(x^2 - \dot{x}^2 t)}{c^2(R^2 + \mathbf{x}^2 - c^2 t^2)^2}\right\}$$

$$+ \frac{m^2 c^4 R^2}{2L_{dS}}\dot{x}^2\left\{\frac{(-2)2x^2[-2R^2\dot{x}^1 - 2\mathbf{x}^2\dot{x}^1 + 2(\mathbf{x}\cdot\dot{\mathbf{x}})x^1 - 2c^2 t(x^1 - \dot{x}^1 t)]}{c^2(R^2 + \mathbf{x}^2 - c^2 t^2)^3}\right.$$

$$\left. + \frac{-4x^2\dot{x}^1 + 2\dot{x}^2 x^1}{c^2(R^2 + \mathbf{x}^2 - c^2 t^2)^2}\right\} \tag{G.6}$$

$$\dot{x}^3\partial_{x^3}(\partial_{\dot{x}^1} L_{dS}) = \frac{-m^4 c^8 R^4}{4L_{dS}^3}\dot{x}^3\left\{\frac{-4x^3[R^2(c^2 - \dot{\mathbf{x}}^2) - \mathbf{x}^2\dot{\mathbf{x}}^2 + (\mathbf{x}\cdot\dot{\mathbf{x}})^2 + c^2(\mathbf{x} - \dot{\mathbf{x}}t)^2]}{(R^2 + \mathbf{x}^2 - c^2 t^2)^3}\right.$$

$$\left. + \frac{-2x^3\dot{\mathbf{x}}^2 + 2(\mathbf{x}\cdot\dot{\mathbf{x}})\dot{x}^3 + 2c^2(x^3 - \dot{x}^3 t)}{c^2(R^2 + \mathbf{x}^2 - c^2 t^2)^2}\right\}$$

$$+ \frac{m^2 c^4 R^2}{2L_{dS}}\dot{x}^3\left\{\frac{(-2)2x^3[-2R^2\dot{x}^1 - 2\mathbf{x}^2\dot{x}^1 + 2(\mathbf{x}\cdot\dot{\mathbf{x}})x^1 - 2c^2 t(x^1 - \dot{x}^1 t)]}{c^2(R^2 + \mathbf{x}^2 - c^2 t^2)^3}\right.$$

$$\left. + \frac{-4x^3\dot{x}^1 + 2\dot{x}^3 x^1}{c^2(R^2 + \mathbf{x}^2 - c^2 t^2)^2}\right\} \tag{G.7}$$

(5) Problem 5:

Solution:

Substituting Eqs.(G.2), (G.4), (G.5), (G.6), (G.7) into (3.34) gives a explicit verify check to the identity (3.34).

(6) Problem 6:

Proof: Setting $\Lambda = 3/R^2$, we have

$$\det\left(\frac{\partial^2 L_{dS}}{\partial \dot{x}^i \partial \dot{x}^j}\right) = \sum_{n=0}^{\infty} \frac{\Lambda^n}{n!}\left[\frac{\partial^n}{\partial \Lambda^n}\det\left(\frac{\partial^2 L_{dS}}{\partial \dot{x}^i \partial \dot{x}^j}\right)\right]_{\Lambda=0}$$

$$= \left[\det\left(\frac{\partial^2 L_{dS}}{\partial \dot{x}^i \partial \dot{x}^j}\right)\right]_{\Lambda=0} + \Lambda\left[\frac{\partial}{\partial\Lambda}\det\left(\frac{\partial^2 L_{dS}}{\partial \dot{x}^i \partial \dot{x}^j}\right)\right]_{\Lambda=0} + \cdots.$$

Since $B_{\mu\nu}(x)|_{\Lambda=0} = \eta_{\mu\nu}$, from (3.24), we have

$$L_{dS}|_{\Lambda=0} = -mc\sqrt{\eta_{\mu\nu}\dot{x}^\mu\dot{x}^\nu} = -mc^2\sqrt{1 + \eta_{ij}\frac{\dot{x}^i\dot{x}^j}{c^2}}$$

$$= -mc^2 - \frac{1}{2}m\eta_{ij}\dot{x}^i\dot{x}^j + \mathcal{O}(\frac{1}{c^2}),$$

and hence

$$\left[\det\left(\frac{\partial^2 L_{dS}}{\partial \dot{x}^i \partial \dot{x}^j}\right)\right]_{\Lambda=0} = m^3 + \mathcal{O}(\frac{1}{c^2}) \neq 0.$$

Therefore

$$\det\left(\frac{\partial^2 L_{dS}}{\partial \dot{x}^i \partial \dot{x}^j}\right) \neq 0.$$

Section 3.4

(1) Problem 1:

Solution: Substituting

$$\begin{pmatrix} A & C \\ B & D \end{pmatrix}^T = \begin{pmatrix} A^T & B^T \\ C^T & D^T \end{pmatrix} \tag{G.8}$$

into Eq.(3.50) gives

$$\begin{pmatrix} AA^T - \lambda CJC^T & AB^T - \lambda CJD^T \\ (AB^T - \lambda CJD^T)^T & BB^T - \lambda DJD^T \end{pmatrix} = \begin{pmatrix} I & 0 \\ 0 & -\lambda J \end{pmatrix}. \tag{G.9}$$

Comparing two sides of above equation, we get Eq.(3.51). Nextly, substituting Eq.(G.8) into Eq.(3.52) gives Eq.(3.53).

(2) Problem 2:

Proof: Start with the definition of the inverse

$$A^{-1}A = I, \tag{G.10}$$

and differentiate, yielding

$$A^{-1}dA + dA^{-1}A = 0, \tag{G.11}$$

rearranging the terms yields

$$dA^{-1} = -A^{-1}dAA^{-1}, \qquad QED. \tag{G.12}$$

which is just Eq.(3.79).

Section 3.6

(1) Problem 1:
Solution: Using Eq.(3.138) and letting $(f,\ g) = (x^i,\ \text{or}\ \pi_j)$, we get Eq.(3.141).

(2) Problem 2:
Solution:

$$\dot{F} \equiv \frac{dF(t,\mathbf{x},\boldsymbol{\pi})}{dt} = \frac{\partial F}{\partial t} + \sum_{i=1}^{3}\left(\frac{\partial F}{\partial x^i}\frac{dx^i}{dt} + \frac{\partial F}{\partial \pi_i}\frac{d\pi_i}{dt}\right)$$

$$= \frac{\partial F}{\partial t} + \sum_{i=1}^{3}\left(\frac{\partial F}{\partial x^i}\frac{\partial H_{dS}}{\partial \pi_i} - \frac{\partial F}{\partial \pi_i}\frac{\partial H_{dS}}{\partial x^i}\right)$$

$$= \frac{\partial F}{\partial t} + \{F, H_{dS}\}_{PB}, \qquad QED$$

where Eqs.(3.135), (3.136) and (3.138) were used.

(3) Problem 3:
Solution: From $f = f(x,y)$, we use notations:

$$u = \frac{\partial f}{\partial x}, \qquad v = \frac{\partial f}{\partial y}.$$

Now, treating u and y as two independent variables, then $x = x(u,y)$, $v = v(u,y)$, we have

$$f(x,y) \to \bar{f}(u,y) = f[x(u,y),y].$$

So, we have

$$\frac{\partial \bar{f}}{\partial y} = \frac{\partial f}{\partial y} + \frac{\partial f}{\partial x}\frac{\partial x}{\partial y} = v + u\frac{\partial x}{\partial y}$$

$$\frac{\partial \bar{f}}{\partial u} = \frac{\partial f}{\partial x}\frac{\partial x}{\partial u} = u\frac{\partial x}{\partial u}. \qquad\qquad (G.13)$$

Rewriting (G.13) as follows:

$$v = -\frac{\partial}{\partial y}(-\bar{f} + ux) \equiv -\frac{\partial g}{\partial y}$$

$$x = \frac{\partial}{\partial u}(-\bar{f} + ux) \equiv \frac{\partial g}{\partial u}, \qquad\qquad (G.14)$$

where $g = -\bar{f} + ux$, then the above equation indicates that the desired function with independent variables u and y is:

$$g(u,y) = -\bar{f} + ux = -f + \frac{\partial f}{\partial x}x.$$

Chapter 5

Section 5.2

(1) Problem 1:
   Solution:

$$\Gamma^0_{11} = a\dot{a}, \quad \Gamma^0_{22} = a\dot{a}r^2, \quad \Gamma^0_{33} = a\dot{a}r^2\sin^2\theta,$$
$$\Gamma^1_{01} = \dot{a}/a, \quad \Gamma^1_{22} = -r,$$
$$\Gamma^1_{33} = -r\sin^2\theta, \tag{G.15}$$
$$\Gamma^2_{02} = \dot{a}/a, \quad \Gamma^2_{12} = 1/r, \quad \Gamma^2_{33} = -\sin\theta\cos\theta,$$
$$\Gamma^3_{03} = \dot{a}/a, \quad \Gamma^3_{13} = 1/r, \quad \Gamma^3_{23} = \cot\theta.$$

(2) Problem 2:
   Solution:

$$\mathcal{R}_{00} = 3\ddot{a}/a$$
$$\mathcal{R}_{11} = -(a\ddot{a} + 2\dot{a}^2)$$
$$\mathcal{R}_{22} = -r^2(a\ddot{a} + 2\dot{a}^2) \tag{G.16}$$
$$\mathcal{R}_{33} = -r^2\sin^2\theta(a\ddot{a} + 2\dot{a}^2)$$
$$\mathcal{R}_{\mu\nu} = 0, \quad for \ \ \mu \neq \nu.$$

$$\mathcal{R} = \frac{6(\dot{a}^2 + a\ddot{a})}{a^2}. \tag{G.17}$$

Chapter 6

Section 6.2

(1) Problem 1:
   Solution: From Eq.(6.27), we have

$$e^a{}_{\mu;\nu} = e^a{}_{\mu,\nu} + \omega^a{}_{b\nu}e^b{}_\mu - \Gamma^\lambda_{\mu\nu}e^a{}_\lambda. \tag{G.18}$$

Substituting Eqs.(6.56), (6.60), (6.64) and (3.8) into Eq.(G.18) gives

$$e^a{}_{\mu;\nu} \equiv e^a{}_{\mu,\nu} + \omega^a{}_{b\nu}e^a{}_\mu - \Gamma^\lambda_{\mu\nu}e^a{}_\lambda$$
$$= \delta^a_\gamma\left(\frac{1}{2\sqrt{\sigma}}\theta^\gamma{}_\mu + \frac{1}{\sigma}\omega^\gamma{}_\mu\right)\frac{2\eta_{\nu\rho}x^\rho}{\sigma R^2}$$
$$+(-\delta^a_\nu\eta_{\mu\rho}x^\rho - \eta_{\mu\nu}\delta^a_\alpha x^\alpha + 2\delta^a_\alpha x^\alpha\omega_{\mu\nu})\frac{1}{R^2(1-\sigma)}\left(\frac{1}{\sqrt{\sigma}} - \frac{1}{\sigma}\right)$$
$$+\frac{1}{R^2(1+\sqrt{\sigma})\sqrt{\sigma}}(\delta^a_\nu\eta_{bd}\delta^d_\gamma x^\gamma - \delta^\beta_b\eta_{\beta\nu}\delta^a_\alpha x^\alpha)\delta^b_\lambda\left(\frac{1}{\sqrt{\sigma}}\theta^\lambda{}_\mu + \frac{1}{\sigma}\omega^\lambda{}_\mu\right)$$
$$-\frac{1}{\sigma R^2}(\eta^\lambda_\mu\eta_{\nu\rho}x^\rho + \eta^\lambda_\nu\eta_{\mu\rho}x^\rho)\delta^a_\alpha\left(\frac{1}{\sqrt{\sigma}}\theta^\alpha{}_\lambda + \frac{1}{\sigma}\omega^\alpha{}_\lambda\right). \tag{G.19}$$

Noting the notations in Eqs.(8.44)–(8.39), the expression of $e^a_{\mu;\nu}$ of (G.19) can be generally rewritten in following form:

$$e^a_{\mu;\nu} = f_1(\sigma, R)\delta^a_\mu \bar{x}_\nu + f_2(\sigma, R)\delta^a_\nu \bar{x}_\mu + f_3(\sigma, R)\eta_{\mu\nu}\delta^a_\alpha x^\alpha$$
$$+ f_4(\sigma, R)\delta^a_\alpha x^\alpha \omega_{\mu\nu}, \tag{G.20}$$

where $f_1(\sigma, R), f_2(\sigma, R), f_3(\sigma, R), f_4(\sigma, R)$ are scalar functions of σ and R, which are determined by Eq.(G.19). Expanding Eq.(G.19) and comparing it with Eq.(G.20), we find that:

$$f_1(\sigma, R) = \frac{1}{2\sqrt{\sigma}}\frac{2}{\sigma R^2} - \frac{1}{R^2\sigma\sqrt{\sigma}} = 0, \tag{G.21}$$

$$f_2(\sigma, R) = \frac{-1}{R^2(1-\sigma)}\left(\frac{1}{\sqrt{\sigma}} - \frac{1}{\sigma}\right) + \frac{1}{R^2(1+\sqrt{\sigma})\sigma^{3/2}} - \frac{1}{R^2\sigma^{3/2}}$$
$$= 0, \tag{G.22}$$

$$f_3(\sigma, R) = \frac{-1}{R^2(1-\sigma)}\left(\frac{1}{\sqrt{\sigma}} - \frac{1}{\sigma}\right) - \frac{1}{R^2(1+\sqrt{\sigma})\sigma} = 0, \tag{G.23}$$

$$f_4(\sigma, R) = \left(\frac{-1}{\sqrt{\sigma}} + \frac{2}{\sigma}\right)\frac{1}{R^2\sigma} + \frac{1}{R^2}\left(\frac{2}{1-\sigma} + \frac{2}{\sigma} + \frac{1}{(1+\sqrt{\sigma})\sqrt{\sigma}}\right)$$
$$\times \left(\frac{1}{\sqrt{\sigma}} - \frac{1}{\sigma}\right)$$
$$= 0. \tag{G.24}$$

Therefore, substituting Eqs.(G.21)–(G.24) into Eq.(G.20) gives

$$e^a_{\mu;\nu} = 0, \tag{G.25}$$

which is the first equation in (6.66). Noting $e^b_\mu e_a^\mu = \delta^b_a$, from Eq.(G.25) we have:

$$e_a^\mu{}_{;\nu} = 0. \tag{G.26}$$

Thus the second equation in (6.66) is also proved.

Chapter 7

Section 7.4

(1) Problem 1:
Answer: Use *Mathematica* :

Parameters defining the eigenfunctions of the one-electron atom for the principal numbers $n = 1$ and $n = 2$.

$In[1] =$ $Z = 1;$

$\kappa = -1;$

$\alpha = 1/137.003599911;$ (*fine − construct constant*)

$c = 2.99792458 * 10^{10};$ (* cm/s *)

$m = 510998.918/c^2;$ (*mass of electron : m $\equiv$ 510998.918eV$/c^2$*)

$\xi = Z * \alpha;$(* $Eq.(A.67)$ *)

$hb = 6.58211899 * 10^{-16};$ (* unit : $eV \cdot s$, hb $= h/(2\pi)$ *)

$kC = m * c/hb;$ (*1/cm*)

$s = (\kappa^2 - Z^2\alpha^2)^{1/2};$

$W = m * (1 + Z^2\alpha^2/(n - Abs[\kappa] + s)^2)^{-1/2} - m$

(*see equation $(A.98)$*)

$\lambda = (kC^2 - WC^2)^{1/2};$ (*1/cm*)

(*$g2s[r] = Bs * r^{(s-1)} * Exp[-\lambda * r] * (bs0 + bs1 * r);$ $Eq.(7.37)$*)

(*$f2s[r] = -As * r^{(s-1)} * Exp[-\lambda * r] * (as0 + as1 * r);$ $Eq.(7.38)$*)

$as0 = WC/kC + 1;$

$bs0 = WC/kC;$

$$Bs = \left(\frac{(2 * \lambda)^{(2*s+1)} * kC * (2s + 1) * (kC + WC)}{2 * WC * (2 * WC + kC) * Gamma[2s + 1]} \right)^{(1/2)};$$

$$As = \left(\frac{(2 * \lambda)^{(2*s+1)} * kC * (2s + 1) * (kC - WC)}{2 * WC * (2 * WC + kC) * Gamma[2s + 1]} \right)^{(1/2)};$$

$Out[21] =$ Null^2

(*To check $\int_0^\infty \left(g2s^2 + f2s^2\right) r^2 dr = 1$, (see footnote :[a]) *)

$In[21] =$ Is $= (Bs^2 * bs0^2 + As^2 * as0^2) * \dfrac{Gamma[2s + 1]}{(2\lambda)^{2s+1}}$

$Out[25] =$ 0.99999

$In[26] =$ IIs $= 2 * (Bs^2 * bs0 * bs1 + As^2 * as0 * as1) * \dfrac{Gamma[2s + 2]}{(2\lambda)^{2s+2}}$

$Out[26] =$ -2.99996

$In[27] =$ IIIs $= (Bs^2 * bs1^2 + As^2 * as1^2) * \dfrac{Gamma[2s + 3]}{(2\lambda)^{2s+3}}$

$Out[27] =$ 2.99997

$In[28] =$ Is + IIs + IIIs

$Out[28] =$ 1. (* QED *)

[a] We try to check $\int_0^\infty \left(g_{(2s1/2)}(r)^2 + f_{(2s1/2)}(r)^2\right) r^2 dr = 1$, where $g_{(2s1/2)}(r)$, $f_{(2s1/2)}(r)$ have been given in (7.37) (7.38) (or (A.104) (A.105) in Appendix A). Note

the calculations and notations used in the program:

$$\int_0^\infty \left(\text{g2s}^2 + \text{f2s}^2 \right) r^2 \, dr$$

$$= \int_0^\infty \left(\text{Bs}^2 * r^{2s-2} e^{-2\lambda r} (\text{bs0} + \text{bs1} * r)^2 + \text{As}^2 * r^{2s-2} e^{-2\lambda r} (\text{as0} + \text{as1} * r)^2 \right) r^2 \, dr$$

$$= \left(\text{Bs}^2 * \text{bs0}^2 + \text{As}^2 * \text{as0}^2 \right) * \frac{\Gamma(2s+1)}{(2\lambda)^{2s+1}} + 2 * \left(\text{Bs}^2 * \text{bs0} * \text{bs1} + \text{As}^2 * \text{as0} * \text{as1} \right)$$

$$* \frac{\Gamma(2s+2)}{(2\lambda)^{2s+2}} + \left(\text{Bs}^2 * \text{bs1}^2 + \text{As}^2 * \text{as1}^2 \right) * \frac{\Gamma(2s+3)}{(2\lambda)^{2s+3}}$$

$$:= \text{Is} + \text{IIs} + \text{IIIs}, \text{ where notations Is, IIs, IIIs are used in program.}$$

Similarly, the normalization of $2p^{1/2}$-radial wave function (7.42) (7.43) can also be verified in similar program.

Chapter 8

Section 8.3

(1) Problem 1:
   Answer:

$$\Delta t \simeq \frac{2\pi a_c}{\alpha c} \simeq 8.6 \times 10^{-19} s, \tag{G.27}$$

where Bohr radius $a_c = \hbar/(m_e c) \simeq 0.3^{-12} m$ were used.

(2) Problem 2:
   Answer:

$$\Delta z = \frac{dz}{dt} \Delta t = H(z)(1+z)\Delta t$$

$$= H_0 \sqrt{\Omega_{m0}(1+z)^3 + 1 - \Omega_{m0}} (1+z)\Delta t.$$

As $z = 3$, one gets

$$\Delta z \simeq 70.5 \times \sqrt{0.274 \times 4^3 + 1 - 0.274} \times 4 \times 8.6 \times 10^{-19} km/Mpc$$

$$\simeq 1.04 \times 10^{-15} km/Mpc$$

$$\simeq 1.04 \times 10^{-15} \times (km/(3.09 \times 10^{19} km))$$

$$\simeq 3.3 \times 10^{-34}, \tag{G.28}$$

which is extremely tiny, and indicates the movements of electrons in the atoms are much faster than the evolutions of the Universe. And hence the adiabatic approximations for the time-dependent QM equations in atom physics in cosmology are legitimate .

Section 8.6

(1) Problem 1:

Answer: The metric of $\mathcal{B}^{\mathcal{M}}$ has been given in (8.35). The corresponding Maxwell equations are (8.64). Noting $\mu = 0$ gives the Coulomb law equation (8.71). Solved this equation, we get answer (8.85).

Bibliography

Bargmann, V., 1932, Sitzung. Preuss. Akad. Wiss. Phys. Math. Kl. **346**.

Bayfield, J.E., 1999, *"Quantum Evolution: An Introduction to Time-Dependent Quantum Mechanics"*, John Wiley & Sons, Inc., New York.

Baym, G., 1967, *"Lectures on Quantum Mechanics"*, (Addison-Wesley, California).

Bekenstein, J.D., 1982, 2002, Phys. Rev. **25**, 1527 (1982); Phys. Rev. **D66** 123514 (2002).

Bennett, C.L., *et al.*, 2003, Astrophys. J. Suppl. **148**, 1, astro-ph/0302207.

Born, M and Fock, V, Z. 1928, Phys., **51**, 165.

Brill, D.R. and Wheeler, J.A., 1957, Rev. Mod. Phys. **29** 465.

Cartan, E., 1966, *"The Theory of Spinors"*, MIT Press, Cambridge, Mass.

Chang, Z, Chen, S.X., Guan, C.B., Huang, C.G., 2005, *"Cosmic ray threshold in an asymptotically dS spacetime"*, Phys. Rev. **D71**, 103007. astro-ph/0505612.

Chen, S.X., Xiao, N.C., Yan, M.L., 2008, Chinese Phys. **C32**, 612.

Dent, T., Stern, S. and Wetterich, C., Phys. Rev. **D78**, 103518 (2008).

de Sitter, W., 1917, Proc. Roy. Acod. Sci. (Amsterdam), **19**, 1217; *ibid.* **20**, 1309; Mon. Not. Roy. Astron. Soc., **78**, 2.

Dirac, P.A.M., 1928, Proc. Roy. Soc.**117**, 610.

Dirac, P.A.M., 1935, Annals of Mathmatics, **36**, No.3, 657.

Duan, I.S., 1958, Soviet Physics JETP **34** 437.

Dzuba, V. A., Flambaum, V. V. and Webb, J. K., 1999a, Phys. Rev. Lett., **82**, 888.

Dzuba, V. A., Flambaum, V. V. and Webb, J. K., 1999b, Phys. Rev., **A 59**, 230.

Ellingsen, S., Voronkov, M., Breen, S., 2011 Phys. Rev. Lett. **107**, 270801.

Filippenko, A. V., 2004, in Measuring and Modeling of the Universe (Carnegie Observatories Astrophysics Series, Vol 2., Cambridge University Press) [astro-ph/0307139]; Lect. Notes Phys. 645, 191 (2004) [astro-ph/0309739].

Fock, V., 1929, Z. Phys. **57** 261.

Gasiorowicz, S., 1974, *"Quantum Mechanics"*, (John Wiley and Sons, New York).

Guo, H.Y., Huang, C.G., Xu, Z. and Zhou, B., 2004, *"On Beltrami Model of de Sitter Spacetime"*, Mod. Phys. Lett. **A19**, 1701, arXiv: hep-th/0311156.

Guo, H.Y., Huang, C.G., Xu, Z. and Zhou, B., 2004, *"On special relativity with cosmological constant"*, Phys. Lett. **A331**, 1.

Guo, H.Y., Huang, C.G., Wu, H.T., Zhou, B., 2010, *"The Principle of Relativity, Kinematics and Algebraic Relations"* Sci. China Phys. Mech. Astron. **53** 591. arXiv:0812.0871 [hep-th].

Guo, H.Y, Huang, C.G., Wu, H.T., 2008, *"Yang's Model as Triply Special Relativity and the Snyder's Model-de Sitter Special Relativity Duality"*, Phys. Lett. **B663**, 270. arXiv:0801.1146 [hep-th].

Guo, H.Y, Huang, C.G., Wu, H.T., 2008, *"Yang's Model as Triply Special Relativity and the Snyder's Model-de Sitter Special Relativity Duality"*, Phys. Lett. **B663**, 270. arXiv:0801.1146 [hep-th].

Guo, H.Y., Huang, C.G., Tian,Y., Wu, H.T., Zhou, B., 2007, *"Snyder's Model—de Sitter Special Relativity Duality and de Sitter Gravity"*, Class. Quant. Grav. **24** 4009, gr-qc/0703078.

Gürsey, F., 1963, *Introduction to Group Theory*, Relativity, Groups and Topology, edt. by DeWitt, C. and DeWitt, B.

Gürsey, F. and Lee, T.D., 1963, *"Spin 1/2 Wave Equation in de Sitter Space"*, Proc. Nat. Sc. Sci., **49**, 179.

Hartle, J.B., 2003, *"Gravity, An Introduction to Einstein's General Relativity"*, Addison Wesley.

Hu, S, Li, S, Yan, M L, 2012, *"Remark on 'Pair Creation Constrains Superluminal Neutrino Propagation' "* arXiv:1203.6026.

Hijazi, O., Hijazi, S. & Xiao, X., 2001, *Dirac Opertor on Embedded Hypersurfaces*, Math. Res. Lett. **8**, 195-208.

Huang, C.G., Tian,Y., Wu, X.N., Xu.Z., Zhou, B., 2012, *"Geometries for Possible Kinematics"* Sci. China **G55** 1978. arXiv:1007.3618 [math-ph].

Inönü, E. and Wigner, E.P., 1953, Proc. Nat. Acad. Sci. **39**, 510-524; Bacry, H. and Lévy-Leblond, J.-M., J. Math. Phys. **9**, 1605-1614 (1968).

Jarosik, N. *et al.*, 2011, Astrophys. J. Suppl. **192**, 14 (2011); Planck Collaboration: P. A. R. Ade, *et al.*, *"Planck 2013 results. XVI. Cosmological parameters"*, arXiv:1303.5076 [astro-ph.CO].

Kleinert, H, 1990, *"Path Integrals in Quantum Mechanics Statistics and Polymer Physics"*, World Scientific, Singapore.

Kibble, B.W.T., 1961, J. Math. Phys. **2**, 212.

King, J. A., *et al.*, 2012, Mon. Not. Roy. Astron. Soc. **422**, 3370-3413.

Komatsu, E., *et al.*, 2009, Astrophys. J. Suppl. **180**, 330.

Kottler, F., 1918, Ann. Physik **56** (361), 401-462.

Landau, L.D. and Lifshitz, E.M., 1987, *The Classical Theory of Fields*, (Translated from Russian by M. Hamermesh), Pergamon Press, Oxford.

Look, K.H.(Lu, Q.K.), 1970, *Why the Minkowski metric must be used ?*, unpublished.

Look, K.H., Tsou, C.L. (Zou, Z.L.) and Kuo, H.Y.(Guo, H.Y.), 1974, *Acta Physica Sinica*, **23** 225 (in Chinese).

Messiah, A., 1970 *"Quantum Mechanics I, II"*, North-Holland Publishing Company.

Murphy, M.T., Webb, J.K., Flambaum, V.V., 2003 *"Further evidence for a variable fine-structure constant from Keck/HIRES QSO absorption spectra"*, Mon. Not. Roy. Astron. Soc. 345, 609.

Murphy, M. T., Webb, J. K., Flambaum, V. V., 2007, Phys. Rev. Lett. **99**, 239001; astro-ph/0612407.

Murphy, M. T., Flambaum, V.V., Webb, J. K., Dzuba, V.V., Prochaska, J. X. and Wolfe, A. M., 2004, Lect. Notes Phys. **648**, 131.

Nieh, H.T., 1980, J. Math. Phys. **21**, 1439.

Nieh, H.T and Yan, M.L., 1982, Ann. Phys. **138**, 237.

Olive, K.A., Peloso, M. and Uzan, J.-P. 2011, Phys. Rev., D **83**, 043509.

Padmanabhan, T., 2009, Phys. Rep. **380**, 235.

Panagia, N., 2005, Nuovo Cimento B **210**, 667. [astro-ph/0502247].

Pauli, W., 1927, Z. Physik, **43**, 601.

Peebles, P.J.E., 2009, Rev. of Mod. Phys. **75**, 559.

Perlmutter, S., *et al.*, 1999, Astrophys. J. 517, 565; astro-ph/9812133.

Perlmutter, S. and Schmidt, B. P., 2003, in *Supernovae & Gamma Ray Bursts*, ed. K.Weiler (Springer, 2003) [astro-ph/0303428].

Riess, A.G. *et al.*, 1998, 1999, Astro. J. **116** 1009 (1998); S. Perlmutter *et al.*, Astrophys. J. **517**, 565 (1999) [astro-ph/9812133].

Roberson, H. P., 1928, Phil. Mag. V, p. 839.

Rose, M.E., 1961, *"Relativistic Electron Theory"*, John Wiley, New York.

Ruiz-Lapuente, P., 2004, *Astrophys. Space Sci.* **290**, 43 [astroph/0304108];

Sandvik, H.B., Barrow, J.D. and Magueijo, J., 2002, Phys. Rev. Lett., **88**, 031302.

Schrödinger, E., 1932, Sitzung. Preuss. Akad. Wiss. Phys. Math. Kl. **105**.

Schmidt, B., *et al.*, 1998, *Astrophys.J.* **507**, 46 , [astro-ph/9805200].

Sobelman, I. I., 1979, *"Atomic spectra and radiative transitions"* Springer-Verlag, Berlin.

Strange, P., 2008, *"Relativistic Quantum Mechanics"*, Cambridge University Press.

Sun, L.F., Yan, M.L., Deng, Y., Huang, W., Hu, S., 2013, *"Schwarzschild-de Sitter Metric and Inertial Beltrami Coordinates"*, Mod. Phys. Lett. **A28** 1350114. arXiv:1308.5222 [gr-qc].

Tegmark, M., *et al.*, 2004, Phys. Rev. **D69** 103501, astro-ph/0310723.

Tian, Y., Guo, H.Y., Huang, C.G., Xu, Z., Zhou, B., 2005, *"Mechanics and Newton-Cartan-like gravity on the Newton-Hooke space-time"*, Phys. Rev. **D71** 044030. hep-th/0411004.

Trefftz, E., 1922, Mathem. Ann. **86**, 317-326.

Utiyama, R., 1956, Phys. Rev. **101**, 1597.

Uzan, J.-P., 2003, Rev. Mod. Phys. **75** 403.

van der Waerden, B.L., 1929, Nachr. Wiss. Gottingen Math. Phys. **100**, *"Die Gruppentheoretische Methode in der Quantenmechanik"*, Springer, Berlin.

van Weerdenburg, F., Murphy, M.T., Malec, A.L., Kaper, L., and Ubachs, W., 2011, Phys. Rev. Lett., **106**, 180802.

Webb, J. K., Flambaum, V. V., Churchill, C. W., Drinkwater, M. J., Barrow, J. D., 1999, Phys. Rev. Lett. **82** 884.

Webb, J. K. *et al.*, 2001, Phys. Rev. Lett. 87, 091301.

Webb, J. K., King, J. A., Murphy, M. T.,Flambaum, V.V., Carswell, R. F., and Bainbridge, M. B., 2011, Phys. Rev. Lett., **107**, 191101.

Weinberg, S, 1972, *"Gravitation and Cosmology: Principles and Applications of the General Theory of Relativity"*, John Wiley & Sons, Inc.

Weinberg, S, 2008, *"Cosmology"*, Oxford University Press.

Weyl, H., 1919, Phys. Z. **20**, 31-34.

Weyl, H., 1929, Proc. Nat. Acad. Sci. **15** 323, Z.Phys. **56** 330, Phys. Rev. **77** (1950) 699.

Witten, E., 2001, *"Quantum Gravity In De Sitter Space"*, hep-th/0106109.

Yan, M.L., Xiao, N.C., Huang, W., Li, S., 2005, *"Hamiltonian Formalism of de-Sitter Invariant Special Relativity"* Commun. Theor. Phys. (Beijing, China) **48**, 27 (2007), hep-th/0512319.

Yan, M.L., Xiao, N.C., Huang, W., Hu, S, 2011, *"Superluminal Neutrinos from Special Relativity with de Sitter Space-time Symmetry"*, Mod. Phys. Lett. **A27** 1250076; e-Print: arXiv:1111.4532.

Yan, M.L., 2012, *"One Electron Atom in Special Relativity with de Sitter Space-Time Symmetry"* Commun. Theor. Phys. (Beijing, China) **57**, 930.

Yan, M.L., 2014, *"One Electron Atom in Special Relativity with de Sitter Space-Time Symmetry-(II)"* Commun. Theor. Phys. (Beijing, China) **62**, 189.

Yan, M.L., Hu, S., Huang, W, 2012, *"On determination of the geometric cosmological constant from the OPERA experiment of superluminal neutrinos"*, Mod. Phys. Lett. **A27**, 1250041, arXiv:1112.6217.

Yang, C.N. and Mills, R.L., 1954, *Conservation of Isotopic Spin and Isotopic Gauge Invariance*, Phys. Rev. **96**, 191-195.

Zhang, Y.Z,. 1997, *"Special Relativity and Its Experimental Foundations"*, World Scientific.

Index

ΛCDM model, 97

α-variation, ix, 158, 195

adiabatic approximation, 121,
    139–141, 154, 167, 174, 256
AdS Quantum Gravity, 106
affine connection, 35, 74
Ampere, 128
Anti-de Sitter, 3
appendix, 213

basic metric, 1–4, 71, 73, 166
Beltrami, 2, 4–6, 51, 55, 56, 63, 70,
    71, 81, 82, 84, 85, 201
Bianchi identity, 28, 37, 77
Bohr radius, 126, 127

canonical momenta, 5, 14, 64, 107
Christoffel symbols, 35, 36, 46, 76, 77,
    120, 172
commutation relations, 23, 108
confluent hypergeometric, 129, 225,
    231
connection field, 111, 112
cosmological constant, vii, viii, 43, 88,
    104–106, 168, 201
Cosmological Principle, 87, 89, 191
Coulomb, 128
covariant derivatives, 28, 34, 35, 113

critic density, 96
curvature tensor, 36, 37

dark energy, viii, 7, 88, 99, 104–106,
    168
de Sitter, vii–ix, 1–5, 7, 8, 43, 44, 56,
    63, 65, 67, 71–76, 78–81, 84, 85, 91,
    96, 105–110, 114, 118, 120–123,
    126, 127, 130, 132, 136, 137, 139,
    150, 157, 165–167, 196, 198, 200,
    201, 203, 210, 211, 239
de Sitter invariance, 211
de Sitter Invariant Special Relativity,
    vii, ix, 1
de Sitter model, 99
de Sitter symmetry, 109
de-Sitter algebra, 110
deceleration parameter, 96, 102
declination, 161, 162, 182, 198
dipole, 161–163, 181, 182, 192, 193,
    196–198
Dirac, vii, viii, 3, 24, 25, 110
Dirac equation, 119, 122
Dirac matrices, 25
Dirac spinor, 111
Dirac-Hartree-Fock method, 159
dispersion relation, 65, 68, 109
dS-GR, 6, 106, 173, 174

dS-SR, 57–59, 61, 63–65, 68, 70, 106,
  136, 138, 139, 141–143, 148–150,
  153, 154, 163, 166–169, 183, 186,
  190, 196–199, 243

E-GR, 6, 38, 41, 42, 172–174
E-SR, 3, 9, 13, 14, 19, 20, 23–25, 29,
  30, 37, 38, 41, 42, 50, 57, 59, 62–65,
  128, 130, 136, 142, 143, 145,
  165–167, 169, 198, 199
eigenvalues, 148
Einstein, vii–ix, 1–7, 9, 10, 25, 29, 30,
  37, 38, 41, 45, 47, 50, 51, 56, 64,
  70, 71, 73–75, 78, 79, 81, 85, 87,
  91–94, 96, 99, 104–106, 108, 120,
  122, 123, 127, 128, 130, 166, 168,
  169, 171–173, 198
Einstein–de Sitter model, 99
Equivalence Principle, 5, 42, 119, 172
Euler-Lagrange, 13, 113

fine-structure constant, 156, 157, 160,
  163, 164, 181, 196, 197
Friedmann equation, 91–93, 95, 98
Friedmann models, 95, 96
Friedmann–Robertson–Walker, 90,
  119, 120, 122, 124, 163, 169, 170
FRW, 119, 122–127, 130, 163–165,
  167, 169–173, 177, 181–183, 196,
  198

Gürsey-Lee, 201
gauge theory, 26, 29–31, 37, 38, 42,
  112
Gauss-Codazzi, 204
geodesic, 37, 43–47, 50, 94
GM-AV, 197–199

Hamiltonian, 3–5, 20, 23, 63–65, 107,
  108, 120–122, 130, 136, 138, 142,
  143, 167, 168, 188, 214, 215

Heisenberg, 23, 105, 108
Homogeneity, 87, 88, 93, 124, 170
Hubble, 95, 96, 98, 99, 103, 122, 157
Hydrogen, viii, 6, 7, 119–122,
  126–130, 132, 136, 137, 142, 143,
  145, 149, 150, 153, 154, 156, 163,
  164, 167, 177, 181, 183, 184, 186,
  190, 197, 198, 213, 220, 226, 239,
  243

inertial reference systems, 2, 49
Isotropy, 87, 93, 124, 170

Keck, ix, 160–162, 181, 182, 191–193,
  195, 197
Klein-Gordon, 24, 107, 109, 110
KWT, 78–81, 84

Lagrangian, 3–5, 13–15, 18–20, 22,
  23, 27, 28, 48–50, 58, 62–65, 80,
  107, 112, 121, 167, 203
Landau-Lifshitz, 2, 48, 79
Levi-Civita, 207
light-cone, 123, 164, 169–172, 177,
  181–183
local inertial coordinate system, 5–7,
  119, 120, 122, 123, 164, 165, 169,
  171–173, 181, 191, 198
localization, 25, 29, 30, 37, 38, 41, 74
Lorentz boost, 21, 22, 61, 62, 74
Lorentz transformation, 10–12, 24,
  51, 62, 111, 113, 201
Lu-Zou-Guo, 3
LZG-transformation, 51, 52, 57, 58,
  60, 74

Many-Multiplet, 158, 199
Maurer-Cartan, 208
Michelson-Morley, 10
Minkowski point, 70, 71, 120, 122,
  166, 167, 170, 174, 182, 190, 198

MM method, 159, 160

Noether charges, 4, 5, 19, 20, 38, 57, 58, 62, 63, 68, 72
Noether's theorem, 3, 15, 20
non-Abelian, 29

orthonormal frame, 202

Poincaré group, viii, 1, 19, 29, 165
Poincaré-Minkowski space-time symmetry, 1
Poisson Bracket, 14, 23, 66, 107
principle of least action, 12, 39, 45, 48, 75
principle of relativity, 1, 9, 10
projection operators, 115, 175, 176
pseudo-sphere, 1

QSO, 121–124, 127, 132, 133, 137, 143, 150, 157, 170, 235
quantization, 5, 7, 23, 107–110
quasar absorption systems, 157

red-shift, viii, 90, 91, 157, 183
Ricci coefficients, 114, 116, 117
Ricci tensor, 37, 40
right ascension, 161, 162, 182, 198
Robertson–Walker metric, 88–93

scalar field, 26, 39, 110, 199, 200
Schwarzschild metric, 78, 79, 81, 84, 85
SM, ix
Standard Model, ix, 163, 196
stress energy tensor, 87
Supernova Cosmology Project, 7

vacuum energy, viii, 99, 100
vielbein, 112

VLT, ix, 160–162, 181, 182, 191, 192, 195, 197

Weyl's approach, 111
WMAP data, 105, 168

Yang-Mills field theory, 29

Printed in the United States
By Bookmasters